Kawasaki KC, KE & KH100 Owners Workshop Manual

by Jeremy Churchill

Models covered
KC100 (Companion). 99cc. UK 1979 to 1986
KE100 A. 99cc. UK 1978 to 1981, US 1975 to 1981
KE100 B. 99cc. UK and US 1981 to 1993
KH100 A (KH100 EL). 99cc. UK 1977 to 1980
KH100 G (KH100 EX). 99cc. UK 1981 to 1992

(1371 – 3S4)

ABCDE
FGHIJ
K

Haynes Publishing Group
Sparkford Nr Yeovil
Somerset BA22 7JJ England

Haynes Publications, Inc
861 Lawrence Drive
Newbury Park
California 91320 USA

Acknowledgements

Our thanks are due to CW Motorcycles of Dorchester who supplied the KH100 G6 featured in the photographs throughout this manual and the KE100 B7 shown on the front cover, and to Kawasaki Motors (UK) Ltd who provided the service literature necessary and gave permission to reproduce many of the line drawings used.

The Avon Rubber Company provided information and technical assistance on tyre care and fitting, NGK Spark Plugs (UK) Ltd provided information on plug maintenance and electrode conditions, and Renold Ltd advised on chain maintenance and renewal.

© **Haynes Publishing Group 1993**

A book in the **Haynes Owners Workshop Manual Series**

Printed by J. H. Haynes & Co. Ltd, Sparkford, Nr Yeovil, Somerset BA22 7JJ, England

All rights reserved. No part of this book may be reproduced or transmitted in any form or by any means, electronic or mechanical, including photocopying, recording or by any information storage or retrieval system, without permission in writing from the copyright holder.

ISBN 1 85010 911 7

Library of Congress Catalog Card Number 93-70493

British Library Cataloguing in Publication Data
A catalogue record for this book is available from the British Library

We take great pride in the accuracy of information given in this manual, but motorcycle manufacturers make alterations and design changes during the production run of a particular motorcycle of which they do not inform us. No liability can be accepted by the authors or publishers for loss, damage or injury caused by any errors in, or omissions from, the information given.

Contents

Introductory pages

About this manual	5
Introduction to the Kawasaki KC, KE and KH100 models	7
Model dimensions and weights	7
Ordering spare parts	8
Safety first!	9
Tools and working facilities	10
Conversion factors	12
Choosing and fitting accessories	13
Fault diagnosis	16
Routine maintenance	25

Chapter 1
Engine, clutch and gearbox — 46

Chapter 2
Fuel system and lubrication — 90

Chapter 3
Ignition system — 109

Chapter 4
Frame and forks — 112

Chapter 5
Wheels, brakes and tyres — 127

Chapter 6
Electrical system — 147

Wiring diagrams — 158

English/American terminology — 169

Index — 170

The KC100 C1 Companion

The KH100 A3 (KH100 EL)

About this manual

The purpose of this manual is to present the owner with a concise and graphic guide which will enable him to tackle any operation from basic routine maintenance to a major overhaul. It has been assumed that any work would be undertaken without the luxury of a well-equipped workshop and a range of manufacturer's service tools.

To this end, the machine featured in the manual was stripped and rebuilt in our own workshop, by a team comprising a mechanic, a photographer and the author. The resulting photographic sequence depicts events as they took place, the hands shown being those of the author and the mechanic.

The use of specialised, and expensive, service tools was avoided unless their use was considered to be essential due to risk of breakage or injury. There is usually some way of improvising a method of removing a stubborn component, providing that a suitable degree of care is exercised.

The author learnt his motorcycle mechanics over a number of years, faced with the same difficulties and using similar facilities to those encountered by most owners. It is hoped that this practical experience can be passed on through the pages of this manual.

Where possible, a well-used example of the machine is chosen for the workshop project, as this highlights any areas which might be particularly prone to giving rise to problems. In this way, any such difficulties are encountered and resolved before the text is written, and the techniques used to deal with them can be incorporated in the relevant section. Armed with a working knowledge of the machine, the author undertakes a considerable amount of research in order that the maximum amount of data can be included in the manual.

A comprehensive section, preceding the main part of the manual, describes procedures for carrying out the routine maintenance of the machine at intervals of time and mileage. This section is included particularly for those owners who wish to ensure the efficient day-to-day running of their motorcycle, but who choose not to undertake overhaul or renovation work.

Each Chapter is divided into numbered sections. Within these sections are numbered paragraphs. Cross reference throughout the manual is quite straightforward and logical. When reference is made 'See Section 6.10' it means Section 6, paragraph 10 in the same Chapter. If another Chapter were intended, the reference would read, for example, 'See Chapter 2, Section 6.10'. All the photographs are captioned with a section/paragraph number to which they refer and are relevant to the Chapter text adjacent.

Figures (usually line illustrations) appear in a logical but numerical order, within a given Chapter. Fig. 1.1 therefore refers to the first figure in Chapter 1.

Left-hand and right-hand descriptions of the machines and their components refer to the left and right of a given machine when the rider is seated normally.

Motorcycle manufacturers continually make changes to specifications and recommendations, and these, when notified, are incorporated into our manuals at the earliest opportunity.

We take great pride in the accuracy of information given in this manual, but motorcycle manufacturers make alterations and design changes during the production run of a particular motorcycle of which they do not inform us. No liability can be accepted by the authors or publishers for loss, damage or injury caused by any errors in, or omissions from, the information given.

The KE100 A8

The KE100 B1

Introduction to the Kawasaki KC, KE and KH100 models

The models covered in this manual, while sharing the same basic engine/gearbox unit, differ widely in design and specification according to the requirements of the market at which each is aimed. Furthermore, all models have undergone some degree of modification during their production; while all such modifications are described in full in this Manual it is essential that the owner can identify exactly his or her machine before starting work.

Given below are the engine and frame numbers with which each variant of a particular model began its production run. Note that Kawasaki's model designation refers to the production year in which that machine left the factory; it is therefore not sufficient to identify a machine by its date of registration or sale. Accurate identification can be made only by reference to the engine and frame numbers.

Year	Model	Engine number	Frame number
KC100 (Companion):			
1980	KC100 C1	KC100CE000001	KC100C-000001
1981	KC100 C2	KC100CE003901	KC100C-003901
1982	KC100 C3	KC100CE006434	KC100C-006433
1983	KC100 C4	KC100CE006601	KC100C-006601
KE100 A			
1976	KE100 A5 (US)	G5E090900	G5-091700
1977	KE100 A6 (US)	G5E105524	G5-105524
1978	KE100 A7 (US and UK)	G5E113701	G5-114201
1979	KE100 A8 (US and UK)	G5E126001	G5-126001
1980	KE100 A9 (US and UK)	G5E134801	G5-134101
1981	KE100 A10 (UK)	G5E146201	G5-144901
1981	KE100 A10 (US)	G5E146201	G5-144901 to 146430, JKAKETA1*BA146431
KE100 B:			
1982	KE100 B1 (UK)	G5E154001	KE100B-000001
1982	KE100 B1 (US)	G5E154001	JKAKETB1*CA000001
1983	KE100 B2 (UK)	G5E160501	KE100B-004001

Year	Model	Engine number	Frame number
KE100 B (continued)			
1983	KE100 B2 (US)	G5E160501	JKAKETB1*DA004001
1984	KE100 B3 (UK)	G5E164601	KE100B-007101
1984	KE100 B3 (US)	G5E164601	JKAKETB1*EA007101
1985	KE100 B4 (UK)	G5E168001	KE100B-009201
1986	KE100 B5 (UK)	G5E170001	KE100B-010501
1986	KE100 B5 (US)	G5E170001	JKAKETB1*GA010501
1987	KE100 B6 (UK)	G5E173001	KE100B-013301
1987	KE100 B6 (US)	G5E173001	JKAKETB1*HA013301
1988-90	KE100 B7 (UK)	G5E176001	KE100B-016001
1988	KE100 B7 (US)	G5E176001	JKAKETB1*JA016001
1989	KE100 B8 (US)	G5E178400	JKAKETB1*KA019001
1990	KE100 B9 (US)	G5E190001	JKAKETB1*LA020501
1991-92	KE100 B10 (UK)	G5E178401	KE100B-024001
1991	KE100 B10 (US)	G5E178401	JKAKETB1*MA024001
1992-93	KE100 B11 (UK)	G5E190001	KE100B-030001
1992	KE100 B11 (US)	G5E190001	JKAKETB1*NC030001
1993	KE100 B12 (US)	G5E192002	JKAKETB1*PC032002

N.B. The digit indicated by the asterisk in the new VIN numbers of US models will vary from machine to machine

Year	Model	Engine number	Frame number
KH100 A (KH100 EL):			
1978	KH100 A2	G7E580367	G7-542610
1979	KH100 A3	G7E627336	G7-577941
1980	KH100 A4	Not available	Not available
KH100 G (KH100 EX):			
1981	KH100 G2	KH100DE018801	KH100E-011465
1982	KH100 G3	KH100DE020351	KH100G-001251
1983	KH100 G4	KH100DE021301	KH100G-002001
1984	KH100 G5	KH100DE022401	KH100G-002801
1985	KH100 G6	KH100DE023301	KH100G-003701
1986	KH100 G7	KH100DE024701	KH100G-005001
1987-92	KH100 G8	Not available	Not available

Model dimensions and weights

	KC100	KH100 A	KH100 G
Overall length	1810 mm (71.3 in)	1890 mm (74.4 in)	1900 mm (74.8 in)
Overall width	740 mm (29.1 in)	680 mm (26.8 in)	775 mm (30.5 in)
Overall height	1020 mm (40.2 in)	970 mm (38.2 in)	1030 mm (40.6 in)
Seat height	765 mm (30.1 in)	765 mm (30.1 in)	765 mm (30.1 in)
Wheelbase	1150 mm (45.3 in)	1215 mm (47.8 in)	1215 mm (47.8 in)
Ground clearance	150 mm (5.9 in)	160 mm (6.3 in)	155 mm (6.1 in)
Dry weight	82 kg (181 lb)	91 kg (201 lb)	93 kg (205 lb)
Kerb weight	92 kg (203 lb)	100 kg (220 lb)	104 kg (229 lb)
Maximum permissible load*	136 kg (300 lb)	136 kg (300 lb)	136 kg (300 lb)

	KE100 A	KE100 B (UK)	KE100 B (US)
Overall length	1980 mm (78.0 in)	2060 mm (81.1 in)	2030 mm (79.9 in)
Overall width	860 mm (33.9 in)	845 mm (33.3 in)	850 mm (33.5 in)
Overall height	1070 mm (42.1 in)	1075 mm (42.3 in)	1080 mm (42.5 in)
Seat height	805 mm (31.7 in)	805 mm (31.7 in)	805 mm (31.7 in)
Wheelbase	1260 mm (49.6 in)	1285 mm (50.6 in)	1285 mm (50.6 in)
Ground clearance	240 mm (9.5 in)	235 mm (9.3 in)	240 mm (9.5 in)
Dry weight	91 kg (201 lb) – A5, A6, A7 92 kg (203 lb) – A8, A9, A10	85 kg (187 lb)	85 kg (187 lb)
Kerb weight	98 kg (216 lb) – A5, A6, A7 99 kg (218 lb) – A8, A9, A10	94 kg (207 lb)	94 kg (207 lb)
Maximum permissible load*	150 kg (331 lb)	150 kg (331 lb)	150 kg (331 lb)

*Loads given are maximum permissible weight of rider, passenger, and accessories or luggage

Ordering spare parts

When ordering spare parts it is advisable to deal direct with an authorized Kawasaki dealer, who will be able to supply many of the items required ex-stock. It is advisable to get acquainted with the local Kawasaki dealer, and to rely on his advice when purchasing spares. He is in a better position to specify exactly the parts required and to identify the relevant spare part numbers so that there is less chance of the wrong part being supplied by the manufacturer due to a vague or incomplete description.

When ordering spares, always quote the frame and engine numbers in full, together with any prefixes or suffixes in the form of letters. The frame number is found stamped on the left or right-hand side of the steering head, in line with the forks. The engine number is stamped on the left-hand side of the crankcase, on the air intake duct.

Use only parts of genuine Kawasaki manufacture. A few pattern parts are available, sometimes at cheaper prices, but there is no guarantee that they will give such good service as the originals they replace. Retain any worn or broken parts until the replacements have been obtained; they are sometimes needed as a pattern to help identify the correct replacement when design changes have been made during a production run.

Some of the more expendable parts such as spark plugs, bulbs, tyres, oils and greases etc., can be obtained from accessory shops and motor factors, who have convenient opening hours, and can often be found not far from home. It is also possible to obtain parts on a mail order basis from a number of specialists who advertise regularly in the motorcycle magazines.

Location of frame number

Location of engine number

Safety first!

Professional motor mechanics are trained in safe working procedures. However enthusiastic you may be about getting on with the job in hand, do take the time to ensure that your safety is not put at risk. A moment's lack of attention can result in an accident, as can failure to observe certain elementary precautions.

There will always be new ways of having accidents, and the following points do not pretend to be a comprehensive list of all dangers; they are intended rather to make you aware of the risks and to encourage a safety-conscious approach to all work you carry out on your vehicle.

Essential DOs and DON'Ts

DON'T start the engine without first ascertaining that the transmission is in neutral.

DON'T suddenly remove the filler cap from a hot cooling system – cover it with a cloth and release the pressure gradually first, or you may get scalded by escaping coolant.

DON'T attempt to drain oil until you are sure it has cooled sufficiently to avoid scalding you.

DON'T grasp any part of the engine, exhaust or silencer without first ascertaining that it is sufficiently cool to avoid burning you.

DON'T allow brake fluid or antifreeze to contact the machine's paintwork or plastic components.

DON'T syphon toxic liquids such as fuel, brake fluid or antifreeze by mouth, or allow them to remain on your skin.

DON'T inhale dust – it may be injurious to health (see *Asbestos* heading).

DON'T allow any spilt oil or grease to remain on the floor – wipe it up straight away, before someone slips on it.

DON'T use ill-fitting spanners or other tools which may slip and cause injury.

DON'T attempt to lift a heavy component which may be beyond your capability – get assistance.

DON'T rush to finish a job, or take unverified short cuts.

DON'T allow children or animals in or around an unattended vehicle.

DON'T inflate a tyre to a pressure above the recommended maximum. Apart from overstressing the carcase and wheel rim, in extreme cases the tyre may blow off forcibly.

DO ensure that the machine is supported securely at all times. This is especially important when the machine is blocked up to aid wheel or fork removal.

DO take care when attempting to slacken a stubborn nut or bolt. It is generally better to pull on a spanner, rather than push, so that if slippage occurs you fall away from the machine rather than on to it.

DO wear eye protection when using power tools such as drill, sander, bench grinder etc.

DO use a barrier cream on your hands prior to undertaking dirty jobs – it will protect your skin from infection as well as making the dirt easier to remove afterwards; but make sure your hands aren't left slippery. Note that long-term contact with used engine oil can be a health hazard.

DO keep loose clothing (cuffs, tie etc) and long hair well out of the way of moving mechanical parts.

DO remove rings, wristwatch etc, before working on the vehicle – especially the electrical system.

DO keep your work area tidy – it is only too easy to fall over articles left lying around.

DO exercise caution when compressing springs for removal or installation. Ensure that the tension is applied and released in a controlled manner, using suitable tools which preclude the possibility of the spring escaping violently.

DO ensure that any lifting tackle used has a safe working load rating adequate for the job.

DO get someone to check periodically that all is well, when working alone on the vehicle.

DO carry out work in a logical sequence and check that everything is correctly assembled and tightened afterwards.

DO remember that your vehicle's safety affects that of yourself and others. If in doubt on any point, get specialist advice.

IF, in spite of following these precautions, you are unfortunate enough to injure yourself, seek medical attention as soon as possible.

Asbestos

Certain friction, insulating, sealing, and other products – such as brake linings, clutch linings, gaskets, etc – contain asbestos. *Extreme care must be taken to avoid inhalation of dust from such products since it is hazardous to health.* If in doubt, assume that they *do* contain asbestos.

Fire

Remember at all times that petrol (gasoline) is highly flammable. Never smoke, or have any kind of naked flame around, when working on the vehicle. But the risk does not end there – a spark caused by an electrical short-circuit, by two metal surfaces contacting each other, by careless use of tools, or even by static electricity built up in your body under certain conditions, can ignite petrol vapour, which in a confined space is highly explosive.

Always disconnect the battery earth (ground) terminal before working on any part of the fuel or electrical system, and never risk spilling fuel on to a hot engine or exhaust.

It is recommended that a fire extinguisher of a type suitable for fuel and electrical fires is kept handy in the garage or workplace at all times. Never try to extinguish a fuel or electrical fire with water.

Note: *Any reference to a 'torch' appearing in this manual should always be taken to mean a hand-held battery-operated electric lamp or flashlight. It does **not** mean a welding/gas torch or blowlamp.*

Fumes

Certain fumes are highly toxic and can quickly cause unconsciousness and even death if inhaled to any extent. Petrol (gasoline) vapour comes into this category, as do the vapours from certain solvents such as trichloroethylene. Any draining or pouring of such volatile fluids should be done in a well ventilated area.

When using cleaning fluids and solvents, read the instructions carefully. Never use materials from unmarked containers – they may give off poisonous vapours.

Never run the engine of a motor vehicle in an enclosed space such as a garage. Exhaust fumes contain carbon monoxide which is extremely poisonous; if you need to run the engine, always do so in the open air or at least have the rear of the vehicle outside the workplace.

The battery

Never cause a spark, or allow a naked light, near the vehicle's battery. It will normally be giving off a certain amount of hydrogen gas, which is highly explosive.

Always disconnect the battery earth (ground) terminal before working on the fuel or electrical systems.

If possible, loosen the filler plugs or cover when charging the battery from an external source. Do not charge at an excessive rate or the battery may burst.

Take care when topping up and when carrying the battery. The acid electrolyte, even when diluted, is very corrosive and should not be allowed to contact the eyes or skin.

If you ever need to prepare electrolyte yourself, always add the acid slowly to the water, and never the other way round. Protect against splashes by wearing rubber gloves and goggles.

Mains electricity and electrical equipment

When using an electric power tool, inspection light etc, always ensure that the appliance is correctly connected to its plug and that, where necessary, it is properly earthed (grounded). Do not use such appliances in damp conditions and, again, beware of creating a spark or applying excessive heat in the vicinity of fuel or fuel vapour. Also ensure that the appliances meet the relevant national safety standards.

Ignition HT voltage

A severe electric shock can result from touching certain parts of the ignition system, such as the HT leads, when the engine is running or being cranked, particularly if components are damp or the insulation is defective. Where an electronic ignition system is fitted, the HT voltage is much higher and could prove fatal.

Tools and working facilities

The first priority when undertaking maintenance or repair work of any sort on a motorcycle is to have a clean, dry, well-lit working area. Work carried out in peace and quiet in the well-ordered atmosphere of a good workshop will give more satisfaction and much better results than can usually be achieved in poor working conditions. A good workshop must have a clean flat workbench or a solidly constructed table of convenient working height. The workbench or table should be equipped with a vice which has a jaw opening of at least 4 in (100 mm). A set of jaw covers should be made from soft metal such as aluminium alloy or copper, or from wood. These covers will minimise the marking or damaging of soft or delicate components which may be clamped in the vice. Some clean, dry, storage space will be required for tools, lubricants and dismantled components. It will be necessary during a major overhaul to lay out engine/gearbox components for examination and to keep them where they will remain undisturbed for as long as is necessary. To this end it is recommended that a supply of metal or plastic containers of suitable size is collected. A supply of clean, lint-free, rags for cleaning purposes and some newspapers, other rags, or paper towels for mopping up spillages should also be kept. If working on a hard concrete floor note that both the floor and one's knees can be protected from oil spillages and wear by cutting open a large cardboard box and spreading it flat on the floor under the machine or workbench. This also helps to provide some warmth in winter and to prevent the loss of nuts, washers, and other tiny components which have a tendency to disappear when dropped on anything other than a perfectly clean, flat, surface.

Unfortunately, such working conditions are not always available to the home mechanic. When working in poor conditions it is essential to take extra time and care to ensure that the components being worked on are kept scrupulously clean and to ensure that no components or tools are lost or damaged.

A selection of good tools is a fundamental requirement for anyone contemplating the maintenance and repair of a motor vehicle. For the owner who does not possess any, their purchase will prove a considerable expense, offsetting some of the savings made by doing-it-yourself. However, provided that the tools purchased meet the relevant national safety standards and are of good quality, they will last for many years and prove an extremely worthwhile investment.

To help the average owner to decide which tools are needed to carry out the various tasks detailed in this manual, we have compiled three lists of tools under the following headings: *Maintenance and minor repair*, *Repair and overhaul*, and *Specialized*. The newcomer to practical mechanics should start off with the simpler jobs around the vehicle. Then, as his confidence and experience grow, he can undertake more difficult tasks, buying extra tools as and when they are needed. In this way, a *Maintenance and minor repair* tool kit can be built-up into a *Repair and overhaul* tool kit over a considerable period of time without any major cash outlays. The experienced home mechanic will have a tool kit good enough for most repair and overhaul procedures and will add tools from the specialized category when he feels the expense is justified by the amount of use these tools will be put to.

It is obviously not possible to cover the subject of tools fully here. For those who wish to learn more about tools and their use there is a book entitled *Motorcycle Workshop Practice Manual* (Book no 1454) available from the publishers of this manual.

As a general rule, it is better to buy the more expensive, good quality tools. Given reasonable use, such tools will last for a very long time, whereas the cheaper, poor quality, item will wear out faster and need to be renewed more often, thus nullifying the original saving. There is also the risk of a poor quality tool breaking while in use, causing personal injury or expensive damage to the component being worked on.

For practically all tools, a tool factor is the best source since he will have a very comprehensive range compared with the average garage or accessory shop. Having said that, accessory shops often offer excellent quality tools at discount prices, so it pays to shop around. There are plenty of tools around at reasonable prices, but always aim to purchase items which meet the relevant national safety standards. If in doubt, seek the advice of the shop proprietor or manager before making a purchase.

The basis of any toolkit is a set of spanners. While open-ended spanners with their slim jaws, are useful for working on awkwardly-positioned nuts, ring spanners have advantages in that they grip the nut far more positively. There is less risk of the spanner slipping off the nut and damaging it, for this reason alone ring spanners are to be preferred. Ideally, the home mechanic should acquire a set of each, but if expense rules this out a set of combination spanners (open-ended at one end and with a ring of the same size at the other) will provide a good compromise. Another item which is so useful it should be

Tools and working facilities

considered an essential requirement for any home mechanic is a set of socket spanners. These are available in a variety of drive sizes. It is recommended that the ½-inch drive type is purchased to begin with as although bulkier and more expensive than the ⅜-inch type, the larger size is far more common and will accept a greater variety of torque wrenches, extension pieces and socket sizes. The socket set should comprise sockets of sizes between 8 and 24 mm, a reversible ratchet drive, an extension bar of about 10 inches in length, a spark plug socket with a rubber insert, and a universal joint. Other attachments can be added to the set at a later date.

Maintenance and minor repair tool kit

    Set of spanners 8 – 24 mm
    Set of sockets and attachments
    Spark plug spanner with rubber insert – 10, 12, or 14 mm as appropriate
    Adjustable spanner
    C-spanner/pin spanner
    Torque wrench (same size drive as sockets)
    Set of screwdrivers (flat blade)
    Set of screwdrivers (cross-head)
    Set of Allen keys 4 – 10 mm
    Impact screwdriver and bits
    Ball pein hammer – 2 lb
    Hacksaw (junior)
    Self-locking pliers – Mole grips or vice grips
    Pliers – combination
    Pliers – needle nose
    Wire brush (small)
    Soft-bristled brush
    Tyre pump
    Tyre pressure gauge
    Tyre tread depth gauge
    Oil can
    Fine emery cloth
    Funnel (medium size)
    Drip tray
    Grease gun
    Set of feeler gauges
    Brake bleeding kit
    Strobe timing light
    Continuity tester (dry battery and bulb)
    Soldering iron and solder
    Wire stripper or craft knife
    PVC insulating tape
    Assortment of split pins, nuts, bolts, and washers

Repair and overhaul toolkit

The tools in this list are virtually essential for anyone undertaking major repairs to a motorcycle and are additional to the tools listed above. Concerning Torx driver bits, Torx screws are encountered on some of the more modern machines where their use is restricted to fastening certain components inside the engine/gearbox unit. It is therefore recommended that if Torx bits cannot be borrowed from a local dealer, they are purchased individually as the need arises. They are not in regular use in the motor trade and will therefore only be available in specialist tool shops.

    Plastic or rubber soft-faced mallet
    Torx driver bits
    Pliers – electrician's side cutters
    Circlip pliers – internal (straight or right-angled tips are available)
    Circlip pliers – external
    Cold chisel
    Centre punch
    Pin punch
    Scriber
    Scraper (made from soft metal such as aluminium or copper)
    Soft metal drift
    Steel rule/straight edge

    Assortment of files
    Electric drill and bits
    Wire brush (large)
    Soft wire brush (similar to those used for cleaning suede shoes)
    Sheet of plate glass
    Hacksaw (large)
    Stud extractor set (E-Z out)

Specialized tools

This is not a list of the tools made by the machine's manufacturer to carry out a specific task on a limited range of models. Occasional references are made to such tools in the text of this manual and, in general, an alternative method of carrying out the task without the manufacturer's tool is given where possible. The tools mentioned in this list are those which are not used regularly and are expensive to buy in view of their infrequent use. Where this is the case it may be possible to hire or borrow the tools against a deposit from a local dealer or tool hire shop. An alternative is for a group of friends or a motorcycle club to join in the purchase.

    Piston ring compressor
    Universal bearing puller
    Cylinder bore honing attachment (for electric drill)
    Micrometer set
    Vernier calipers
    Dial gauge set
    Cylinder compression gauge
    Multimeter
    Dwell meter/tachometer

Care and maintenance of tools

Whatever the quality of the tools purchased, they will last much longer if cared for. This means in practice ensuring that a tool is used for its intended purpose; for example screwdrivers should not be used as a substitute for a centre punch, or as chisels. Always remove dirt or grease and any metal particles but remember that a light film of oil will prevent rusting if the tools are infrequently used. The common tools can be kept together in a large box or tray but the more delicate, and more expensive, items should be stored separately where they cannot be damaged. When a tool is damaged or worn out, be sure to renew it immediately. It is false economy to continue to use a worn spanner or screwdriver which may slip and cause expensive damage to the component being worked on.

Fastening systems

Fasteners, basically, are nuts, bolts and screws used to hold two or more parts together. There are a few things to keep in mind when working with fasteners. Almost all of them use a locking device of some type; either a lock washer, lock nut, locking tab or thread adhesive. All threaded fasteners should be clean, straight, have undamaged threads and undamaged corners on the hexagon head where the spanner fits. Develop the habit of replacing all damaged nuts and bolts with new ones.

Rusted nuts and bolts should be treated with a rust penetrating fluid to ease removal and prevent breakage. After applying the rust penetrant, let it 'work' for a few minutes before trying to loosen the nut or bolt. Badly rusted fasteners may have to be chiseled off or removed with a special nut breaker, available at tool shops.

Flat washers and lock washers, when removed from an assembly should always be replaced exactly as removed. Replace any damaged washers with new ones. Always use a flat washer between a lock washer and any soft metal surface (such as aluminium), thin sheet metal or plastic. Special lock nuts can only be used once or twice before they lose their locking ability and must be renewed.

If a bolt or stud breaks off in an assembly, it can be drilled out and removed with a special tool called an E-Z out. Most dealer service departments and motorcycle repair shops can perform this task, as well as others (such as the repair of threaded holes that have been stripped out).

Conversion factors

Length (distance)

Inches (in)	X	25.4	= Millimetres (mm)	X 0.0394	= Inches (in)
Feet (ft)	X	0.305	= Metres (m)	X 3.281	= Feet (ft)
Miles	X	1.609	= Kilometres (km)	X 0.621	= Miles

Volume (capacity)

Cubic inches (cu in; in^3)	X	16.387	= Cubic centimetres (cc; cm^3)	X 0.061	= Cubic inches (cu in; in^3)
Imperial pints (Imp pt)	X	0.568	= Litres (l)	X 1.76	= Imperial pints (Imp pt)
Imperial quarts (Imp qt)	X	1.137	= Litres (l)	X 0.88	= Imperial quarts (Imp qt)
Imperial quarts (Imp qt)	X	1.201	= US quarts (US qt)	X 0.833	= Imperial quarts (Imp qt)
US quarts (US qt)	X	0.946	= Litres (l)	X 1.057	= US quarts (US qt)
Imperial gallons (Imp gal)	X	4.546	= Litres (l)	X 0.22	= Imperial gallons (Imp gal)
Imperial gallons (Imp gal)	X	1.201	= US gallons (US gal)	X 0.833	= Imperial gallons (Imp gal)
US gallons (US gal)	X	3.785	= Litres (l)	X 0.264	= US gallons (US gal)

Mass (weight)

Ounces (oz)	X	28.35	= Grams (g)	X 0.035	= Ounces (oz)
Pounds (lb)	X	0.454	= Kilograms (kg)	X 2.205	= Pounds (lb)

Force

Ounces-force (ozf; oz)	X	0.278	= Newtons (N)	X 3.6	= Ounces-force (ozf; oz)
Pounds-force (lbf; lb)	X	4.448	= Newtons (N)	X 0.225	= Pounds-force (lbf; lb)
Newtons (N)	X	0.1	= Kilograms-force (kgf; kg)	X 9.81	= Newtons (N)

Pressure

Pounds-force per square inch (psi; lbf/in^2; lb/in^2)	X	0.070	= Kilograms-force per square centimetre (kgf/cm^2; kg/cm^2)	X 14.223	= Pounds-force per square inch (psi; lbf/in^2; lb/in^2)
Pounds-force per square inch (psi; lbf/in^2; lb/in^2)	X	0.068	= Atmospheres (atm)	X 14.696	= Pounds-force per square inch (psi; lbf/in^2; lb/in^2)
Pounds-force per square inch (psi; lbf/in^2; lb/in^2)	X	0.069	= Bars	X 14.5	= Pounds-force per square inch (psi; lbf/in^2; lb/in^2)
Pounds-force per square inch (psi; lbf/in^2; lb/in^2)	X	6.895	= Kilopascals (kPa)	X 0.145	= Pounds-force per square inch (psi; lbf/in^2; lb/in^2)
Kilopascals (kPa)	X	0.01	= Kilograms-force per square centimetre (kgf/cm^2; kg/cm^2)	X 98.1	= Kilopascals (kPa)

Torque (moment of force)

Pounds-force inches (lbf in; lb in)	X	1.152	= Kilograms-force centimetre (kgf cm; kg cm)	X 0.868	= Pounds-force inches (lbf in; lb in)
Pounds-force inches (lbf in; lb in)	X	0.113	= Newton metres (Nm)	X 8.85	= Pounds-force inches (lbf in; lb in)
Pounds-force inches (lbf in; lb in)	X	0.083	= Pounds-force feet (lbf ft; lb ft)	X 12	= Pounds-force inches (lbf in; lb in)
Pounds-force feet (lbf ft; lb ft)	X	0.138	= Kilograms-force metres (kgf m; kg m)	X 7.233	= Pounds-force feet (lbf ft; lb ft)
Pounds-force feet (lbf ft; lb ft)	X	1.356	= Newton metres (Nm)	X 0.738	= Pounds-force feet (lbf ft; lb ft)
Newton metres (Nm)	X	0.102	= Kilograms-force metres (kgf m; kg m)	X 9.804	= Newton metres (Nm)

Power

Horsepower (hp)	X	745.7	= Watts (W)	X 0.0013	= Horsepower (hp)

Velocity (speed)

Miles per hour (miles/hr; mph)	X	1.609	= Kilometres per hour (km/hr; kph)	X 0.621	= Miles per hour (miles/hr; mph)

*Fuel consumption**

Miles per gallon, Imperial (mpg)	X	0.354	= Kilometres per litre (km/l)	X 2.825	= Miles per gallon, Imperial (mpg)
Miles per gallon, US (mpg)	X	0.425	= Kilometres per litre (km/l)	X 2.352	= Miles per gallon, US (mpg)

Temperature

Degrees Fahrenheit = (°C x 1.8) + 32   Degrees Celsius (Degrees Centigrade; °C) = (°F − 32) x 0.56

*It is common practice to convert from miles per gallon (mpg) to litres/100 kilometres (l/100km), where mpg (Imperial) x l/100 km = 282 and mpg (US) x l/100 km = 235

Choosing and fitting accessories

The range of accessories available to the modern motorcyclist is almost as varied and bewildering as the range of motorcycles. This Section is intended to help the owner in choosing the correct equipment for his needs and to avoid some of the mistakes made by many riders when adding accessories to their machines. It will be evident that the Section can only cover the subject in the most general terms and so it is recommended that the owner, having decided that he wants to fit, for example, a luggage rack or carrier, seeks the advice of several local dealers and the owners of similar machines. This will give a good idea of what makes of carrier are easily available, and at what price. Talking to other owners will give some insight into the drawbacks or good points of any one make. A walk round the motorcycles in car parks or outside a dealer will often reveal the same sort of information.

The first priority when choosing accessories is to assess exactly what one needs. It is, for example, pointless to buy a large heavy-duty carrier which is designed to take the weight of fully laden panniers and topbox when all you need is a place to strap on a set of waterproofs and a lunchbox when going to work. Many accessory manufacturers have ranges of equipment to cater for the individual needs of different riders and this point should be borne in mind when looking through a dealer's catalogues. Having decided exactly what is required and the use to which the accessories are going to be put, the owner will need a few hints on what to look for when making the final choice. To this end the Section is now sub-divided to cover the more popular accessories fitted. Note that it is in no way a customizing guide, but merely seeks to outline the practical considerations to be taken into account when adding aftermarket equipment to a motorcycle.

Fairings and windscreens

A fairing is possibly the single, most expensive, aftermarket item to be fitted to any motorcycle and, therefore, requires the most thought before purchase. Fairings can be divided into two main groups: front fork mounted handlebar fairings and windscreens, and frame mounted fairings.

The first group, the front fork mounted fairings, are becoming far more popular than was once the case, as they offer several advantages over the second group. Front fork mounted fairings generally are much easier and quicker to fit, involve less modification to the motorcycle, do not as a rule restrict the steering lock, permit a wider selection of handlebar styles to be used, and offer adequate protection for much less money than the frame mounted type. They are also lighter, can be swapped easily between different motorcycles, and are available in a much greater variety of styles. Their main disadvantages are that they do not offer as much weather protection as the frame mounted types, rarely offer any storage space, and, if poorly fitted or naturally incompatible, can have an adverse effect on the stability of the motorcycle.

The second group, the frame mounted fairings, are secured so rigidly to the main frame of the motorcycle that they can offer a substantial amount of protection to motorcycle and rider in the event of a crash. They offer almost complete protection from the weather and, if double-skinned in construction, can provide a great deal of useful storage space. The feeling of peace, quiet and complete relaxation encountered when riding behind a good full fairing has to be experienced to be believed. For this reason full fairings are considered essential by most touring motorcyclists and by many people who ride all year round. The main disadvantages of this type are that fitting can take a long time, often involving removal or modification of standard motorcycle components, they restrict the steering lock and they can add up to about 40 lb to the weight of the machine. They do not usually affect the stability of the machine to any great extent once the front tyre pressure and suspension have been adjusted to compensate for the extra weight, but can be affected by sidewinds.

The first thing to look for when purchasing a fairing is the quality of the fittings. A good fairing will have strong, substantial brackets constructed from heavy-gauge tubing; the brackets must be shaped to fit the frame or forks evenly so that the minimum of stress is imposed on the assembly when it is bolted down. The brackets should be properly painted or finished – a nylon coating being the favourite of the better manufacturers – the nuts and bolts provided should be of the same thread and size standard as is used on the motorcycle and be properly plated. Look also for shakeproof locking nuts or locking washers to ensure that everything remains securely tightened down. The fairing shell is generally made from one of two materials: fibreglass or ABS plastic. Both have their advantages and disadvantages, but the main consideration for the owner is that fibreglass is much easier to repair in the event of damage occurring to the fairing. Whichever material is used, check that it is properly finished inside as well as out, that the edges are protected by beading and that the fairing shell is insulated from vibration by the use of rubber grommets at all mounting points. Also be careful to check that the windscreen is retained by plastic bolts which will snap on impact so that the windscreen will break away and not cause personal injury in the event of an accident.

Having purchased your fairing or windscreen, read the manufacturer's fitting instructions very carefully and check that you have all the necessary brackets and fittings. Ensure that the mounting brackets are located correctly and bolted down securely. Note that some manufacturers use hose clamps to retain the mounting brackets; these should be discarded as they are convenient to use but not strong enough for the task. Stronger clamps should be substituted; car exhaust pipe clamps of suitable size would be a good alternative. Ensure that the front forks can turn through the full steering lock available without fouling the fairing. With many types of frame-mounted fairing the handlebars will have to be altered or a different type fitted and the steering lock will be restricted by stops provided with the fittings. Also check that the fairing does not foul the front wheel or mudguard, in any steering position, under full fork compression. Re-route any cables, brake pipes or electrical wiring which may snag on the fairing and take great care to protect all electrical connections, using insulating tape. If the manufacturer's instructions are followed carefully at every stage no serious problems should be encountered. Remember that hydraulic pipes that have been disconnected must be carefully re-tightened and the hydraulic system purged of air bubbles by bleeding.

Two things will become immediately apparent when taking a motorcycle on the road for the first time with a fairing – the first is the tendency to underestimate the road speed because of the lack of wind pressure on the body. This must be very carefully watched until one has grown accustomed to riding behind the fairing. The second thing is the alarming increase in engine noise which is an unfortunate but inevitable by-product of fitting any type of fairing or windscreen, and is caused by normal engine noise being reflected, and in some cases amplified, by the flat surface of the fairing.

Luggage racks or carriers

Carriers are possibly the commonest item to be fitted to modern motorcycles. They vary enormously in size, carrying capacity, and durability. When selecting a carrier, always look for one which is made specifically for your machine and which is bolted on with as few separate brackets as possible. The universal-type carrier, with its mass of brackets and adaptor pieces, will generally prove too weak to be of any real use. A good carrier should bolt to the main frame, generally using the two suspension unit top mountings and a mudguard mounting bolt as attachment points, and have its luggage platform as low and as far forward as possible to minimise the effect of any load on the machine's stability. Look for good quality, heavy gauge tubing, good welding and good finish. Also ensure that the carrier does not prevent opening of the seat, sidepanels or tail compartment, as appropriate. When using a carrier, be very careful not to overload it. Excessive weight placed so high and so far to the rear of any motorcycle will have an adverse effect on the machine's steering and stability.

Luggage

Motorcycle luggage can be grouped under two headings: soft and hard. Both types are available in many sizes and styles and have advantages and disadvantages in use.

Soft luggage is now becoming very popular because of its lower cost and its versatility. Whether in the form of tankbags, panniers, or strap-on bags, soft luggage requires in general no brackets and no modification to the motorcycle. Equipment can be swapped easily from one motorcycle to another and can be fitted and removed in seconds. Awkwardly shaped loads can easily be carried. The disadvantages of soft luggage are that the contents cannot be secure against the casual thief, very little protection is afforded in the event of a crash, and waterproofing is generally poor. Also, in the case of panniers, carrying capacity is restricted to approximately 10 lb, although this amount will vary considerably depending on the manufacturer's recommendation. When purchasing soft luggage, look for good quality material, generally vinyl or nylon, with strong, well-stitched attachment points. It is always useful to have separate pockets, especially on tank bags, for items which will be needed on the journey. When purchasing a tank bag, look for one which has a separate, well-padded, base. This will protect the tank's paintwork and permit easy access to the filler cap at petrol stations.

Hard luggage is confined to two types: panniers, and top boxes or tail trunks. Most hard luggage manufacturers produce matching sets of these items, the basis of which is generally that manufacturer's own heavy-duty luggage rack. Variations on this theme occur in the form of separate frames for the better quality panniers, fixed or quickly-detachable luggage, and in size and carrying capacity. Hard luggage offers a reasonable degree of security against theft and good protection against weather and accident damage. Carrying capacity is greater than that of soft luggage, around 15 – 20 lb in the case of panniers, although top boxes should never be loaded as much as their apparent capacity might imply. A top box should only be used for lightweight items, because one that is heavily laden can have a serious effect on the stability of the machine. When purchasing hard luggage look for the same good points as mentioned under fairings and windscreens, ie good quality mounting brackets and fittings, and well-finished fibreglass or ABS plastic cases. Again as with fairings, always purchase luggage made specifically for your motorcycle, using as few separate brackets as possible, to ensure that everything remains securely bolted in place. When fitting hard luggage, be careful to check that the rear suspension and brake operation will not be impaired in any way and remember that many pannier kits require re-siting of the indicators. Remember also that a non-standard exhaust system may make fitting extremely difficult.

Handlebars

The occupation of fitting alternative types of handlebar is extremely popular with modern motorcyclists, whose motives may vary from the purely practical, wishing to improve the comfort of their machines, to the purely aesthetic, where form is more important than function. Whatever the reason, there are several considerations to be borne in mind when changing the handlebars of your machine. If fitting lower bars, check carefully that the switches and cables do not foul the petrol tank on full lock and that the surplus length of cable, brake pipe, and electrical wiring are smoothly and tidily disposed of. Avoid tight kinks in cable or brake pipes which will produce stiff controls or the premature and disastrous failure of an overstressed component. If necessary, remove the petrol tank and re-route the cable from the engine/gearbox unit upwards, ensuring smooth gentle curves are produced. In extreme cases, it will be necessary to purchase a shorter brake pipe to overcome this problem. In the case of higher handlebars than standard it will almost certainly be necessary to purchase extended cables and brake pipes. Fortunately, many standard motorcycles have a custom version which will be equipped with higher handlebars and, therefore, factory-built extended components will be available from your local dealer. It is not usually necessary to extend electrical wiring, as switch clusters may be used on several different motorcycles, some being custom versions. This point should be borne in mind however when fitting extremely high or wide handlebars.

When fitting different types of handlebar, ensure that the mounting clamps are correctly tightened to the manufacturer's specifications and that cables and wiring, as previously mentioned, have smooth easy runs and do not snag on any part of the motorcycle throughout the full steering lock. Ensure that the fluid level in the front brake master cylinder remains level to avoid any chance of air entering the hydraulic system. Also check that the cables are adjusted correctly and that all handlebar controls operate correctly and can be easily reached when riding.

Crashbars

Crashbars, also known as engine protector bars, engine guards, or case savers, are extremely useful items of equipment which can contribute protection to the machine's structure if a crash occurs. They do not, as has been inferred in the US, prevent the rider from crashing, or necessarily prevent rider injury should a crash occur.

It is recommended that only the smaller, neater, engine protector type of crashbar is considered. This type will offer protection while restricting, as little as is possible, access to the engine and the machine's ground clearance. The crashbars should be designed for use specifically on your machine, and should be constructed of heavy-gauge tubing with strong, integral mounting brackets. Where possible, they should bolt to a strong lug on the frame, usually at the engine mounting bolts.

The alternative type of crashbar is the larger cage type. This type is not recommended in spite of their appearance which promises some protection to the rider as well as to the machine. The larger amount of leverage imposed by the size of this type of crashbar increases the risk of severe frame damage in the event of an accident. This type also decreases the machine's ground clearance and restricts access to the engine. The amount of protection afforded the rider is open to some doubt as the design is based on the premise that the rider will stay in the normally seated position during an accident, and the crash bar structure will not itself fail. Neither result can in any way be guaranteed.

As a general rule, always purchase the best, ie usually the most expensive, set of crashbars you an afford. The investment will be repaid by minimising the amount of damage incurred, should the machine be involved in an accident. Finally, avoid the universal type of crashbar. This should be regarded only as a last resort to be used if no alternative exists. With its usual multitude of separate brackets and spacers, the universal crashbar is far too weak in design and construction to be of any practical value.

Exhaust systems

The fitting of aftermarket exhaust systems is another extremely popular pastime amongst motorcyclists. The usual motive is to gain more performance from the engine but other considerations are to gain more ground clearance, to lose weight from the motorcycle, to obtain a more distinctive exhaust note or to find a cheaper alternative to the manufacturer's original equipment exhaust system. Original equipment exhaust systems often cost more and may well have a relatively short life. It should be noted that it is rare for an aftermarket exhaust system alone to give a noticeable increase in the engine's power output. Modern motorcycles are designed to give the highest power output possible allowing for factors such as quietness, fuel economy, spread of power, and long-term reliability. If there were a magic formula which allowed the exhaust system to produce more power without affecting these other considerations you can be sure that the manufacturers, with their large research and development facilities, would have found it and made use of it. Performance increases of a worthwhile and noticeable nature only come from well-tried and properly matched modifications to the entire engine, from the air filter, through the carburettors, port timing or camshaft and valve design, combustion chamber shape, compression ratio, and the exhaust system. Such modifications are well outside the scope of this manual but interested owners might refer to specialist books produced by the publisher of this manual which go into the whole subject in great detail.

Whatever your motive for wishing to fit an alternative exhaust system, be sure to seek expert advice before doing so. Changes to the carburettor jetting will almost certainly be required for which you must consult the exhaust system manufacturer. If he cannot supply adequately specific information it is reasonable to assume that insufficient development work has been carried out, and that particular make should be avoided. Other factors to be borne in mind are whether the exhaust system allows the use of both centre and side stands, whether it allows sufficient access to permit oil and filter changing and whether modifications are necessary to the standard exhaust system. Many two-stroke expansion chamber systems require the use of the standard exhaust pipe; this is all very well if the standard exhaust pipe and silencer are separate units but can cause problems if the two, as with so many modern two-strokes, are a one-piece unit. While the exhaust pipe can be removed easily by means of a hacksaw it is not so easy to refit the original silencer should you at any time wish to return the machine to standard trim. The same applies to several four-stroke systems.

On the subject of the finish of aftermarket exhausts, avoid black-painted systems unless you enjoy painting. As any trail-bike owner will tell you, rust has a great affinity for black exhausts and re-painting or rust removal becomes a task which must be carried out with monotonous regularity. A bright chrome finish is, as a general rule, a far better proposition as it is much easier to keep clean and to prevent rusting. Although the general finish of aftermarket exhaust systems is not always up to the standard of the original equipment the lower cost of such systems does at least reflect this fact.

When fitting an alternative system always purchase a full set of new exhaust gaskets, to prevent leaks. Fit the exhaust first to the cylinder head or barrel, as appropriate, tightening the retaining nuts or bolts by hand only and then line up the exhaust rear mountings. If the new system is a one-piece unit and the rear mountings do not line up exactly, spacers must be fabricated to take up the difference. Do not force the system into place as the stress thus imposed will rapidly cause cracks and splits to appear. Once all the mountings are loosely fixed, tighten the retaining nuts or bolts securely, being careful not to overtighten them. Where the motorcycle manufacturer's torque settings are available, these should be used. Do not forget to carry out any carburation changes recommended by the exhaust system's manufacturer.

Electrical equipment

The vast range of electrical equipment available to motorcyclists is so large and so diverse that only the most general outline can be given here. Electrical accessories vary from electric ignition kits fitted to replace contact breaker points, to additional lighting at the front and rear, more powerful horns, various instruments and gauges, clocks, anti-theft systems, heated clothing, CB radios, radio-cassette players, and intercom systems, to name but a few of the more popular items of equipment.

As will be evident, it would require a separate manual to cover this subject alone and this section is therefore restricted to outlining a few basic rules which must be borne in mind when fitting electrical equipment. The first consideration is whether your machine's electrical system has enough reserve capacity to cope with the added demand of the accessories you wish to fit. The motorcycle's manufacturer or importer should be able to furnish this sort of information and may also be able to offer advice on uprating the electrical system. Failing this, a good dealer or the accessory manufacturer may be able to help. In some cases, more powerful generator components may be available, perhaps from another motorcycle in the manufacturer's range. The second consideration is the legal requirements in force in your area. The local police may be prepared to help with this point. In the UK for example, there are strict regulations governing the position and use of auxiliary riding lamps and fog lamps.

When fitting electrical equipment always disconnect the battery first to prevent the risk of a short-circuit, and be careful to ensure that all connections are properly made and that they are waterproof. Remember that many electrical accessories are designed primarily for use in cars and that they cannot easily withstand the exposure to vibration and to the weather. Delicate components must be rubber-mounted to insulate them from vibration, and sealed carefully to prevent the entry of rainwater and dirt. Be careful to follow exactly the accessory manufacturer's instructions in conjunction with the wiring diagram at the back of this manual.

Accessories – general

Accessories fitted to your motorcycle will rapidly deteriorate if not cared for. Regular washing and polishing will maintain the finish and will provide an opportunity to check that all mounting bolts and nuts are securely fastened. Any signs of chafing or wear should be watched for, and the cause cured as soon as possible before serious damage occurs.

As a general rule, do not expect the re-sale value of your motorcycle to increase by an amount proportional to the amount of money and effort put into fitting accessories. It is usually the case that an absolutely standard motorcycle will sell more easily at a better price than one that has been modified. If you are in the habit of exchanging your machine for another at frequent intervals, this factor should be borne in mind to avoid loss of money.

Fault diagnosis

Contents

Introduction	1

Engine does not start when turned over
No fuel flow to carburettor	2
Fuel not reaching cylinder	3
Engine flooding	4
No spark at plug	5
Weak spark at plug	6
Compression low	7

Engine stalls after starting
General causes	8

Poor running at idle and low speed
Weak spark at plug or erratic firing	9
Fuel/air mixture incorrect	10
Compression low	11

Acceleration poor
General causes	12

Poor running or lack of power at high speeds
Weak spark at plug or erratic firing	13
Fuel/air mixture incorrect	14
Compression low	15

Knocking or pinking
General causes	16

Overheating
Firing incorrect	17
Fuel/air mixture incorrect	18
Lubrication inadequate	19
Miscellaneous causes	20

Clutch operating problems
Clutch slip	21
Clutch drag	22

Gear selection problems
Gear lever does not return	23
Gear selection difficult or impossible	24
Jumping out of gear	25
Overselection	26

Abnormal engine noise
Knocking or pinking	27
Piston slap or rattling from cylinder	28
Other noises	29

Abnormal transmission noise
Clutch noise	30
Transmission noise	31

Exhaust smokes excessively
White/blue smoke (caused by oil burning)	32
Black smoke (caused by over-rich mixture)	33

Poor handling or roadholding
Directional instability	34
Steering bias to left or right	35
Handlebar vibrates or oscillates	36
Poor front fork performance	37
Front fork judder when braking	38
Poor rear suspension performance	39

Abnormal frame and suspension noise
Front end noise	40
Rear suspension noise	41

Brake problems
Brakes are spongy or ineffective – disc brakes	42
Brakes drag – disc brakes	43
Brake lever or pedal pulsates in operation – disc brakes	44
Disc brake noise	45
Brakes are spongy or ineffective – drum brakes	46
Brake drag – drum brakes	47
Brake lever or pedal pulsates in operation – drum brakes	48
Drum brake noise	49
Brake induced fork judder	50

Electrical problems
Battery dead or weak	51
Battery overcharged	52
Total electrical failure	53
Circuit failure	54
Bulbs blowing repeatedly	55

Fault diagnosis

1 Introduction

This Section provides an easy reference-guide to the more common faults that are likely to afflict your machine. Obviously, the opportunities are almost limitless for faults to occur as a result of obscure failures, and to try and cover all eventualities would require a book. Indeed, a number have been written on the subject.

Successful fault diagnosis is not a mysterious 'black art' but the application of a bit of knowledge combined with a systematic and logical approach to the problem. Approach any fault diagnosis by first accurately identifying the symptom and then checking through the list of possible causes, starting with the simplest or most obvious and progressing in stages to the most complex. Take nothing for granted, but above all apply liberal quantities of common sense.

The main symptom of a fault is given in the text as a major heading below which are listed, as Section headings, the various systems or areas which may contain the fault. Details of each possible cause for a fault and the remedial action to be taken are given, in brief, in the paragraphs below each Section heading. Further information should be sought in the relevant Chapter.

Engine does not start when turned over

2 No fuel flow to carburettor

- Fuel tank empty or level too low. Check that the tap is turned to 'On' or 'Reserve' position as required. If in doubt, prise off the fuel feed pipe at the carburettor end and check that fuel runs from the pipe when the tap is turned on.
- Tank filler cap vent obstructed. This can prevent fuel from flowing into the carburettor float bowl because air cannot enter the fuel tank to replace it. The problem is more likely to appear when the machine is being ridden. Check by listening close to the filler cap and releasing it. A hissing noise indicates that a blockage is present. Remove the cap and clear the vent hole with wire or by using an air line from the inside of the cap.
- Fuel tap or filter blocked. Blockage may be due to accumulation of rust or paint flakes from the tank's inner surface or of foreign matter from contaminated fuel. Remove the tap and clean it and the filter. Look also for water droplets in the fuel.
- Fuel line blocked. Blockage of the fuel line is more likely to result from a kink in the line rather than the accumulation of debris.

3 Fuel not reaching cylinder

- Float chamber not filling. Caused by float needle or floats sticking in up position. This may occur after the machine has been left standing for an extended length of time allowing the fuel to evaporate. When this occurs a gummy residue is often left which hardens to a varnish-like substance. This condition may be worsened by corrosion and crystalline deposits produced prior to the total evaporation of contaminated fuel. Sticking of the float needle may also be caused by wear. In any case removal of the float chamber will be necessary for inspection and cleaning.
- Blockage in starting circuit, slow running circuit or jets. Blockage of these items may be attributable to debris from the fuel tank by-passing the filter system or to gumming up as described in paragraph 1. Water droplets in the fuel will also block jets and passages. The carburettor should be dismantled for cleaning.
- Fuel level too low. The fuel level in the float chamber is controlled by float height. The fuel level may increase with wear or damage but will never reduce, thus a low fuel level is an inherent rather than developing condition. Check the float height, renewing the float or needle if required.

4 Engine flooding

- Float valve needle worn or stuck open. A piece of rust or other debris can prevent correct seating of the needle against the valve seat thereby permitting an uncontrolled flow of fuel. Similarly, a worn needle or needle seat will prevent valve closure. Dismantle the carburettor float bowl for cleaning and, if necessary, renewal of the worn components.
- Fuel level too high. The fuel level is controlled by the float height which may increase due to wear of the float needle, pivot pin or operating tang. Check the float height, and make any necessary adjustments. A leaking float will cause an increase in fuel level, and thus should be renewed.
- Cold starting mechanism. Check the choke (starter mechanism) for correct operation. If the mechanism jams in the 'On' position subsequent starting of a hot engine will be difficult.
- Blocked air filter. A badly restricted air filter will cause flooding. Check the filter and clean or renew as required. A collapsed inlet hose will have a similar effect. Check that the air filter inlet has not become blocked by a rag or similar item.

5 No spark at plug

- Ignition switch not on.
- Engine stop switch off.
- Spark plug dirty, oiled or 'whiskered'. Because the induction mixture of a two-stroke engine is inclined to be of a rather oily nature it is comparatively easy to foul the plug electrodes, especially where there have been repeated attempts to start the engine. A machine used for short journeys will be more prone to fouling because the engine may never reach full operating temperature, and the deposits will not burn off. On rare occasions a change of plug grade may be required but the advice of a dealer should be sought before making such a change. 'Whiskering' is a comparatively rare occurrence on modern machines but may be encountered where pre-mixed petrol and oil (petroil) lubrication is employed. An electrode deposit in the form of a barely visible filament across the plug electrodes can short circuit the plug and prevent its sparking. On all two-stroke machines it is a sound precaution to carry a new spare spark plug for substitution in the event of fouling problems.
- Spark plug failure. Clean the spark plug thoroughly and reset the electrode gap. Refer to the spark plug section and the colour condition guide in Routine Maintenance. If the spark plug shorts internally or has sustained visible damage to the electrodes, core or ceramic insulator it should be renewed. On rare occasions a plug that appears to spark vigorously will fail to do so when refitted to the engine and subjected to the compression pressure in the cylinder.
- Spark plug cap or high tension (HT) lead faulty. Check condition and security. Replace if deterioration is evident. Most spark plug caps have an internal resistor designed to inhibit electrical interference with radio and television sets. On rare occasions the resistor may break down, thus preventing sparking. If this is suspected, fit a new cap as a precaution.
- Spark plug cap loose. Check that the spark plug cap fits securely over the plug and, where fitted, the screwed terminal on the plug end is secure.
- Shorting due to moisture. Certain parts of the ignition system are susceptible to shorting when the machine is ridden or parked in wet weather. Check particularly the area from the spark plug cap back to the ignition coil. A water dispersant spray may be used to dry out waterlogged components. Recurrence of the problem can be prevented by using an ignition sealant spray after drying out and cleaning.
- Ignition or stop switch shorted. May be caused by water corrosion or wear. Water dispersant and contact cleaning sprays may be used. If this fails to overcome the problem dismantling and visual inspection of the switches will be required.
- Shorting or open circuit in wiring. Failure in any wire connecting any of the ignition components will cause ignition malfunction. Check also that all connections are clean, dry and tight.
- Ignition coil failure. Check the coil, referring to Chapter 3.
- Capacitor (condenser) failure. The capacitor may be checked most easily by substitution with a replacement item. Blackened contact breaker points indicate capacitor malfunction but this may not always occur.
- Contact breaker points pitted, burned or closed up. Check the contact breaker points, referring to Routine Maintenance. Check also that the low tension leads at the contact breaker are secure and not shorting out.

6 Weak spark at plug

● Feeble sparking at the plug may be caused by any of the faults mentioned in the preceding Section other than those items in the first three paragraphs. Check first the contact breaker assembly and the spark plug, these being the most likely culprits.

7 Compression low

● Spark plug loose. This will be self-evident on inspection, and may be accompanied by a hissing noise when the engine is turned over. Remove the plug and check that the threads in the cylinder head are not damaged. Check also that the plug sealing washer is in good condition.
● Cylinder head gasket leaking. This condition is often accompanied by a high pitched squeak from around the cylinder head and oil loss, and may be caused by insufficiently tightened cylinder head fasteners, a warped cylinder head or mechanical failure of the gasket material. Re-torqueing the fasteners to the correct specification may seal the leak in some instances but if damage has occurred this course of action will provide, at best, only a temporary cure.
● Low crankcase compression. This can be caused by worn main bearings and seals and will upset the incoming fuel/air mixture. A good seal in these areas is essential on any two-stroke engine.
● Worn disc valve. Disc valve wear is not common, but will cause similar symptoms to those described above. Overhaul will be necessary.
● Piston rings sticking or broken. Sticking of the piston rings may be caused by seizure due to lack of lubrication or overheating as a result of poor carburation or incorrect fuel type. Gumming of the rings may result from lack of use, or carbon deposits in the ring grooves. Broken rings result from over-revving, over-heating or general wear. In either case a top-end overhaul will be required.

Engine stalls after starting

8 General causes

● Improper cold start mechanism operation. Check that the operating controls function smoothly and, where applicable, are correctly adjusted. A cold engine may not require application of an enriched mixture to start initially but may baulk without choke once firing. Likewise a hot engine may start with an enriched mixture but will stop almost immediately if the choke is inadvertently in operation.
● Ignition malfunction. See Section 9. Weak spark at plug.
● Carburettor incorrectly adjusted. Maladjustment of the mixture strength or idle speed may cause the engine to stop immediately after starting. See Chapter 2.
● Fuel contamination. Check for filter blockage by debris or water which reduces, but does not completely stop, fuel flow, or blockage of the slow speed circuit in the carburettor by the same agents. If water is present it can often be seen as droplets in the bottom of the float bowl. Clean the filter and, where water is in evidence, drain and flush the fuel tank and float bowl.
● Intake air leak. Check for security of the carburettor mounting and hose connections, and for cracks or splits in the hoses. Check also that the carburettor cover is secure.
● Air filter blocked or omitted. A blocked filter will cause an over-rich mixture; the omission of a filter will cause an excessively weak mixture. Both conditions will have a detrimental effect on carburation. Clean or renew the filter as necessary.
● Fuel filler cap air vent blocked. Usually caused by dirt or water. Clean the vent orifice.
● Choked exhaust system. Caused by excessive carbon build-up in the system, particularly around the silencer baffles. Refer to Routine Maintenance for further information.
● Excessive carbon build-up in the engine. This can result from failure to decarbonise the engine at the specified interval or through excessive oil consumption. Check pump adjustment.

Poor running at idle and low speed

9 Weak spark at plug or erratic firing

● Spark plug fouled, faulty or incorrectly adjusted. See Section 4 or refer to Routine Maintenance.
● Spark plug cap or high tension lead shorting. Check the condition of both these items ensuring that they are in good condition and dry and that the cap is fitted correctly.
● Spark plug type incorrect. Fit plug of correct type and heat range as given in Specifications. In certain conditions a plug of hotter or colder type may be required for normal running.
● Contact breaker points pitted, burned or closed-up. Check the contact breaker assembly, referring to Routine Maintenance.
● Ignition timing incorrect. Check the ignition timing and reset it if necessary. See Routine Maintenance.
● Faulty ignition coil. Partial failure of the coil internal insulation will diminish the performance of the coil. No repair is possible, a new component must be fitted.
● Faulty capacitor (condenser). A failure of the capacitor will cause blackening of the contact breaker point faces and will allow excessive sparking at the points. A faulty capacitor may best be checked by substitution of a serviceable replacement item.
● Defective flywheel generator ignition source. Refer to Chapter 3 for further details on test procedures.

10 Fuel/air mixture incorrect

● Intake air leak. Check carburettor mountings and air cleaner hoses for security and signs of splitting. Ensure that carburettor cover is fully sealed.
● Mixture strength incorrect. Adjust slow running mixture strength using pilot adjustment screw.
● Pilot jet or slow running circuit blocked. The carburettor should be removed and dismantled for thorough cleaning. Blow through all jets and air passages with compressed air to clear obstructions.
● Air cleaner clogged or omitted. Clean or fit air cleaner element as necessary. Check also that the element and air filter cover are correctly seated.
● Cold start mechanism in operation. Check that the choke has not been left on inadvertently and the operation is correct. Where applicable check the operating cable free play.
● Fuel level too high or too low. Check the float height, renewing float or needle if required. See Section 3 or 4.
● Fuel tank air vent obstructed. Obstructions usually caused by dirt or water. Clean vent orifice.

11 Compression low

● See Section 7.

Acceleration poor

12 General causes

● All items as for previous Section.
● Choked air filter. Failure to keep the air filter element clean will allow the build-up of dirt with proportional loss of performance. In extreme cases of neglect acceleration will suffer.
● Choked exhaust system. This can result from failure to remove accumulations of carbon from the silencer baffles at the prescribed intervals. The increased back pressure will make the machine noticeably sluggish. Refer to Routine Maintenance for further information on decarbonisation.
● Excessive carbon build-up in the engine. This can result from failure to decarbonise the engine at the specified interval or through excessive oil consumption. Check pump adjustment.

Fault diagnosis

● Ignition timing incorrect. Check the contact breaker gap and set within the prescribed range ensuring that the ignition timing is correct. If the contact breaker assembly is worn it may prove impossible to get the gap and timing settings to coincide, necessitating renewal.
● Ignition timing incorrect. Check the ignition timing as described in Routine Maintenance.
● Carburation fault. See Section 10.
● Mechanical resistance. Check that the brakes are not binding. On small machines in particular note that the increased rolling resistance caused by under-inflated tyres may impede acceleration.

Poor running or lack of power at high speeds

13 Weak spark at plug or erratic firing

● All items as for Section 9.
● HT lead insulation failure. Insulation failure of the HT lead and spark plug cap due to old age or damage can cause shorting when the engine is driven hard. This condition may be less noticeable, or not noticeable at all at lower engine speeds.

14 Fuel/air mixture incorrect

● All items as for Section 10, with the exception of items relative exclusively to low speed running.
● Main jet blocked. Debris from contaminated fuel, or from the fuel tank, and water in the fuel can block the main jet. Clean the fuel filter, the float bowl area, and if water is present, flush and refill the fuel tank.
● Main jet is the wrong size. The standard carburettor jetting is for sea level atmospheric pressure. For high altitudes, usually above 5000 ft, a smaller main jet will be required.
● Jet needle and needle jet worn. These can be renewed individually but should be renewed as a pair. Renewal of both items requires partial dismantling of the carburettor.
● Air bleed holes blocked. Dismantle carburettor and use compressed air to blow out all air passages.
● Reduced fuel flow. A reduction in the maximum fuel flow from the fuel tank to the carburettor will cause fuel starvation, proportionate to the engine speed. Check for blockages through debris or a kinked fuel line.

15 Compression low

● See Section 7.

Knocking or pinking

16 General causes

● Carbon build-up in combustion chamber. After a high mileage has been covered large accumulations of carbon may occur. These may glow red hot and cause premature ignition of the fuel/air mixture, in advance of normal firing by the spark plug. Cylinder head removal will be required to allow inspection and cleaning.
● Fuel incorrect. A low grade fuel, or one of poor quality may result in compression induced detonation of the fuel resulting in knocking and pinking noises. Old fuel can cause similar problems. A too highly leaded fuel will reduce detonation but will accelerate deposit formation in the combustion chamber and may lead to early pre-ignition as described in item 1.
● Spark plug heat range incorrect. Uncontrolled pre-ignition can result from the use of a spark plug the heat range of which is too hot.
● Weak mixture. Overheating of the engine due to a weak mixture can result in pre-ignition occurring where it would not occur when engine temperature was within normal limits. Maladjustment, blocked jets or passages and air leaks can cause this condition.

Overheating

17 Firing incorrect

● Spark plug fouled, defective or maladjusted. See Section 5.
● Spark plug type incorrect. Refer to the Specifications and ensure that the correct plug type is fitted.
● Incorrect ignition timing. Timing that is far too much advanced or far too much retarded will cause overheating. Check the ignition timing is correct.

18 Fuel/air mixture incorrect

● Slow speed mixture strength incorrect. Adjust pilot air screw.
● Main jet wrong size. The carburettor is jetted for sea level atmospheric conditions. For high altitudes, usually above 5000 ft, a smaller main jet will be required.
● Air filter badly fitted or omitted. Check that the filter element is in place and that it and the air filter box cover are sealing correctly. Any leaks will cause a weak mixture.
● Induction air leaks. Check the security of the carburettor mountings and hose connections, and for cracks and splits in the hoses. Check also that the carburettor cover is secure.
● Fuel level too low. See Section 3.
● Fuel tank filler cap air vent obstructed. Clear blockage.

19 Lubrication inadequate

● Oil pump settings incorrect. The oil pump settings are of great importance since the quantities of oil being injected are very small. Any variation in oil delivery will have a significant effect on the engine. Refer to Routine Maintenance for further information.
● Oil tank empty or low. This will have disastrous consequences if left unnoticed. Check and replenish tank regularly.
● Transmission oil low or worn out. Check the level regularly and investigate any loss of oil. If the oil level drops with no sign of external leakage it is likely that the crankshaft main bearing oil seals are worn, allowing transmission oil to be drawn into the crankcase during induction.

20 Miscellaneous causes

● Engine fins clogged. A build-up of mud in the cylinder head and cylinder barrel cooling fins will decrease the cooling capabilities of the fins. Clean the fins as required.

Clutch operating problems

21 Clutch slip

● No clutch lever play. Adjust clutch lever end play according to the procedure in Routine Maintenance.
● Friction plates worn or warped. Overhaul clutch assembly, replacing plates out of specification.
● Steel plates worn or warped. Overhaul clutch assembly, replacing plates out of specification.
● Clutch spring broken or worn. Old or heat-damaged (from slipping clutch) springs should be replaced with new ones.
● Clutch inner cable snagging. Caused by a frayed cable or kinked outer cable. Replace the cable with a new one. Repair of a frayed cable is not advised.
● Clutch release mechanism defective. Worn or damaged parts in the clutch release mechanism could include the thrust bearing, actuating arm or worm. Replace parts as necessary.
● Clutch centre and outer drum worn. Severe indentation by the clutch plate tangs of the channels in the centre and drum will cause

snagging of the plates preventing correct engagement. If this damage occurs, renewal of the worn components is required.
● Lubricant incorrect. Use of a transmission lubricant other than that specified may allow the plates to slip.

22 Clutch drag

● Clutch lever play excessive. Adjust lever at bars or at cable end if necessary.
● Clutch plates warped or damaged. This will cause a drag on the clutch, causing the machine to creep. Overhaul clutch assembly.
● Clutch spring tension uneven. Usually caused by a sagged or broken spring. Check and replace springs.
● Transmission oil deteriorated. Badly contaminated transmission oil and a heavy deposit of oil sludge on the plates will cause plate sticking. The oil recommended for this machine is of the detergent type, therefore it is unlikely that this problem will arise unless regular oil changes are neglected.
● Transmission oil viscosity too high. Drag in the plates will result from the use of an oil with too high a viscosity. In very cold weather clutch drag may occur until the engine has reached operating temperature.
● Clutch centre and outer drum worn. Indentation by the clutch plate tangs of the channels in the centre and drum will prevent easy plate disengagement. If the damage is light the affected areas may be dressed with a fine file. More pronounced damage will necessitate renewal of the components.
● Clutch housing seized to shaft. Lack of lubrication, severe wear or damage can cause the housing to seize to the shaft. Overhaul of the clutch, and perhaps the transmission, may be necessary to repair damage.
● Clutch release mechanism defective. Worn or damaged release mechanism parts can stick and fail to provide leverage. Overhaul clutch cover components.

Gear selection problems

23 Gear lever does not return

● Weak or broken return spring. Renew the spring.
● Gearchange shaft bent or seized. Distortion of the gearchange shaft often occurs if the machine is dropped heavily on the gear lever. Provided that damage is not severe straightening of the shaft is permissible.

24 Gear selection difficult or impossible

● Clutch not disengaging fully. See Section 22.
● Gearchange shaft bent. This often occurs if the machine is dropped heavily on the gear lever. Straightening of the shaft is permissible if the damage is not too great.
● Gearchange arms, pawls or pins worn or damaged. Wear or breakage of any of these items may cause difficulty in selecting one or more gears. Overhaul the selector mechanism.
● Gearchange shaft return spring maladjusted (where applicable). This is often characterised by difficulties in changing up or down, but rarely in both directions. Adjust the spring anchor bolt as described in Chapter 1.
● Selector drum detent roller arm or plunger damaged. Failure, rather than wear of these items may jam the drum thereby preventing gearchanging or causing false selection at high speed.
● Selector forks bent or seized. This can be caused by dropping the machine heavily on the gearchange lever or as a result of lack of lubrication. Though rare, bending of a shaft can result from a missed gearchange or false selection at high speed.
● Selector fork end and pin wear. Pronounced wear of these items and the grooves in the selector drum can lead to imprecise selection and, eventually, no selection. Renewal of the worn components will be required.
● Structural failure. Failure of any one component of the selector mechanism will result in improper or fouled gear selection.

25 Jumping out of gear

● Detent assembly worn or damaged. Wear of the plunger or roller arm and the cam with which it locates and breakage of the detent spring can cause imprecise gear selection resulting in jumping out of gear. Renew the damaged components.
● Gear pinion dogs worn or damaged. Rounding off the dog edges and the mating recesses in adjacent pinion can lead to jumping out of gear when under load. The gears should be inspected and renewed. Attempting to reprofile the dogs is not recommended.
● Selector forks, drum and pinion grooves worn. Extreme wear of these interconnected items can occur after high mileages especially when lubrication has been neglected. The worn components must be renewed.
● Gear pinions, bushes and shafts worn. Renew the worn components.
● Bent gearchange shaft. Often caused by dropping the machine on the gear lever.
● Gear pinion tooth broken. Chipped teeth are unlikely to cause jumping out of gear once the gear has been selected fully; a tooth which is completely broken off, however, may cause problems in this respect and in any event will cause transmission noise.

26 Overselection

● Pawl spring weak or broken. Renew the spring.
● Detent plunger worn or broken. Renew the damaged items.
● Stopper arm spring worn or broken. Renew the spring.
● Gearchange arm stop pads worn. Repairs can be made by welding and reprofiling with a file.

Abnormal engine noise

27 Knocking or pinking

● See Section 16.

28 Piston slap or rattling from cylinder

● Cylinder bore/piston clearance excessive. Resulting from wear, or partial seizure. This condition can often be heard as a high, rapid tapping noise when the engine is under little or no load, particularly when power is just beginning to be applied. If possible, the cylinder must be rebored and an oversize piston fitted, but in some cases the cylinder barrel and piston must be renewed.
● Connecting rod bent. This can be caused by over-revving, trying to start a very badly flooded engine (resulting in a hydraulic lock in the cylinder) or by earlier mechanical failure. Attempts at straightening a bent connecting rod are not recommended. Careful inspection of the crankshaft should be made before renewing the damaged connecting rod.
● Gudgeon pin, piston boss bore or small-end bearing wear or seizure. Excess clearance or partial seizure between normal moving parts of these items can cause continuous or intermittent tapping noises. Rapid wear or seizure is caused by lubrication starvation.
● Piston rings worn, broken or sticking. Renew the rings after careful inspection of the piston and bore.

29 Other noises

● Big-end bearing wear. A pronounced knock from within the crankcase which worsens rapidly is indicative of big-end bearing failure as a result of extreme normal wear or lubrication failure. Remedial action in the form of a bottom end overhaul should be taken; continuing to run the engine will lead to further damage including the possibility of connecting rod breakage.
● Main bearing failure. Extreme normal wear or failure of the main

Fault diagnosis

bearings is characteristically accompanied by a rumble from the crankcase and vibration felt through the frame and footrests. Renew the worn bearings and carry out a very careful examination of the crankshaft.
● Crankshaft excessively out of true. A bent crank may result from over-revving or damage from an upper cylinder component or gearbox failure. Damage can also result from dropping the machine on either crankshaft end. Straightening of the crankshaft is not possible in normal circumstances; a replacement item should be fitted.
● Engine mounting loose. Tighten all the engine mounting nuts and bolts.
● Cylinder head gasket leaking. The noise most often associated with a leaking head gasket is a high pitched squeaking, although any other noise consistent with gas being forced out under pressure from a small orifice can also be emitted. Gasket leakage is often accompanied by oil seepage from around the mating joint or from the cylinder head holding down bolts and nuts. Leakage results from insufficient or uneven tightening of the cylinder head fasteners, or from random mechanical failure. Retightening to the correct torque figure will, at best, only provide a temporary cure. The gasket should be renewed at the earliest opportunity.
● Exhaust system leakage. Popping or crackling in the exhaust system, particularly when it occurs with the engine on the overrun, indicates a poor joint either at the cylinder port or at the exhaust pipe/silencer connection. Failure of the gasket or looseness of the clamp should be looked for.

Abnormal transmission noise

30 Clutch noise

● Clutch outer drum/friction plate tang clearance excessive.
● Clutch outer drum/spacer clearance excessive.
● Clutch outer drum/thrust washer clearance excessive.
● Primary drive gear teeth worn or damaged.
● Clutch shock absorber assembly worn or damaged.

31 Transmission noise

● Bearing or bushes worn or damaged. Renew the affected components.
● Gear pinions worn or chipped. Renew the gear pinions.
● Metal chips jammed in gear teeth. This can occur when pieces of metal from any failed component are picked up by a meshing pinion. The condition will lead to rapid bearing wear or early gear failure.
● Gearbox oil level too low. Top up immediately to prevent damage to gearbox and engine.
● Gearchange mechanism worn or damaged. Wear or failure of certain items in the selection and change components can induce mis-selection of gears (see Section 24) where incipient engagement of more than one gear set is promoted. Remedial action, by the overhaul of the gearbox, should be taken without delay.
● Chain snagging on cases or cycle parts. A badly worn chain or one that is excessively loose may snag or smack against adjacent components.

Exhaust smokes excessively

32 White/blue smoke (caused by oil burning)

● Oil pump settings incorrect. Check and reset the oil pump as described in Routine Maintenance.
● Crankshaft main bearing oil seals worn. Wear in the main bearing oil seals, often in conjunction with wear in the bearings themselves, can allow transmission oil to find its way into the crankcase and thence to the combustion chamber. This condition is often indicated by a mysterious drop in the transmission oil level with no sign of external leakage.
● Accumulated oil deposits in exhaust system. If the machine is used for short journeys only it is possible for the oil residue in the exhaust gases to condense in the relatively cool silencer. If the machine is then taken for a longer run in hot weather, the accumulated oil will burn off producing ominous smoke from the exhaust.

33 Black smoke (caused by over-rich mixture)

● Air filter element clogged. Clean or renew the element.
● Main jet loose or too large. Remove the float chamber to check for tightness of the jet. If the machine is used at high altitudes rejetting will be required to compensate for the lower atmospheric pressure.
● Cold start mechanism jammed on. Check that the mechanism works smoothly and correctly and that, where fitted, the operating cable is lubricated and not snagged.
● Fuel level too high. The fuel level is controlled by the float height which can increase as a result of wear or damage. Remove the float bowl and check the float height. Check also that floats have not punctured; a punctured float will lose buoyancy and allow an increased fuel level.
● Float valve needle stuck open. Caused by dirt or a worn valve. Clean the float chamber or renew the needle and, if necessary, the valve seat.

Poor handling or roadholding

34 Directional instability

● Steering head bearing adjustment too tight. This will cause rolling or weaving at low speeds. Re-adjust the bearings.
● Steering head bearing worn or damaged. Correct adjustment of the bearing will prove impossible to achieve if wear or damage has occurred. Inconsistent handling will occur including rolling or weaving at low speed and poor directional control at indeterminate higher speeds. The steering head bearing should be dismantled for inspection and renewed if required. Lubrication should also be carried out.
● Bearing races pitted or dented. Impact damage caused, perhaps, by an accident or riding over a pot-hole can cause indentation of the bearing, usually in one position. This should be noted as notchiness when the handlebars are turned. Renew and lubricate the bearings.
● Steering stem bent. This will occur only if the machine is subjected to a high impact such as hitting a curb or a pot-hole. The lower yoke/stem should be renewed; do not attempt to straighten the stem.
● Front or rear tyre pressures too low.
● Front or rear tyre worn. General instability, high speed wobbles and skipping over white lines indicates that tyre renewal may be required. Tyre induced problems, in some machine/tyre combinations, can occur even when the tyre in question is by no means fully worn.
● Swinging arm bearings worn. Difficulty in holding line, particularly when cornering or when changing power settings indicates wear in the swinging arm bearings. The swinging arm should be removed from the machine and the bearings renewed.
● Swinging arm flexing. The symptoms given in the preceding paragraph will also occur if the swinging arm fork flexes badly. This can be caused by structural weakness as a result of corrosion, fatigue or impact damage, or because the rear wheel spindle is slack.
● Wheel bearings worn. Renew the worn bearings.
● Loose wheel spokes. The spokes should be tightened evenly to maintain tension and trueness of the rim.
● Tyres unsuitable for machine. Not all available tyres will suit the characteristics of the frame and suspension, indeed, some tyres or tyre combinations may cause a transformation in the handling characteristics. If handling problems occur immediately after changing to a new tyre type or make, revert to the original tyres to see whether an improvement can be noted. In some instances a change to what are, in fact, suitable tyres may give rise to handling deficiencies. In this case a thorough check should be made of all frame and suspension items which affect stability.

35 Steering bias to left or right

● Rear wheel out of alignment. Caused by uneven adjustment of chain tensioner adjusters allowing the wheel to be askew in the fork

ends. A bent rear wheel spindle will also misalign the wheel in the swinging arm.
● Wheels out of alignment. This can be caused by impact damage to the frame, swinging arm, wheel spindles or front forks. Although occasionally a result of material failure or corrosion it is usually as a result of a crash.
● Front forks twisted in the steering yokes. A light impact, for instance with a pot-hole or low curb, can twist the fork legs in the steering yokes without causing structural damage to the fork legs or the yokes themselves. Re-alignment can be made by loosening the yoke pinch bolts, wheel spindle and mudguard bolts. Re-align the wheel with the handlebars and tighten the bolts working upwards from the wheel spindle. This action should be carried out only when there is no chance that structural damage has occurred.

36 Handlebar vibrates or oscillates

● Tyres worn or out of balance. Either condition, particularly in the front tyre, will promote shaking of the fork assembly and thus the handlebars. A sudden onset of shaking can result if a balance weight is displaced during use.
● Tyres badly positioned on the wheel rims. A moulded line on each wall of a tyre is provided to allow visual verification that the tyre is correctly positioned on the rim. A check can be made by rotating the tyre; any misalignment will be immediately obvious.
● Wheel rims warped or damaged. Inspect the wheels for runout as described in Chapter 5.
● Swinging arm bearings worn. Renew the bearings.
● Wheel bearings worn. Renew the bearings.
● Steering head bearings incorrectly adjusted. Vibration is more likely to result from bearings which are too loose rather than too tight. Re-adjust the bearings.
● Loose fork component fasteners. Loose nuts and bolts holding the fork legs, wheel spindle, mudguards or steering stem can promote shaking at the handlebars. Fasteners on running gear such as the forks and suspension should be check tightened occasionally to prevent dangerous looseness of components occurring.
● Engine mounting bolts loose. Tighten all fasteners.

37 Poor front fork performance

● Damping fluid level incorrect. If the fluid level is too low poor suspension control will occur resulting in a general impairment of roadholding and early loss of tyre adhesion when cornering and braking. Too much oil is unlikely to change the fork characteristics unless severe overfilling occurs when the fork action will become stiffer and oil seal failure may occur.
● Damping oil viscosity incorrect. The damping action of the fork is directly related to the viscosity of the damping oil. The lighter the oil used, the less will be the damping action imparted. For general use, use the recommended viscosity of oil, changing to a slightly higher or heavier oil only when a change in damping characteristic is required. Overworked oil, or oil contaminated with water which has found its way past the seals, should be renewed to restore the correct damping performance and to prevent bottoming of the forks.
● Damping components worn or corroded. Advanced normal wear of the fork internals is unlikely to occur until a very high mileage has been covered. Continual use of the machine with damaged oil seals which allows the ingress of water, or neglect, will lead to rapid corrosion and wear. Dismantle the forks for inspection and overhaul.
● Weak fork springs. Progressive fatigue of the fork springs, resulting in a reduced spring free length, will occur after extensive use. This condition will promote excessive fork dive under braking, and in its advanced form will reduce the at-rest extended length of the forks and thus the fork geometry. Renewal of the springs as a pair is the only satisfactory course of action.
● Bent stanchions or corroded stanchions. Both conditions will prevent correct telescoping of the fork legs, and in an advanced state can cause sticking of the fork in one position. In a mild form corrosion will cause stiction of the fork thereby increasing the time the suspension takes to react to an uneven road surface. Bent fork stanchions should be attended to immediately because they indicate

that impact damage has occurred, and there is a danger that the forks will fail with disastrous consequences.

38 Front fork judder when braking (see also Section 50)

● Wear between the fork stanchions and the fork legs. Renewal of the affected components is required.
● Slack steering head bearings. Re-adjust the bearings.
● Warped brake disc or drum. If irregular braking action occurs fork judder can be induced in what are normally serviceable forks. Renew the damaged brake components.

39 Poor rear suspension performance

● Rear suspension unit damper worn out or leaking. The damping performance of most rear suspension units falls off with age. This is a gradual process, and thus may not be immediately obvious. Indications of poor damping include hopping of the rear end when cornering or braking, and a general loss of positive stability.
● Weak rear springs. If the suspension unit springs fatigue they will promote excessive pitching of the machine and reduce the ground clearance when cornering. The units must be renewed complete, and as a matched pair.
● Swinging arm flexing or bearings worn. See Sections 34 and 36.
● Bent suspension unit damper rod. This is likely to occur only if the machine is dropped or if seizure of the piston occurs. If either happens the suspension units should be renewed as a pair.

Abnormal frame and suspension noise

40 Front end noise

● Oil level low or too thin. This can cause a 'spurting' sound and is usually accompanied by irregular fork action.
● Spring weak or broken. Makes a clicking or scraping sound. Fork oil will have a lot of metal particles in it.
● Steering head bearings loose or damaged. Clicks when braking. Check, adjust or replace.
● Fork clamps loose. Make sure all fork clamp pinch bolts are tight.
● Fork stanchion bent. Good possibility if machine has been dropped. Repair or replace tube.

41 Rear suspension noise

● Fluid level too low. Leakage of a suspension unit, usually evident by oil on the outer surfaces, can cause a spurting noise. The suspension units should be renewed as a pair.
● Defective rear suspension unit with internal damage. Renew the suspension units as a pair.

Brake problems

42 Brakes are spongy or ineffective – disc brakes

● Air in brake circuit. This is only likely to happen in service due to neglect in checking the fluid level or because a leak has developed. The problem should be identified and the brake system bled of air.
● Pad worn. Check the pad wear against the wear lines provided and renew the pads if necessary.
● Contaminated pads. Cleaning pads which have been contaminated with oil, grease or brake fluid is unlikely to prove successful; the pads should be renewed.
● Pads glazed. This is usually caused by overheating. The surface of the pads may be roughened using glass-paper or a fine file.
● Brake fluid deterioration. A brake which on initial operation is firm but rapidly becomes spongy in use may be failing due to water

contamination of the fluid. The fluid should be drained and then the system refilled and bled.
● Master cylinder seal failure. Wear or damage of master cylinder internal parts will prevent pressurisation of the brake fluid. Overhaul the master cylinder unit.
● Caliper seal failure. This will almost certainly be obvious by loss of fluid, a lowering of fluid in the master cylinder reservoir and contamination of the brake pads and caliper. Overhaul the caliper assembly.

43 Brakes drag – disc brakes

● Disc warped. The disc must be renewed.
● Caliper piston, caliper or pads corroded. The brake caliper assembly is vulnerable to corrosion due to water and dirt, and unless cleaned at regular intervals and lubricated in the recommended manner, will become sticky in operation.
● Piston seal deteriorated. The seal is designed to return the piston in the caliper to the retracted position when the brake is released. Wear or old age can affect this function. The caliper should be overhauled if this occurs.
● Brake pad damaged. Pad material separating from the backing plate due to wear or faulty manufacture. Renew the pads. Faulty installation of a pad also will cause dragging.
● Wheel spindle bent. The spindle may be straightened if no structural damage has occurred.
● Brake lever or pedal not returning. Check that the lever or pedal works smoothly throughout its operating range and does not snag on any adjacent cycle parts. Lubricate the pivot if necessary.
● Twisted caliper support bracket. This is likely to occur only after impact in an accident. No attempt should be made to re-align the caliper; the bracket should be renewed.

44 Brake lever or pedal pulsates in operation – disc brakes

● Disc warped or irregularly worn. The disc must be renewed.
● Wheel spindle bent. The spindle may be straightened provided no structural damage has occurred.

45 Disc brake noise

● Brake squeal. This can be caused by the omission or incorrect installation of the anti-squeal shim fitted to the rear of one pad. The arrow on the shim should face the direction of wheel normal rotation. Squealing can also be caused by dust on the pads, usually in combination with glazed pads, or other contamination from oil, grease, brake fluid or corrosion. Persistent squealing which cannot be traced to any of the normal causes can often be cured by applying a thin layer of high temperature silicone grease to the rear of the pads. Make absolutely certain that no grease is allowed to contaminate the braking surface of the pads.
● Glazed pads. This is usually caused by high temperatures or contamination. The pad surfaces may be roughened using glass-paper or a fine file. If this approach does not effect a cure the pads should be renewed.
● Disc warped. This can cause a chattering, clicking or intermittent squeal and is usually accompanied by a pulsating brake lever or pedal or uneven braking. The disc must be renewed.
● Brake pads fitted incorrectly or undersize. Longitudinal play in the pads due to omission of the locating springs (where fitted) or because pads of the wrong size have been fitted will cause a single tapping noise every time the brake is operated. Inspect the pads for correct installation and security.

46 Brakes are spongy or ineffective – drum brakes

● Brake cable deterioration. Damage to the outer cable by stretching or being trapped will give a spongy feel to the brake lever. The cable should be renewed. A cable which has become corroded due to old age or neglect of lubrication will partially seize making operation very heavy. Lubrication at this stage may overcome the problem but the fitting of a new cable is recommended.
● Worn brake linings. Determine lining wear using the external brake wear indicator on the brake backplate, or by removing the wheel and withdrawing the brake backplate. Renew the shoe/lining units as a pair if the linings are worn below the recommended limit.
● Worn brake camshaft. Wear between the camshaft and the bearing surface will reduce brake feel and reduce operating efficiency. Renewal of one or both items will be required to rectify the fault.
● Worn brake cam and shoe ends. Renew the worn components.
● Linings contaminated with dust or grease. Any accumulations of dust should be cleaned from the brake assembly and drum using a petrol dampened cloth. Do not blow or brush off the dust because it is asbestos based and thus harmful if inhaled. Light contamination from grease can be removed from the surface of the brake linings using a solvent; attempts at removing heavier contamination are less likely to be successful because some of the lubricant will have been absorbed by the lining material which will severely reduce the braking performance.

47 Brake drag – drum brakes

● Incorrect adjustment. Re-adjust the brake operating mechanism.
● Drum warped or oval. This can result from overheating or impact or uneven tension of the wheel spokes. The condition is difficult to correct, although if slight ovality only occurs, skimming the surface of the brake drum can provide a cure. This is work for a specialist engineer. Renewal of the complete wheel hub is normally the only satisfactory solution.
● Weak brake shoe return springs. This will prevent the brake lining/shoe units from pulling away from the drum surface once the brake is released. The springs should be renewed.
● Brake camshaft, lever pivot or cable poorly lubricated. Failure to attend to regular lubrication of these areas will increase operating resistance which, when compounded, may cause tardy operation and poor release movement.

48 Brake lever or pedal pulsates in operation – drum brakes

● Drums warped or oval. This can result from overheating or impact or uneven spoke tension. This condition is difficult to correct, although if slight ovality only occurs skimming the surface of the drum can provide a cure. This is work for a specialist engineer. Renewal of the hub is normally the only satisfactory solution.

49 Drum brake noise

● Drum warped or oval. This can cause intermittent rubbing of the brake linings against the drum. See the preceding Section.
● Brake linings glazed. This condition, usually accompanied by heavy lining dust contamination, often induces brake squeal. The surface of the linings may be roughened using glass-paper or a fine file.

50 Brake induced fork judder

● Worn front fork stanchions and legs, or worn or badly adjusted steering head bearings. These conditions, combined with uneven or pulsating braking as described in Sections 44 and 48 will induce more or less judder when the brakes are applied, dependent on the degree of wear and poor brake operation. Attention should be given to both areas of malfunction. See the relevant Sections.

Electrical problems

51 Battery dead or weak

● Battery faulty. Battery life should not be expected to exceed 3 to 4 years. Gradual sulphation of the plates and sediment deposits will

reduce the battery performance. Plate and insulator damage can often occur as a result of vibration. Complete power failure, or intermittent failure, may be due to a broken battery terminal. Lack of electrolyte will prevent the battery maintaining charge.
● Battery leads making poor contact. Remove the battery leads and clean them and the terminals, removing all traces of corrosion and tarnish. Reconnect the leads and apply a coating of petroleum jelly to the terminals.
● Load excessive. If additional items such as spot lamps, are fitted, which increase the total electrical load above the maximum generator output, the battery will fail to maintain full charge. Reduce the electrical load to suit the electrical capacity.
● Rectifier or ballast resistor failure.
● Generator coils open-circuit or shorted.
● Charging circuit shorting or open circuit. This may be caused by frayed or broken wiring, dirty connectors or a faulty ignition switch. The system should be tested in a logical manner. See Section 54.

52 Battery overcharged

● Regulator (where applicable) faulty. Overcharging is indicated if the battery becomes hot or it is noticed that the electrolyte level falls repeatedly between checks. In extreme cases the battery will boil causing corrosive gases and electrolyte to be emitted through the vent pipes.
● Battery wrongly matched to the electrical circuit. Ensure that the specified battery is fitted to the machine.

53 Total electrical failure

● Fuse blown. Check the main fuse. If a fault has occurred, it must be rectified before a new fuse is fitted.
● Battery faulty. See Section 51.
● Earth failure. Check that the frame main earth strap from the battery is securely affixed to the frame and is making a good contact.

● Ignition switch or power circuit failure. Check for current flow through the battery positive lead (white or white/black) to the ignition switch. Check the ignition switch for continuity.

54 Circuit failure

● Cable failure. Refer to the machine's wiring diagram and check the circuit for continuity. Open circuits are a result of loose or corroded connections, either at terminals or in-line connectors, or because of broken wires. Occasionally, the core of a wire will break without there being any apparent damage to the outer plastic cover.
● Switch failure. All switches may be checked for continuity in each switch position, after referring to the switch position boxes incorporated in the wiring diagram for the machine. Switch failure may be a result of mechanical breakage, corrosion or water.
● Fuse blown. Refer to the wiring diagram to check whether or not a circuit fuse is fitted. Replace the fuse, if blown, only after the fault has been identified and rectified.

55 Bulbs blowing repeatedly

● Vibration failure. This is often an inherent fault related to the natural vibration characteristics of the engine and frame and is, thus, difficult to resolve. Modifications of the lamp mounting, to change the damping characteristics, may help.
● Intermittent earth. Repeated failure of one bulb, particularly where the bulb is fed directly from the generator, indicates that a poor earth exists somewhere in the circuit. Check that a good contact is available at each earthing point in the circuit.
● Reduced voltage. Where a quartz-halogen bulb is fitted the voltage to the bulb should be maintained or early failure of the bulb will occur. Do not overload the system with additional electrical equipment in excess of the system's power capacity and ensure that all circuit connections are maintained clean and tight.

Routine maintenance

Specifications

Engine
	KE100 A5, A6	All other models
Spark plug:		
Make	NGK	NGK
Type	B8HS	B8ES
Gap	0.6 – 0.7 mm (0.024 – 0.028 in)	0.7 – 0.8 mm (0.028 – 0.032 in)

	KE100 A8 to A10, B1 to B12 (US)	All other models
Ignition timing:		
Piston position BTDC	2.58 mm (0.1016 in)	1.96 mm (0.0772 in)

Contact breaker gap .... 0.35 mm (0.014 in)
Idle speed .... 1300 ± 100 rpm
Throttle cable free play at twistgrip .... 2 – 3 mm (0.08 – 0.12 in)
Choke cable free play – KH100 A, KH100 G .... 1 – 2 mm (0.04 – 0.08 in)
Clutch cable free play – at lever clamp .... 2 – 3 mm (0.08 – 0.12 in)
Oil pump cable adjustment .... Marks should align with throttle closed but all slack removed from cables

Cycle parts
Front (drum) brake adjustment:
  KE100 A5, A6 – brake lever ball-end to twistgrip, brake firmly applied .... 55 – 65 mm (2.17 – 2.56 in)
  All other KE100 models, KC100, KH100 A – at lever clamp, brake lightly applied .... 4 – 5 mm (0.16 – 0.20 in)
Rear brake – at pedal tip .... 20 – 30 mm (0.8 – 1.2 in)
Drive chain free play .... 15 – 20 mm (0.6 – 0.8 in)

Tyre pressures-tyres cold:	Front	Rear
KE100 A up to 97.5 kg (215 lb)	25 psi (1.75 kg/cm²)	25 psi (1.75 kg/cm²)
KE100 A 97.5 kg (215 lb) to maximum permissible load	25 psi (1.75 kg/cm²)	32 psi (2.25 kg/cm²)
All other models – up to 97.5 kg (215 lb)	21 psi (1.50 kg/cm²)	28 psi (2.00 kg/cm²)
All other models – 97.5 kg (215 lb) to maximum permissible load	21 psi (1.50 kg/cm²)	32 psi (2.25 kg/cm²)

Note: Loads given represent total weight of rider, passenger and any accessories or luggage. Refer to Model Dimensions and Weights for maximum permissible loads. Pressures apply to OE tyres only – if different tyres are fitted check with tyre manufacturer or supplier whether different pressures are necessary.

Recommended lubricants
Engine:
  Tank capacity – KH100 models .... 1.4 lit (2.46 Imp pint)
  Tank capacity – all other models .... 1.2 lit (2.11 Imp pint, 1.27 US qt)
  Recommended oil .... Good quality air-cooled 2-stroke engine oil
Transmission:
  Capacity .... 0.6 lit (1.06 Imp pint, 0.63 US qt)
  Recommended oil .... Good quality engine oil SAE 10W/30 or 10W/40 SE or SF
Air filter .... See text

Front forks:
	Capacity – per leg	Oil level*
KC100	130cc (4.58 Imp fl oz)	315 ± 2 mm (12.40 ± 0.08 in)
KE100 A	162 ± 4cc (5.70 ± 0.14 Imp fl oz, 5.48 ± 0.13 US fl oz)	408 ± 2 mm (16.06 ± 0.08 in)
KE100 B	182 ± 4cc (6.41 ± 0.14 Imp fl oz, 6.15 ± 0.13 US fl oz)	408.5 ± 2 mm (16.08 ± 0.08 in)
KH100 A	140 ± 4cc (4.93 ± 0.14 Imp fl oz)	355 ± 2 mm (13.98 ± 0.08 in)
KH100 G	72 ± 4cc (2.53 ± 0.141 Imp fl oz)	396 ± 2 mm (15.59 ± 0.08 in)

Recommended fork oil:
  KC100, KH100 A .... SAE 30 fork oil or SAE 10W/20 engine oil
  KE100 A .... SAE 10 fork oil or SAE 10W/20 engine oil
  KE100 B, KH100 G .... SAE 10W/20 engine oil

*Fork oil level is measured from top of stanchion to top of oil, forks fully extended and (internal) springs removed

Final drive chain .... Commercial chain lubricant
Brake fluid – KH100 G .... SAE J1703 (UK) or DOT 3 (US) hydraulic fluid, extra heavy duty
Brake caliper .... Silicone-or PBC-based brake caliper grease
Brake camshafts and wheel bearings .... High melting-point grease
Speedometer drive .... High melting-point grease
Control cables .... Engine oil or light machine oil
Control pivots, stand pivots, rear suspension pivot, steering head bearings, throttle twistgrip .... General purpose grease

Routine maintenance

Introduction

Periodic routine maintenance is a continuous process which should commence immediately the machine is used. The object to maintain all adjustments and to diagnose and rectify minor defects before they develop into more extensive, and often more expensive, problems.

It follows that if the machine is maintained properly, it will both run and perform with maximum efficiency, and be less prone to unexpected breakdowns. Regular inspection of the machine will show up any parts which are wearing, and with a little experience, it is possible to obtain the maximum life from any one component, renewing it when it becomes so worn that it is liable to fail.

All tasks are grouped under various mileage headings, all of which are also given calendar-based intervals; if the machine only covers a low mileage maintenance should be carried out according to the calendar headings instead. All intervals are intended as a guide only. As a machine gets older it develops faults and requires more frequent attention. If it is used under particularly arduous conditions it is advisable to reduce the period between each check.

Daily (pre-ride check)

It is recommended that the following items are checked whenever the machine is about to be used. This is important to prevent the risk of unexpected failure of any component while riding the machine and, with experience, can be reduced to a simple checklist which will only take a few moments to complete. For those owners who are not inclined to check all items with such frequency, it is suggested that the best course is to carry out the checks in the form of a service which can be undertaken each week or before any long journey. It is essential that all items are checked and serviced with reasonable frequency.

1 Engine oil level check

Although it is safe to use the machine as long as oil is visible in the tank sight glass in the left-hand side panel on KC100 models and in the (right-hand) side panel on all others, it is recommended that the level is maintained to within approximately an inch of the filler hole to allow a good reserve. On all KH100 models, remove the right-hand side panel, unclip the tank and allow it to swing out for topping up. On all other models unlock and raise the seat to expose the tank filler cap. Use only good quality two-stroke engine oil suitable for air-cooled engines when topping up. It is useful to keep a spare container of oil with the machine so that a supply is always available. If the oil tank is ever allowed to run dry do not start the engine until it has been refilled and the oil feed pipes have been primed with the oil and bled to remove any air bubbles. If engine damage is suspected, the engine must be dismantled for checking and repair.

2 Check the transmission oil level

Start the engine and allow it to idle for a few minutes to warm the oil and to distribute it fully so that the true level is recorded; if this is not possible switch off the ignition and turn the engine over 3 – 4 times on the kickstart. Wait for 2 – 3 minutes to allow the level to settle, then remove the hexagon-headed level screw from the crankcase right-hand cover, immediately in front of the kickstart shaft.

With the machine standing absolutely upright on its wheels on level ground, oil should slowly trickle out if the level is correct. If not, check that the machine is upright; if no oil appears, add oil of the specified type through the filler plug orifice in the top of the crankcase right-hand cover. If oil pours out, check that the machine is upright and allow the surplus to drain off until the flow slows to a trickle.

Renew the level screw's sealing washer if damaged or worn and refit the screw, tightening it securely, then check that the filler plug is securely refitted.

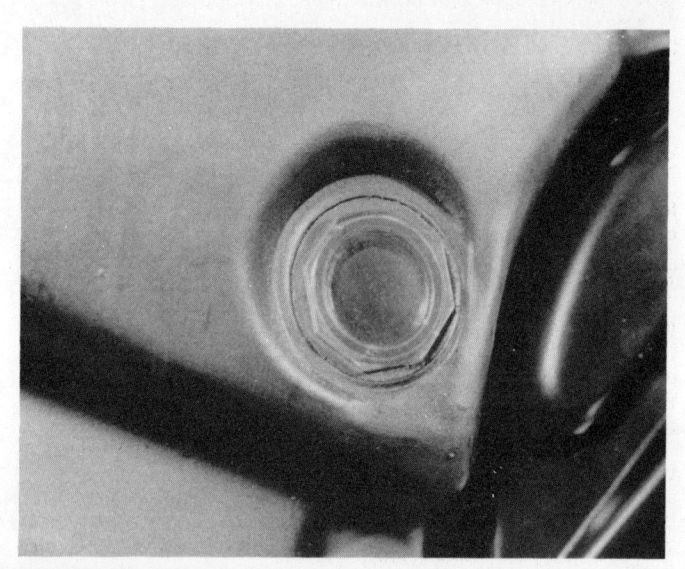

Engine oil level is visible through oil tank sight glass

Use only good quality oil of specified type when topping up

With machine upright, oil should just trickle from level plug hole if level is correct

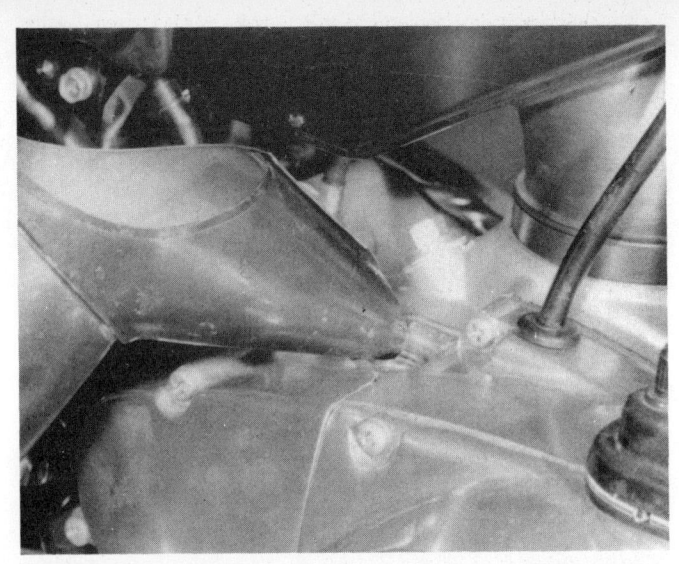

Routine maintenance

over three-quarters of the tread width around the entire circumference with no sign of bald patches. Many riders consider nearer 2 mm to be the limit for secure roadholding, traction, and braking, especially in adverse weather conditions and it should be noted that this is the limit specified by Kawasaki themselves, all measurements being taken at the centre of the tread.

5 Check the brakes

Check that the front and rear brakes work effectively and without binding. Ensure that the cable or the rod linkage is lubricated and properly adjusted. Check the fluid level in the master cylinder reservoir (where appropriate) and ensure that there are no fluid leaks. Should topping-up be required, use only the recommended hydraulic fluid.

6 Final drive chain

Check that the final drive chain is correctly adjusted and well lubricated. Remember that if the machine is used in adverse conditions the chain will require frequent, even daily, lubrication. Refer to the monthly/400 mile service interval.

7 Safety check

Check the throttle and clutch cables and levers, the gear lever and the footrests to ensure that they are adjusted correctly, functioning correctly, and that they are securely fastened. If a bolt is going to work loose, or a cable snap, it is better that it is discovered at this stage with the machine at a standstill, rather than when it is being ridden. Work methodically around the machine checking that all nuts and bolts are securely fastened, that the steering is smooth but without free play throughout its full movement and that it is not hindered by poorly-routed control cables.

Check that the centre and/or side stand (as appropriate) are correctly lubricated and securely retained in the retracted position by their return springs. Where fitted, check that the engine kill switch is working properly.

8 Legal check

Check that all lights, turn signals, horn and speedometer are working correctly to ensure that the machine complies with all local legislation in this respect. Check also that the headlamp is aimed correctly to suit legal requirements; on all machines vertical aim is altered by slackening the headlamp mounting bolts and tilting the shell to the required position. On US models only, horizontal aim can also be altered by rotating the spring-loaded screw set in the side of the headlamp rim.

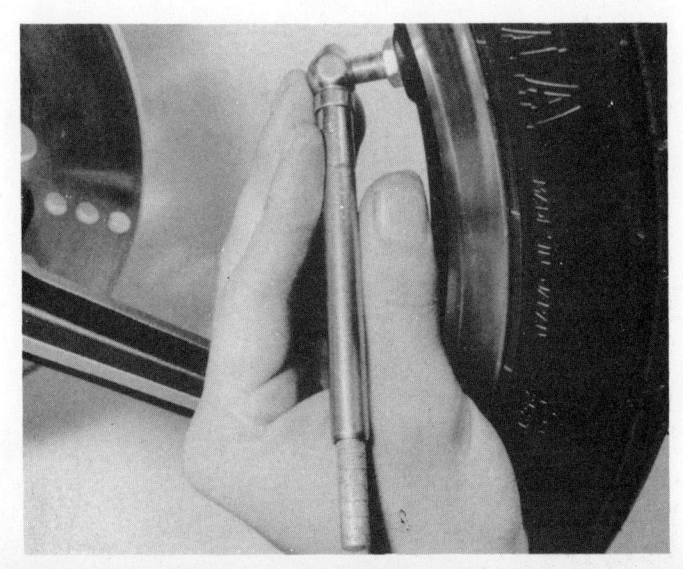

Use only good quality oil of specified grade when topping up

3 Petrol level

Checking the petrol level may seem obvious, but it is all too easy to forget. Ensure that you have enough petrol to complete your journey, or at least to get you to the nearest petrol station.

4 Tyres

Check the tyre pressure with a gauge that is known to be accurate. It is worthwhile purchasing a pocket gauge for this purpose because the gauges on garage forecount airlines are notoriously inaccurate. The pressures should be checked with the tyres **cold**. Even a few miles travelled will warm up the tyres to a point where pressures increase and an inaccurate reading will result.

At the same time as the tyre pressures are checked, examine the tyres themselves. Check them for damage, especially for splitting of the sidewalls. Remove any small stones or other road debris caught between the treads. This is particularly important on the rear tyre, where rapid deflation due to penetration of the inner tube will almost certainly cause total loss of control. When checking the tyres for damage, they should be examined for tread depth. For UK machines, it is vital to keep the tread depth within the legal limits of 1 mm of depth

Use an accurate gauge to check tyre pressures

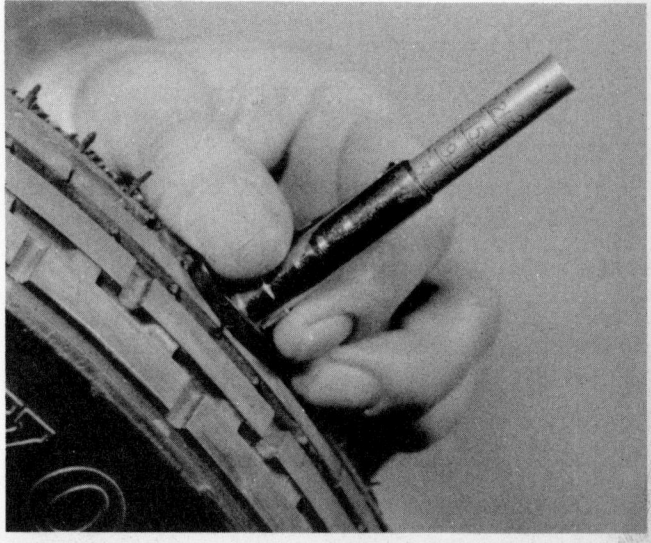

Check tyre tread wear regularly – renew tyres if excessively worn

Routine maintenance

Monthly, or every 400 miles (650 km)

Although this is where the regular routine maintenance procedure begins, the monthly check is really only to emphasise the essential nature of the various checks listed under the daily heading. When carrying out this check start, therefore, by repeating all tasks listed under that heading, then carry out the following:

1 Check the front brake hydraulic fluid level – KH100 G only

To check the fluid level, turn the handlebars until the reservoir is horizontal and check that the fluid level, as seen through the translucent reservoir body on KH100 G2, G3 models or through the sight glass in the rear face of the reservoir body on later models, is not below the lower level mark on the body. Remember that while the fluid level will fall steadily as the pad friction material is used up, if the level falls below the lower level mark there is a risk of air entering the system; it is therefore sufficient to maintain the fluid level above the lower level mark, by topping-up if necessary. Do not top up to the higher level mark (formed by a cast line on the inside of the reservoir on later models) unless this is necessary after new pads have been fitted. If topping up is necessary, wipe any dirt off the reservoir, remove the retaining screws and lift away the reservoir cover and diaphragm. Use only good quality brake fluid of the recommended type and ensure that it comes from a freshly opened sealed container; brake fluid is hygroscopic, which means that it absorbs moisture from the air, therefore old fluid may become contaminated to such an extent that its boiling point has been lowered to an unsafe level. Remember also that brake fluid is an excellent paint stripper and will attack plastic components; wash away any spilled fluid immediately with copious quantities of water. When the level is correct, clean and dry the diaphragm, fold it into its compressed state and fit it to the reservoir. Refit the reservoir cover (and gasket, where fitted) and tighten securely, but do not overtighten, the retaining screws.

2 Check the battery

The battery is located under the seat on all KE100 models; to check it, raise the seat and lift the battery up as far as possible, releasing the spring-loaded retainer on early models. If it is to be removed, disconnect the terminals and vent tube, then lift it out. On all KC100 models the battery is behind the right-hand side panel; remove the securing screws(s), withdraw the retaining strap and lift the battery out until the terminals and vent tube can be disconnected. On all KH100 models remove the left-hand side panel and unhook the rubber strap to release the battery.

On all models, whenever the battery is disconnected, remember to disconnect the negative (−) terminal first, to prevent the possibility of short circuits. The electrolyte level, visible through the translucent casing, must be between the two level marks. If necessary remove the cell caps and top up to the upper level using only distilled water. Check that the terminals are clean and apply a thin smear of petroleum jelly (not grease) to each to prevent corrosion. On refitting, check that the vent hose is not blocked and that it is correctly routed with no kinks, also that it hangs well below any other component, particularly the chain or exhaust system. Remember always to connect the negative (−) terminal last when refitting the battery.

Always check that the terminals are tight and that the fuse connections are clean and tight, that each fuse is of the correct rating and in good condition, and that a spare is available on the machine should the need arise.

At regular intervals remove the battery and check that there is no pale grey sediment deposited at the bottom of the casing. This is caused by sulphation of the plates as a result of re-charging at too high a rate or as a result of the battery being left discharged for long periods. A good battery should have little or no sediment visible and its plates should be straight and pale grey or brown in colour. If sediment

Brake fluid level is checked via sight glass in reservoir body – note lower level mark (arrowed)

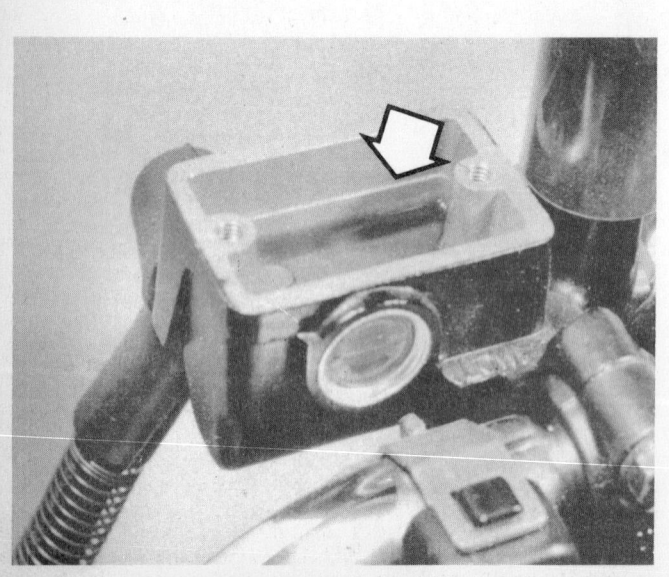

Do not allow fluid level above higher level line (arrowed) when topping up

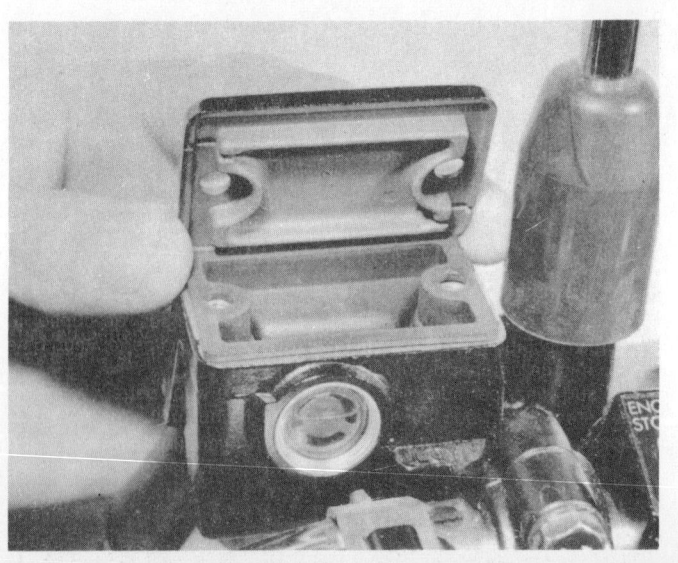

Reservoir diaphragm must be clean, dry and folded as shown on refitting

Routine maintenance 29

Electrolyte level should be between level marks on battery casing

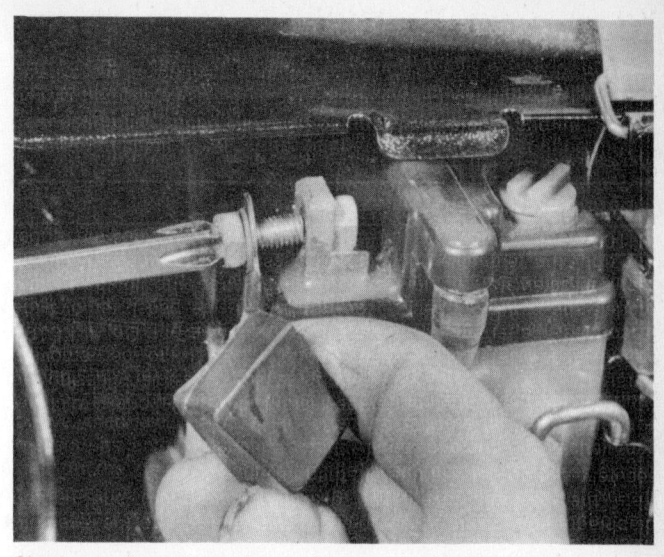

Check security and cleanliness of battery terminals at regular intervals

deposits are deep enough to reach the bottom of the plates, or if the plates are buckled and have whitish deposits on them, the battery is faulty and must be renewed. Remember that a poor battery will give rise to a large number of minor electrical faults.

If the machine is not in regular use, disconnect the battery and give it a refresher charge every month to six weeks, as described in Chapter 6.

3 Check, adjust and lubricate the final drive chain

The exact interval at which the final drive chain will require lubrication, adjustment and renewal is entirely dependent on the usage to which the machine is put and on the amount of care devoted to chain maintenance. In some cases the chain will require daily lubrication, in other cases it need only be lubricated at weekly intervals. The best rule to follow is that if the chain rollers look dry, then the chain needs lubrication immediately. Do not allow the chain to run dry until the links start to kink, or until traces of reddish-brown deposit can be seen on the sideplates. For ease of reference the full procedure is given here but it must be up to the owner to decide how often the chain needs attention. Owners of KC100 models should note that the presence of the full enclosure chainguard means only that service intervals can be extended compared with other machines; the chain must not be neglected.

Cleaning

To clean the chain disconnect it at its connecting link and remove it from the machine. Note that refitting the chain is greatly simplified if a worn out length is temporarily connected to it. As the original chain is pulled off the sprockets, the worn-out chain will follow it and remain in place while the work is carried out. On reassembly, the process is repeated, pulling the worn-out chain over the sprockets so that the new chain, or the freshly cleaned and lubricated chain, is pulled easily into place.

Immerse the chain in a bath containing a mixture of petrol and paraffin and use a stiff-bristled brush to scrub away all the traces of dirt and old lubricant. Take the necessary fire precautions when using this flammable solvent. Swill the chain around to ensure that the solvent penetrates fully into the bushes and rollers and removes any lubricant which may still be present. When the chain is completely clean, remove it from the bath and hang it up to dry.

On refitting, ensure that the connecting link spring clip is fitted with its closed end facing the normal direction of chain travel.

Checking for wear

The chain consists of a multitude of small bearing surfaces which will wear rapidly if the chain is not regularly lubricated and adjusted. A simple check for wear is as follows. With the chain fully lubricated and

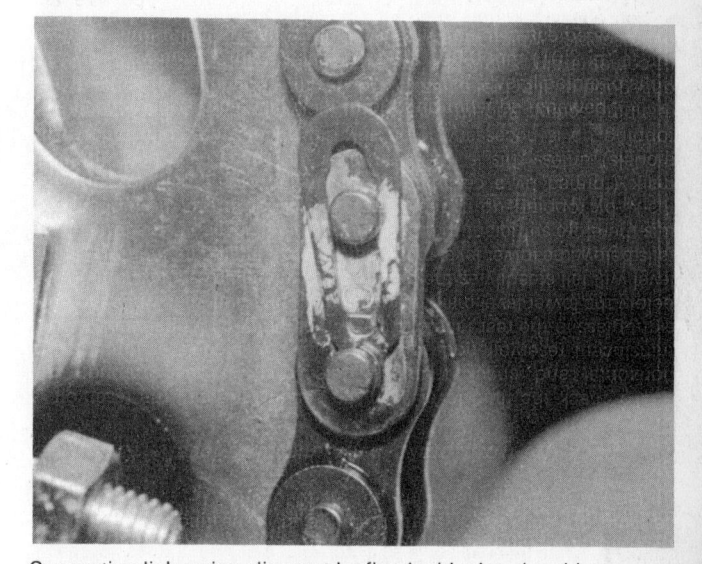

Connecting link spring clip must be fitted with closed end in direction of chain travel

correctly adjusted as described below, attempt to pull the chain backwards off the rear sprocket. If the chain can be pulled clear of the sprocket teeth it must be considered worn out and renewed.

A more accurate measurement of chain wear involves the removal of the chain from the machine as described above.

To check the chain, it must be cleaned and dried as described above, then laid out on a flat surface. Compress the chain fully and measure its length from end to end. Anchor one end of the chain and pull on the other end, drawing the chain out to its fullest extent. Measure the stretched length. If the stretched measurement exceeds the compressed measurement by more than 2% or 1/4 in per foot, the chain must be considered worn out and be renewed. Kawasaki's own version of this test, which can be carried out with the chain in place on the machine if the chainguard is completely removed and if the chain is pulled taut by hanging a 10 kg (20 lb) weight on its lower run, is as follows. Mark any rivet head and count off 21 rivets (pins) along the chains's length ie a 20-link length. Measure the distance between the 1st and 21st pins: if the distance exceeds 259.0 mm (10.2 in) at any point along the chain's length (repeat the test at several points to allow for uneven wear) the chain is worn out and must be renewed.

30 Routine maintenance

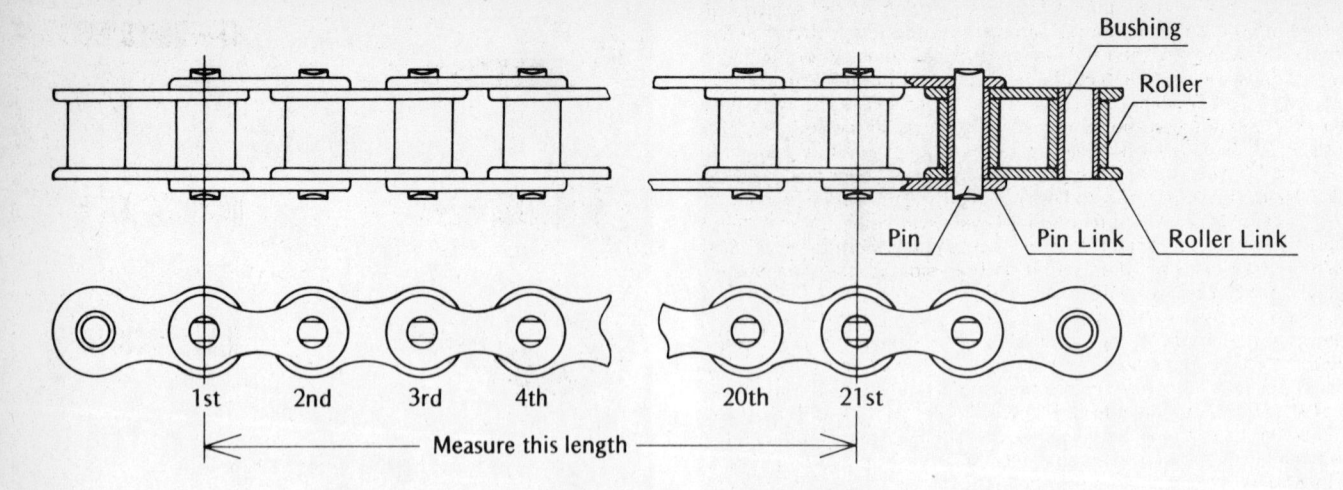

Checking the final drive chain for wear

Renewal

If the chain is found to be worn out, or if any links are kinked or stiff through lack of lubrication, or if damage such as split or missing rollers or side plates is found, the chain must be renewed. This should be done always in conjunction with both sprockets since the running together of new and part-worn components greatly increases the rate of wear of all three items, resulting in even greater expenditure than necessary.

Note that replacement chains are available in standard metric sizes from Renold Ltd, the British chain manufacturer. When purchasing a new chain always quote the size, the number of links required and the machine to which the chain is to be fitted. **Standard** chain sizes are given in the Specifications Section of Chapter 1, but note that the gearing may well have been altered on some machines which means the chain will have been shortened or lengthened.

Lubrication

For the purpose of daily or weekly lubrication one of the many proprietary aerosol-applied chain lubricants can be applied while the chain is in place on the machine. It should be applied at least once a week, and daily if the machine is used in wet weather conditions. Engine oil can be used for this task, but remember that it is flung off the chain far more easily than grease, thus making the rear end of the machine unnecessarily dirty, and requires more frequent application if it is to perform its task adequately. Also remember that surplus oil will eventually find its way onto the tyre, with quite disastrous consequences. While this will serve as a stop-gap measure, it does not reach the inner bearing surfaces of the chain. These can be lubricated correctly only by removing and cleaning the chain as described above, and by immersing it in a molten bath of special chain lubricant such as Chainguard or Linklyfe; which should be done at intervals of 500 – 1000 miles for normal road use and more frequently if the machine is used in wet weather or poor conditions. Follow carefully the manufacturer's instructions when using Chainguard or Linklyfe and take great care to swill the chain gently in the molten lubricant to ensure that it penetrates to all bearing surfaces. Wipe off the surplus before refitting the chain to the machine.

Adjustment

It is necessary to check the chain tension at regular intervals to compensate for wear. Since this wear does not take place evenly along the length of the chain, tight spots will appear which must be compensated for when adjustment is made. On KC100 and KH100 models place the machine on its centre stand and rotate the rear wheel; on KE100 models the machine must be standing on its wheels, supported by the side stand, and must be pushed along to check the chain tension at various points.

On all models, check that the transmission is in neutral and find the tightest spot on the chain by feeling the amount of free play present on the bottom run of the chain, midway between the sprockets, testing along the entire length of the chain. When the tightest spot has been found, measure the total amount of up and down movement available; this should be between 15 – 25 mm (0.6 – 1.0 in).

If adjustment is necessary, remove the securing split pin and

Adjusting the chain

Ensure alignment marks in adjusters are matched with same fixed reference marks to ensure correct wheel alignment

Routine maintenance

slacken the rear wheel spindle nut, the sprocket carrier sleeve nut (where fitted), and brake torque arm nut by just enough to permit the spindle to be moved, then tighten by an equal amount both adjuster nuts to draw the spindle back the necessary amount until free play is between 15 – 20 mm (0.6 – 0.8 in). To preserve accurate wheel alignment, ensure that the same notches stamped in the adjusters line up with the same lines stamped on each swinging arm fork end.

On all models, a final check of accurate wheel alignment can be made by laying a plank of wood or drawing a length of string parallel to the machine so that it touches both walls of the rear tyre. Wheel alignment is correct when the plank or string is equidistant from both walls of the front tyre when tested on both sides of the machine, as shown in the accompanying illustration.

When the chain is correctly tensioned, apply the rear brake firmly to centralise the shoes on the drum and tighten first the sleeve nut (if fitted), then the spndle retaining nut and brake torque arm nut securely. Use the recommended torque settings where possible and do not forget to secure the nuts with new split pins. Note that if the chain tension has been altered significantly, the rear brake and stop lamp rear switch adjustment will also require resetting; these should be checked before taking the machine out on the road.

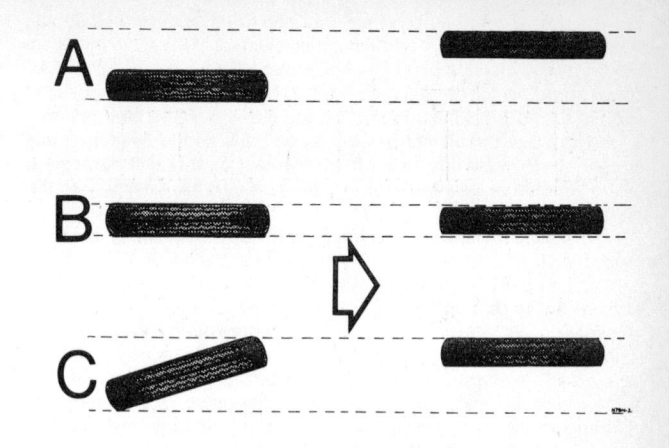

Checking the wheel alignment

A and C – Incorrect
B – Correct

Six monthly, or every 2500 miles (4000 km)

Repeat the tasks listed under the monthly heading, then carry out the following:

1 Clean the air filter element

KC100

Slacken the clamp screw at the base of the filter assembly and withdraw it from the machine. Unhook the wire clamps and separate the two halves of the assembly, then unscrew the wing nut or bolt and remove the element and plated cap. On reassembly apply a smear of SAE 30 engine oil to the sealing strips on the element and inside the body.

KH100

Withdraw the left-hand side panel and remove the two screws which fasten the cover to the side of the air filter casing. Remove the cover, unscrew the wing nut (noting the presence of any washers beneath it) and remove the element. On reassembly, ensure that the element is fitted with the 'up' mark facing upwards and that it is seated correctly.

KE100

Remove the two screws securing the cover to the filter assembly right-hand end, withdraw the cover and pull out the element, noting which way round it was fitted. On all UK models and US KE100 A5, A6, A7 models carefully peel the foam element off the wire frame. On all other US models separate the outer foam element from the inner paper element. On reassembly ensure that the foam element is correctly installed so that no unfiltered air can leak past and fit the element assembly into the filter casing following the instructions given by the warning label on the inside of the cover; the element must be fitted 'diagonally across' the casing rather than 'along' it, even though the latter might seem correct at first glance. Remember that air must flow in through the entry duct and must pass through the element (not merely around it) to get into the carburettor chamber.

All models

Kawasaki recommend that all elements be cleaned in a bath of non-oily high flash-point solvent such as Stoddard solvent (white spirit). This is somewhat unusual in the case of dry paper elements and great care should be taken. When the element is clean, dry it by shaking it or by blowing from the inside outwards with compressed air. When dry, all paper elements can be refitted. On all later US KE100 models

Air filter cleaning, KH100 – remove two screws (arrowed) ...

... to expose air filter element. Note retaining wing nut

the foam outer element must be cleaned and dried as described above, then refitted dry around the inner paper element. However the foam element fitted to all UK KE100 models and to US KE100 A5, A6 and A7 models should be soaked in clean SAE 30 SE engine oil and squeezed to remove the surplus, finishing off by squeezing it when wrapped in a clean rag to leave the element as dry as possible and only slightly oily to the touch. Never wring out a foam element as this will damage it. Refit the element to its supporting frame then refit the assembly to the machine.

If any element is severely clogged, worn out, or has any holes or splits in it, it should be renewed. Check also that the foam sealing band is intact on all paper elements and that the element is seated correctly in its housing so that unfiltered air cannot leak past.

Warning: *Solvents of the type recommended are not only inflammable, they are toxic if inhaled for prolonged periods. Always wear hand and eye protection, work in a well-ventilated area and take all precautions to prevent the risk of fire when using such solvents.*

If riding in a particularly dusty or moist atmosphere, increase the frequency of cleaning the element. Never run the engine without the element fitted; the carburettor is jetted to compensate for its being fitted and the resulting weak mixture will cause overheating of the engine.

2 Check the spark plug

The spark plug supplied as original equipment will prove satisfactory in most operating conditions; alternatives are available to allow for varying altitudes, climatic conditions and the use to which the machine is put. If the spark plug is suspected of being faulty it can be tested only by the substitution of a brand new (not second-hand) plug of the correct make, type, and heat range; always carry a spare on the machine.

Note that the advice of an authorized Kawasaki dealer or similar expert should be sought before the plug heat range is altered from standard. The use of too cold, or hard, a grade of plug will result in fouling and the use of too hot, or soft, a grade of plug will result in engine damage due to the excess heat being generated. If the correct grade of plug is fitted, however, it will be possible to use the condition of the spark plug electrodes to diagnose a fault in the engine or to decide whether the engine is operating efficiently or not. The accompanying series of colour photographs will show this clearly.

It is advisable to carry a new spare spark plug on the machine, having first set the electrodes to the correct gap. Whilst spark plugs do not often fail, a new replacement is well worth having if a breakdown does occur. Ensure that the spare is of the correct heat range and type.

The electrode gap can be assessed using feeler gauges. If necessary, alter the gap by bending the outer electrode, preferably using a proper electrode tool. **Never** bend the centre electrode, otherwise the porcelain insulator will crack, and may cause damage to the engine if particles break away whilst the engine is running. If the outer electrode is seriously eroded as shown in the photographs, or if the spark plug is heavily fouled, it should be renewed. Renew the spark plug at least once annually, regardless of its apparent condition, as it will have passed peak efficiency. Clean the electrodes using a wire brush or a sharp-pointed knife, followed by rubbing a strip of fine emery across the electrodes. If a sand-blaster is used, check carefully that there are no particles of sand trapped inside the plug body to fall into the engine at a later date. For this reason such cleaning methods are no longer recommended; if the plug is so heavily fouled it should be renewed.

Before refitting a spark plug into the cylinder head, coat the threads sparingly with a graphited grease to aid future removal. Use the correct size spanner when tightening the plug, otherwise the spanner may slip and damage the ceramic insulator. The plug should be tightened by hand only at first and then secured with a quarter turn of the spanner so that it seats firmly on its sealing ring. If a torque wrench is available, tighten the plug to the specified torque setting.

Never overtighten a spark plug otherwise there is risk of stripping the threads from the cylinder head, especially as it is cast in light alloy. A stripped thread can be repaired without having to scrap the cylinder head by using a 'Helicoil' thread insert. This is a low-cost service, operated by a number of dealers.

3 Check the contact breaker points and ignition timing

Note: since the generator stator plate is located by its two countersunk retaining screws, the ignition timing can only be altered by opening or closing the contact breaker gap; therefore both operations are described as one. The full procedure is given here for ease of reference, but if the points are found to be in good condition and if the gap has not altered or is within the tolerance, then the ignition timing will be sufficiently accurate and there will be no need to carry out the full check.

If the points are to be checked, it is only necessary to remove the outer inspection cover on KC100 and KH100 models. If the points are to be renewed or the ignition timing is to be adjusted, it will be necessary to remove the complete crankcase left-hand cover. On KE100 models the crankcase left-hand cover must first be removed to permit work of any sort to the generator.

Checking the condition of the contact breaker points

Remove the cover. The contact breaker assembly can be viewed through one of the generator rotor aperture slots. Remove the spark plug and turn the rotor until the points begin to open. Use a small screwdriver to push the moving point open against its spring. Examine the point contact faces. If they are burnt or pitted, remove the points for cleaning or renewal. Light surface deposits can be removed with crocus paper or a piece of stiff card.

To remove the points, lock the crankshaft by selecting top gear and applying the rear brake hard or by applying a holding tool to the rotor itself, as shown in Fig. 1.7, then remove the rotor retaining nut (and washers) and withdraw the rotor using a flywheel puller as described in Chapter 1. Remove the contact breaker retaining screw and lift away the assembly until the nut can be slackened to release the low tension lead terminal. Note carefully the arrangement of the insulating washers on the small bolt which secures the moving contact spring blade and the lead terminal to the fixed contact.

If the contact faces are badly burnt or pitted, or if the moving contact fibre heel shows signs of wear or damage, renew the assembly. It is essential that the points are in good condition if the ignition timing is to be correct; use only genuine Kawasaki parts when renewing. If the faces are only mildly marked, clean them using an oilstone or fine emery but be careful to keep them square. If it is necessary to separate the moving contact from the fixed one, carefully remove the circlip fitted to the pivot post and note carefully the arrangement of washers at both the pivot post and spring blade fixing. On reassembly, the moving contact must be able to move freely; apply a smear of grease to the pivot post. Note also that the low tension lead terminal and the moving contact spring blade must be connected to each other via the small bolt, but that both must be completely insulated from the fixed contact. **The engine will not run if a short-circuit occurs at this point.**

Refit the points to the stator plate and the rotor to the crankshaft. Tighten the rotor retaining nut securely, to the recommended torque setting (where given) as described in Chapter 1, then apply a few drops of oil to the cam lubricating wick.

Checking the ignition timing

Certain items of specialised equipment are needed to check the ignition timing, these being rather expensive and infrequently required. It is therefore recommended that the machine be taken to a Kawasaki dealer for the ignition timing to be checked. The items required are; a dial gauge set (Kawasaki Part Number 56019-029 or 57001-402), comprising an accurate dial gauge, a suitable extension rod with a ball-point tip and an adaptor suitable for a 14 mm spark plug thread, and a self-powered points checker. This last item can be replaced by a proprietary multimeter set to the x 1 ohm resistance scale or by a battery and bulb test circuit. For those who have the necessary equipment, proceed as follows.

Remove the spark plug and the flywheel generator cover. Remove the left-hand side panel or raise the seat and disconnect the black or black/white wire which leads from the points to the ignition switch and the HT coil. Fit the dial gauge adaptor in the spark plug thread and tighten it securely, then screw the extension rod into the gauge, and insert the gauge assembly into the adaptor, securing it with the grub screw. Rotate the flywheel until the piston is at top dead centre (TDC); as the piston approaches TDC the gauge reading will decrease, stop momentarily as TDC is reached, then increase again as the piston descends. Set the gauge to zero as TDC is reached, then rock the flywheel to and fro to make sure that the needle does not go past zero.

The exact time at which the contact breaker points open is determined by the use of a points checker or a multimeter; the gauge needle will swing from 'Closed' to 'Open' (points checker) or will flicker to indicate increased resistance (multimeter). A battery and bulb test circuit can be used so that the bulb lights when the points are

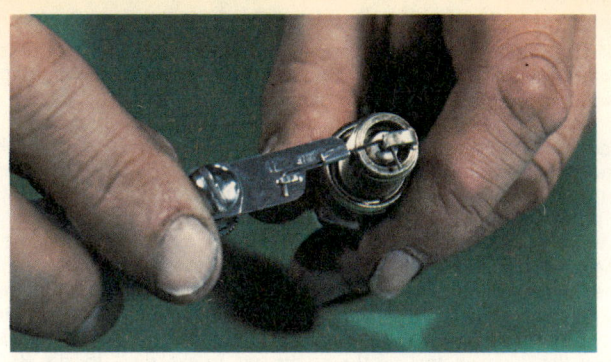

Spark plug maintenance: Checking plug gap with feeler gauges

Altering the plug gap. Note use of correct tool

Spark plug conditions: A brown, tan or grey firing end is indicative of correct engine running conditions and the selection of the appropriate heat rating plug

White deposits have accumulated from excessive amounts of oil in the combustion chamber or through the use of low quality oil. Remove deposits or a hot spot may form

Black sooty deposits indicate an over-rich fuel/air mixture, or a malfunctioning ignition system. If no improvement is obtained, try one grade hotter plug

Wet, oily carbon deposits form an electrical leakage path along the insulator nose, resulting in a misfire. The cause may be a badly worn engine or a malfunctioning ignition system

A blistered white insulator or melted electrode indicates over-advanced ignition timing or a malfunctioning cooling system. If correction does not prove effective, try a colder grade plug

A worn spark plug not only wastes fuel but also overloads the whole ignition system because the increased gap requires higher voltage to initiate the spark. This condition can also affect air pollution

Routine maintenance

closed; note, however, that the bulb will not go out, but will merely glow dimmer as the points open. To make this more obvious to the eye, a high-wattage bulb must be used.

Connect the meter positive (+) terminal to the wire leading from the points and the meter negative (−) terminal to a good earth point on the engine. If a battery and bulb are used, obtain three lengths of wire, connect the battery negative (−) terminal to a good earth point, the battery positive (+) terminal to the bulb contact, and the bulb body to the wire leading from the points.

Rotate the flywheel clockwise until a reading of 2 or 3 mm (0.08 or 0.12 in), as appropriate, is shown on the gauge, then rotate it slowly anticlockwise, until the reading is correct for the model being serviced (ie 1.96 mm/0.0772 in or 2.58 mm/0.1016 in BTDC); at this piston position, the contact breaker points should just be opening. **Note:** for future reference check that the timing mark on the generator rotor is aligned exactly with the fixed index mark on the crankcase or cover; if the marks do not align, as is possible with over-generous production tolerances, make a new fixed mark on the crankcase for use in future adjustments.

The setting is adjusted by opening or closing the contact breaker points gap to advance or retard respectively the ignition timing. Repeat the procedure to check that the timing is now correct.

When the timing is found to be correct, measure very carefully the points gap. Check the points gap as follows, to ensure that the dwell angle is correct for the maximum spark intensity. If the gap is found to be outside the permitted tolerance of 0.30 – 0.40 mm (0.012 – 0.016 in) the contact breaker points are excessively worn, either on the contact faces or on the heel of the moving contact, and must be renewed. It is essential that both the contact breaker points gap and the ignition timing setting are kept within the limits specified for each.

Working as described above, fit a new set of contact breaker points; note that it is essential that only genuine Kawasaki points are used. Refit the flywheel and set the points gap to exactly 0.35 mm (0.014 in), then repeat the procedure given above. The ignition timing should be correct.

Note: The above procedure is described in full as it is the most accurate means of setting the ignition timing. In practice, once the timing marks have been checked with a dial gauge and verified or corrected there is no need to repeat the full procedure at every service

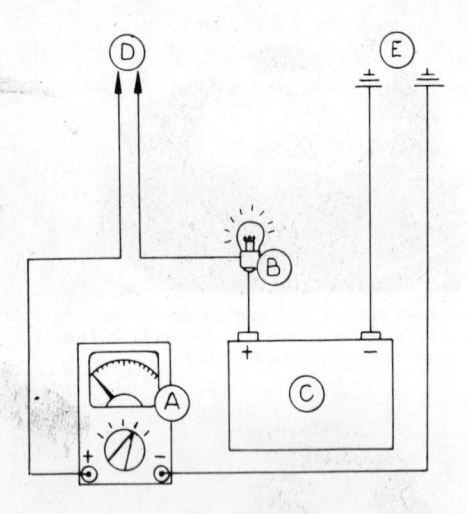

Simple circuit testing apparatus for determining contact breaker point separation

- A Multimeter
- B Bulb
- C Battery
- D Black or black/white wire terminal from points
- E Earth (crankcase)

If ignition timing is correct, points should be opening as rotor mark aligns with fixed mark on crankcase cover ...

... or on crankcase. Check accuracy of marks using dial gauge before checking timing

Ignition timing is adjusted by opening or closing the contact breaker gap

Routine maintenance

interval. Instead it is sufficient to check that the points open precisely as the timing marks align.

While it may appear quicker and easier to use a strobe timing light to check the ignition timing it must be stressed that such a method should not be used until the timing marks have been checked (and corrected, if necessary) with a dial gauge. If this is the case, obtain one of the better quality xenon-tube lights which require a separate power source; the cheaper type of timing lamp may produce an inaccurate reading. Connect the light following its makers's instructions, start the engine and check that the timing marks align at idle speed. There is no point in checking at higher engine speeds as no advance mechanism is fitted. Adjustment of the timing is carried out as described above, requiring that the engine is stopped and re-started at frequent intervals until the setting is correct.

4 Check the carburettor and oil pump settings
Carburettor

If rough running of the engine has developed, some adjustment of the carburettor pilot setting and tick over speed may be required. If this is the case refer to Chapter 2 for details. Do not make these adjustments unless they are obviously required; there is little to be gained by unwarranted attention to the carburettor. On KH100 models the choke (starter) cable should have 1 – 2 mm (0.04 – 0.08 in) of free play; if adjustment is necessary, use the adjuster on the carburettor top.

Once the carburettor has been checked and reset if necessary, the throttle cable free play can be checked. Open and close the throttle several times, allowing it to snap shut under its own pressure. Ensure that it is able to shut off quickly and fully at all handlebar positions, then check that there is 2 – 3 mm (0.08 – 0.12 in) of cable free play, measured in terms of twistgrip rotation. If adjustment is necessary, use first the cable upper end adjuster which is set below or near the twistgrip itself. If there is an insufficient range of adjustment the surplus free play can be eliminated by peeling back the rubber cover over the carburettor and by using the adjuster on the carburettor top. Note that the securing clip fitted to the adjuster on KH100 models must be removed for the free play to be checked accurately, also that disturbing the carburettor/junction box cable on any model will mean that the oil pump adjustment must be checked and reset.

When adjustment is correct, open and close the throttle again to settle the cable and to check that adjustment is not disturbed. Do not forget to check the oil pump cable.

Give the pipe which connects the fuel tap and carburettor a close visual examination, checking for cracks or any signs of leakage. In time, the synthetic rubber pipe will tend to deteriorate, and will eventually leak. Apart from the obvious fire risk, the leaking fuel will affect fuel economy. If the pipe is to be renewed, always use the correct replacement type to ensure a good leak-proof fit. Never use natural tubing because this will tend to break up when in contact with petrol and will obstruct the carburettor jets.

KH100 – Check free play of choke cable at handlebar lever ...

... and adjust if necessary at carburettor adjuster

Throttle cable free play is adjusted first at twistgrip adjuster (where fitted) ...

... then at carburettor adjuster. Note securing clip on throttle cable – KH100 (arrowed)

Note: whenever the carburettor covers are disturbed, always ensure that they are sealed airtight on reassembly. Check the security of the retaining screws, of the rubber grommet set in the front of the crankcase cover, and of the air filter assembly. If unfiltered air is allowed to enter the carburettor chamber it could promote excess engine wear or cause severe engine damage due to the weakened mixture.

Oil pump

The oil pump setting must be checked at regular intervals and whenever the throttle cable is disturbed to ensure that the pump remains correctly synchronised with the carburettor and the engine receives its correct supply of oil.

Remove the retaining screws and withdraw the oil pump cover from behind the kickstart lever. Having first completed carburettor maintenance so that the throttle cable and engine idle speed are now known to be correctly set, start the engine and allow it to tick over.

Very carefully open the throttle until the engine just starts to increase speed, ie so that the throttle is closed but so that all free play has just been eliminated from the cables. At this point the scribed index mark on the pump control lever should align exactly with the reference mark stamped in the lever stop boss on the pump body (see the accompanying photograph).

If adjustment is necessary switch off the engine and first check that the cable outer is seated correctly at each end and that the lower end nipple is securely clamped by the tang on the pump control lever, before using the cable adjuster to achieve the correct setting.

When adjustment is complete, fully open and close the throttle several times to settle the cables then carefully re-check the setting. If all is well, refit the pump cover.

5 Check the brakes

Although a full overhaul will not always be required at each service interval, the full procedure is given here for each type of brake that is fitted to the models described in this Manual.

Drum brakes – adjustment and checking for wear

All drum brakes described in this Manual require regular adjustment to take up wear in the friction material and operating mechanism. Front brakes are adjusted at the cable lower end adjusting nut, reserving the handlebar adjuster for quick roadside adjustments. Rear brakes are adjusted at the nut at the rear end of the operating rod. The adjusting procedure is the same for both, but for front brakes first slacken its locknut and screw fully in the cable handlebar adjuster.

Support the machine securely so that the wheel to be adjusted is raised just clear of the ground, check that the machine cannot fall, then spin the wheel. Tighten the adjuster nut until a rubbing sound is heard as the shoes begin to contact the drum. Unscrew the nut by one or two turns until the sound stops. Spin the wheel hard, apply the brake firmly to settle all disturbed components, then recheck the setting.

This should approximate the specified setting which is that for front brakes of most models there should be 4 – 5 mm (0.16 – 0.20 in) of free play, measured between the handlebar lever butt end and its clamp, when the brake is lightly applied. The exceptions are the KE100 A5 and A6, where it is specified that there must be a distance of 55 – 65 mm (2.17 – 2.56 in) between the twistgrip end and the brake lever ball end when the brake is firmly applied. For all rear brakes there should be 20 – 30 mm (0.8 – 1.2 in) free play at the pedal tip.

On all models, adjust the stop lamp rear switch by means of its two metal nuts or single plastic sleeve nut (whichever is fitted) so that the lamp lights as soon as the pedal has taken up its free play and the shoes can be felt to be engaging the drum; this is usually after about 15 mm (0.6 in) of pedal movement. **Note:** Do not rotate the switch body during adjustment; this will damage the lead wire terminals necessitating switch renewal.

On all types of drum brake it is essential that the operating mechanism is arranged at all times to give the rider the greatest possible leverage. To achieve this the angle between the brake cable or rod and the operating arm must be less than 90° when the brake is fully applied. If the angle exceeds 90°, disconnect the brake rod from the

Oil pump control lever mark must align exactly with fixed mark on pump body (throttle closed, slack just removed from cables)

Rear brake is adjusted at rear end of operating rod

Do not allow switch body to rotate when adjusting stop lamp rear switch

Routine maintenance

operating arm, remove the pinch bolt securing the arm to the brake camshaft and pull the arm off the camshaft, taking care not to disturb the wear indicator pointer. Reposition the arm on the shaft so that with the brake correctly adjusted and fully applied, the angle between arm and rod is 80 – 90°; this may require some experimentation, especially if the brake shoes are nearing the end of their useful life.

The amount of friction material remaining on the brake shoes can be checked without dismantling any component. On all drum brakes a wear indicator pointer is set on the camshaft to align with an arc marked 'Usable range' stamped on the backplate: if, when the brake is fully applied, the pointer indicates outside the arc, the shoes are worn out and must be renewed. This will involve the removal of the wheel and the dismantling of the brake; all work necessary is described in Chapter 5.

Note: If a drum brake is spongy and imprecise in operation at any time, particularly if the wheel has just been disturbed, the shoes will probably need to be centralised on the drum. Slacken lightly the spindle nut (and torque arm fastener, on rear brakes), then spin the wheel and apply the brake hard. Maintain full pressure while the spindle nut is tightened securely, then check the brake adjustment.

Disc brake – general check

Hydraulic disc brakes require no adjustment; all that is necessary is to maintain a check on the fluid level and pad wear, the fluid level check being described under the monthly/400 mile heading. To check the amount of friction material remaining on each pad, look carefully at the caliper from in front of the fork lower leg or upwards from behind the leg; it should be possible to see the notches or steps in the friction material which form the wear limit marks. **Note:** If the pads are so fouled with road dirt that the friction material cannot be clearly seen, they must be removed immediately for cleaning and checking.

On KH100 G2 and G3 models the wear limit marks are in the form of a deep notch cut in the top and bottom edges of the fixed pad 'B' (that furthest away from the caliper piston); if these notches are found to be erased by wear the pads are worn out and must be renewed.

On all later KH100 models the friction material must be at least 1.0 mm (0.04 in) thick, this being indicated by a stepped portion around the outside of the friction material; if either pad is found to be worn down to the level of this step or beyond at any point, both must be renewed.

Pad removal and refitting – KH100 G2 and G3

Remove the two bolts securing the caliper to the fork leg and withdraw the caliper from the disc, taking care not to twist the brake hose. Press the fixed pad 'B' towards the piston (if may be necessary to press the piston back into the caliper bore to gain sufficient clearance) then withdraw it from the caliper noting carefully the way in which the spring and anti-squeal shim are fitted. Similarly remove the moving pad 'A'. Remove the shim and spring from each pad, noting carefully the way in which each is fitted.

On refitting carefully clean all components, and check that the caliper body pivots smoothly and easily about its axis; if any undue stiffness is encountered the caliper must be dismantled and checked as described in Chapter 5. Fit the anti-squeal shims to their respective pads, matching the longer edges of the shims to the longer base sides of the pads, do not forget to refit the springs. Refer to the illustration in Chapter 5 if in doubt as to the correct position.

Noting that pad 'A' is thicker and does not have the wear limit notches of pad 'B', fit pad 'A' against the piston in the caliper body, then refit pad 'B' and install the caliper on the fork leg. Tighten the two bolts to a torque setting of 3.0 kgf m (22 lbf ft).

Pad removal and refitting – all later KH100 G models

Remove the two bolts securing the caliper mounting bracket to the fork lower leg and withdraw the caliper from the disc, taking care not to twist the brake hose. Press the mounting bracket as far as possible towards the piston and disengage the fixed pad, then displace the moving pad. Check that the caliper slides across smoothly and easily on its axle bolts. On reassembly, first see the general notes below. Check that the anti-rattle spring is correctly located then refit the moving pad ensuring that the insulating cap is firmly installed in the

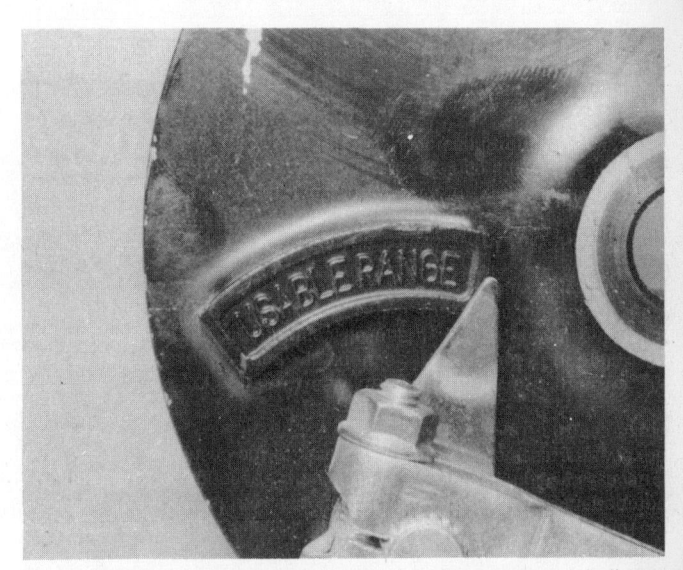

Wear of brake shoes can be checked by observing wear indicator pointer

Measuring pad wear – later KH100 G models – note stepped portion acting as wear indicator

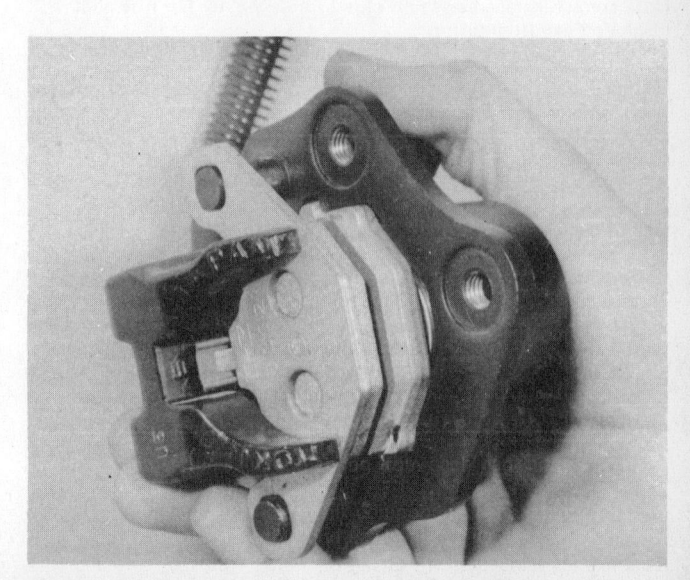

Pad removal – later KH100 G models – press mounting bracket towards piston and release fixed pad from axle pins ...

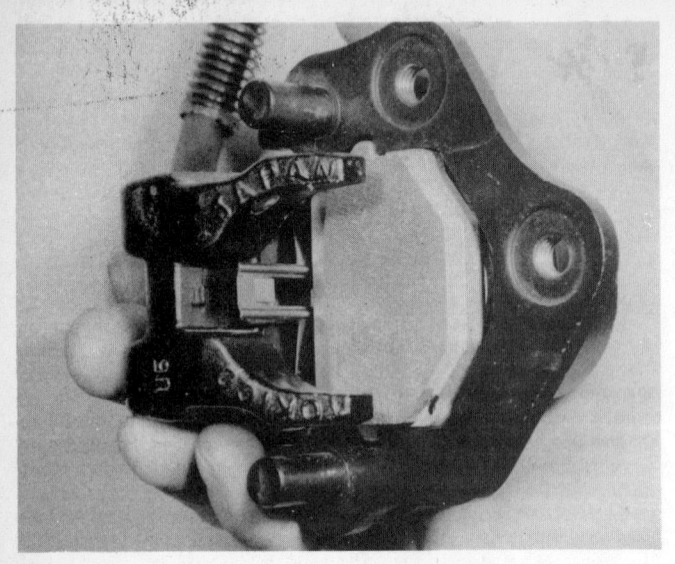

... then withdraw moving pad from mounting bracket

Check, clean and grease caliper axle pins if necessary

Note correct installed position of anti-rattle spring. **Do not** omit insulating cap from piston

Tighten caliper mounting bolts to specified torque setting

piston. Press the mounting bracket towards the piston and refit the fixed pad, noting that its locating pins should also be greased to prevent corrosion. Refit the caliper, tightening its mounting bolts to a torque setting of 2.3 kgf m (16.5 lbf ft).

Brake pads and calipers – general
If the friction material of any brake pad is excessively worn at any point as described above, or if the friction material is fouled with oil or grease, both pads must be renewed as a set. There is no satisfactory way of degreasing friction material. If the pads are not excessively worn, clean them thoroughly with a wire brush, paying careful attention to the sides and rear of the pad metal backing. Dig out any embedded particles from the friction material and clean out any grooves or wear indicator shoulders. Areas of glazing can be removed by rubbing the pad with emery paper. Moving pads (those next to the caliper piston) must slide in the caliper body; thoroughly clean the edges of the pad backing and the matching surface in the caliper body and apply a thin smear of grease to both. Also apply a thin smear of grease to the backs of each pad to prevent brake squeal. Kawasaki recommended the use of PBC (Poly Butyl Cuprysil) based grease which is easily obtainable from auto accessory shops; **do not** use ordinary high-melting point grease. This will melt and foul the friction

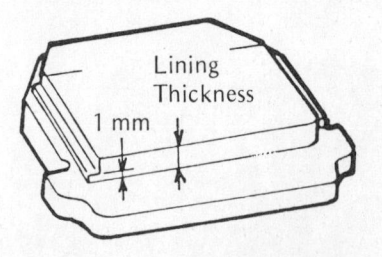

Brake pad friction material wear limits – KH100 G4, G5, G6, G7 and G8 models

material, rendering the brake ineffective. Check that the caliper moves smoothly and easily on its axle bolt or pivot if not, the caliper must be checked as described in the relevant Section of Chapter 5.
If new pads are to be fitted, the piston must be pushed back to make room. Remove the master cylinder reservoir cover and diaphragm and pack clean rag around the reservoir to catch any spilt fluid. Using

Routine maintenance

hand pressure only, push the piston fully back into the caliper body, watching carefully the fluid level in the reservoir; if it rises above the 'Upper' level mark soak up the surplus fluid using a clean rag. If the piston cannot be pushed back using hand pressure, the caliper must be dismantled as described in Chapter 5 for cleaning and examination. On reassembly apply a thin smear of PBC grease to all moving parts, but take great care to keep it away from the friction material. All single-piston calipers are dependent for their efficiency on being able to move smoothly so that equal pressure is applied to both pads. When the caliper is refitted, apply the brake lever or pedal several times to bring the pads back into contact with the disc. Check the hydraulic fluid level; if the original pads have been refitted, it will suffice that the level is above the 'Lower' level mark on the reservoir body, next to the sight glass. If new pads have been fitted, the level must be topped up to the 'Upper' level mark; if this is not marked on the reservoir as described above, it is in the form of a line cast on the inside of the reservoir. Dry the diaphragm with a clean cloth, fold it into its compressed state, and refit it. Refit the cover, tightening it or its retaining screws securely. **Before using the machine on the road, check that the brakes and stop lamp work properly and that full lever (or pedal) pressure is restored.** New pads should be used lightly but firmly, avoiding excessively light or heavy pressures, to enable them to bed in properly for the first 50 miles.

Note: *brake fluid is an extremely effective paint stripper; do not spill any onto any plastic or painted component, and use copious quantities of fresh water to flush away any fluid that is spilt.*

6 Check the clutch adjustment

If the clutch is correctly adjusted there should be 2 – 3 mm (0.08 – 0.12 in) of free play in the cable, measured between the handlebar lever butt end and the clamp, and it should operate smoothly with no sign of slip or drag. Minor adjustments may be made using the handlebar adjuster, but if this does not rectify the problem the release mechanism must be adjusted correctly as follows.

KH100 G, KE100 B

Slacken both their locknuts and screw each cable adjuster fully in to gain the maximum cable free play. Remove the idle adjusting screw extension and, if fitted, the choke knob. Peel off the retaining spring and slide the rubber carburettor cover up the cables clear of the crankcase. Remove the metal carburettor cover.

Slacken the clutch release adjusting screw locknut and turn the screw in or out until the marks on the release lever and crankcase cover are aligned when the handlebar lever is lightly applied. Tighten the locknut without disturbing the screw setting. Using the cable middle adjuster first, so as to reserve the handlebar adjuster for quick roadside alterations, adjust the cable to the specified free play.

Checking clutch cable free play

Slacken cable adjuster locknuts and screw in adjuster

Slacken release mechanism adjuster locknut and set adjuster screw ...

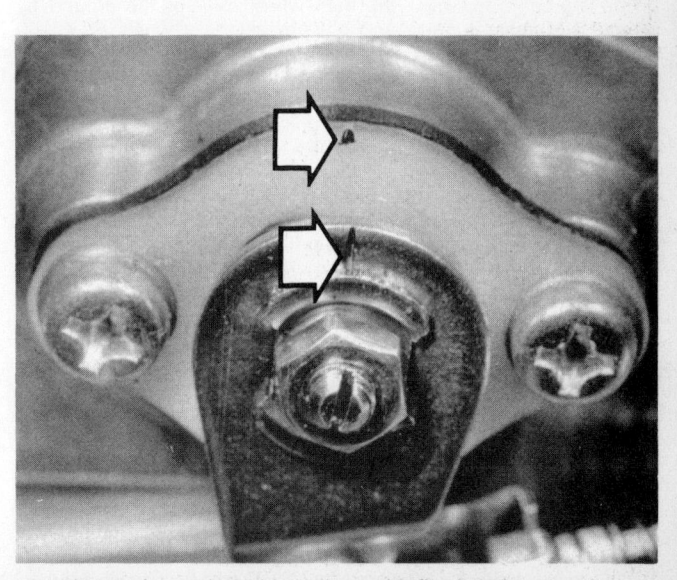

... so that mark on release lever aligns with fixed mark on crankcase cover or adjuster body (arrowed)

All other models

Slacken both cable adjuster locknuts and turn the handlebar adjuster until there is a gap of 5 – 6 mm (0.20 – 0.24 in) between the locknut and the inner edge of the adjuster cable abutment, ie the exposed length of adjuster thread. Remove the idle adjusting screw extension and, if fitted, the choke knob. Peel off the retaining spring and slide the rubber carburettor cover up the cables clear of the crankcase. Remove the metal carburettor cover.

Check that the mark on the clutch release lever is aligned with the fixed mark on the crankcase cover or release mechanism body; if not use the cable mid-way adjuster to bring the two into line. On earlier models which may not have these marks, the release lever must be at an angle of 80 – 90° to the cable inner. Slacken the locknut and rotate the adjusting screw anticlockwise by one or two turns to check that there is no pressure on it, then rotate it clockwise until it seats firmly. Do not apply excessive pressure; this is the point at which the release mechanism is starting to lift the pressure plate. Hold the screw steady in that position while the locknut is tightened.

Check that the lower end of the clutch outer cable is securely fitted in its recess in the crankcase cover, then unscrew the sleeve nut on the cable mid-way adjuster until all cable free play is eliminated. Tighten the mid-way adjuster locknut and use the handlebar adjuster to set the specified free play at the lever.

All models – general

When adjustment is correct apply a few drops of oil to all exposed lengths of inner cable, all adjuster threads and to all bearing surfaces or pivots. Use a new gasket if necessary when refitting the metal carburettor cover and ensure that the carburettor chamber is sealed airtight. Finally, check that all adjuster locknuts have been securely fastened and that any rubber adjuster sleeves have been properly refitted. Check the clutch adjustment carefully before taking the machine out on the road.

Side-stand retracting mechanism

Certain models are fitted with an automatic side-stand retracting mechanism which is operated by a short cable branching off the main clutch cable. Whenever the clutch is adjusted, check the adjustment of this mechanism, which should be set so that all cable free play is just eliminated when the stand is in the down position. Secure the cable adjuster locknut and apply a few drops of oil to the mechanism pivots, cable nipples and bearing surfaces.

As a check of the stand retracting mechanism position the machine upright on its wheels with the stand in the down position. If correctly adjusted, the stand should retract as soon as the clutch lever is pulled in.

7 Check the steering and suspension

For KC and KH100 models, place the machine on its centre stand on level ground and raise the front wheel clear of the ground by placing a suitable support under the crankcase. Owners of KE100 models must procure a stout wooden box or purpose built stand to hold the machine securely upright with the front wheel clear of the ground; old milk bottle crates are very effective.

Check the bearing adjustment by grasping the bottom of both fork lower legs, then pulling and pushing in a fore and aft direction; any free play should be felt between the fork bottom yoke and the frame head lug. Check for overtightened bearings by placing the forks in the straight ahead position and tapping lightly on one handlebar end; the forks should fall away smoothly and easily to the opposite lock, taking into account the effect of cables and wiring, with no trace of notchiness. If adjustment is required, proceed as follows:

Slacken fully the bottom yoke pinch bolts and apply a few drops of penetrating fluid to the stanchions so that they are free to slide in the bottom yoke. Slacken fully the steering stem head clamp bolt (where fitted), then the large steering stem bolt at the centre of the top fork yoke, and use a C-spanner to rotate the slotted adjuster nut immediately below the top yoke. As a guide to adjustment, tighten the slotted nut until a light resistance is felt, then back it off by $1/8$ turn. The object is to remove all discernible play without applying any appreciable preload but, somewhat unusually, Kawasaki specify that the nut should be tightened to a torque setting of 1.8 – 2.2 kgf m (13 – 16 lbf ft) on KE100 machines, and 2.0 kgf m (14.5 lbf ft) on KH100 machines. No settings are available for KC100 models. Owners who do not have the equipment to apply a torque setting to this type of nut should note that it can be approximated by tightening the nut until light resistance is felt then by tightening it for another $1/16$ turn (ie a further 20° rotation.

Note: This latter method of adjustment will require extreme care if overtightened bearings are not to result. It is possible to apply a

Check adjustment of side stand retracting mechanism (where fitted) whenever clutch is adjusted ...

... and oil linkage pivots frequently

Adjusting the steering head bearings

loading of several tons on the small steering head bearings without this being obvious when turning the handlebars. This will cause an accelerated rate of wear, and make the machine weave from side to side at low speeds. When adjustment is correct, tighten the steering stem bolt securely, to the recommended torque setting where specified, and recheck the adjustment. If all is well, tighten the steering stem clamp bolt (if applicable) and the bottom yoke pinch bolts; again use the recommended torque settings where possible.

Ensure that the front forks work smoothly and progressively by pumping them up and down whilst the front brake is held on. Any faults revealed by this check should be investigated immediately; refer to Chapter 4, particularly if the oil seals are leaking. Inspect the stanchions, looking for signs of chips or other damage, then lift the dust excluder at the top of each fork lower leg and wipe away any dirt from its sealing lips and above the fork oil seal. Pack grease above the seal and refit the dust excluder. Note that none of this would be necessary, and fork stiction would be reduced, if gaiters are fitted; they are available from any good motorcycle dealer.

The rear suspension can be checked with the machine on the centre stand or, for KE100 models, the stand described above. Check that all the suspension components are securely attached to the frame. Check for free play in the swinging arm by pushing and pulling it horizontally. If free play is found, refer to the relevant Sections of Chapter 4; note that while actual wear is very rare with this type of pivot bearing, it is advisable to remove the swinging arm pivot bolt at regular intervals so that it can be cleaned and greased to minimise corrosion. Refer to Chapter 4, but note that if care is taken to support the machine securely so that the rear suspension is wedged in place, then the bolt can be removed with no need for extensive dismantling.

8 Check the wheels and tyres

Raise the wheel to be examined clear of the ground, using blocks positioned beneath the engine. Check the wheel spins freely; if necessary, slacken the brake adjuster or push back the brake pads, and in the case of the rear wheel, detach the final drive chain.

KC100, KE100 and KH100 A models

Examine the rim for serious corrosion or impact damage. Slight deformities can often be corrected by adjusting spoke tension. Serious damage and corrosion will necessitate renewal, which is best left to an expert. A light alloy rim will prove more corrosion resistant.

Place a wire pointer close to the rim and rotate the wheel to check it for runout. If the rim is more than 2.0 mm (0.08 in) out of true in the radial or axial planes, check spoke tension by tapping them with a screwdriver. A loose spoke will sound quite different to those around it. Worn bearings will also cause rim runout.

Adjust spoke tension by turning the square-headed nipples with the appropriate spoke key which can be purchased from a dealer. With the spokes evenly tensioned, remaining distortion can be pulled out by tightening the spokes on one side of the wheel and slackening those directly opposite. This will pull the rim across whilst maintaining spoke tension.

More than slight adjustment will cause the spoke ends to protrude through the nipple and chafe the inner tube, causing a puncture. Remove the tyre and tube and file off the protruding ends. The rim band protects the tube against chafing, check it is in good condition before fitting.

Check spoke tension and general wheel condition regularly. Frequent cleaning will help prevent corrosion. Replace a spoke immediately because the load taken by it will be transferred to adjacent spokes which may fail in turn.

KH100 G models

Carefully check the complete wheel for cracks and chipping, particularly at the spoke roots and the edge of the rim. As a general rule a damaged wheel must be renewed as cracks will cause stress points which may lead to sudden failure under heavy load. Small nicks may be radiused carefully with a fine file and emery paper (No 600 – No 1000) to relieve the stress. If there is any doubt as to the condition of a wheel, advice should be sought from a reputable dealer or specialist repairer.

Each wheel is covered with a coating of lacquer, to prevent corrosion. If damage occurs to the wheel and the lacquer finish is penetrated, the bared aluminium alloy will soon start to corrode. A whitish grey oxide will form over the damaged area, which in itself is a protective coating. This deposit, however, should be removed carefully as soon as possible and a new protective coating of lacquer applied.

Check the lateral runout at the rim by spinning the wheel and placing a fixed pointer close to the rim edge. If the maximum runout is greater than 0.5 mm (0.020 in), the manufacturer recommends that the wheel be renewed. This is, however, a counsel of perfection; a runout somewhat greater than this can probably be accommodated without noticeable effect on steering. No means is available for straightening a warped wheel without resorting to the expense of having the wheel skimmed on both faces. If warpage was caused by impact during an accident, the safest measure is to renew the wheel complete. Worn wheel bearings may cause rim runout. These should be renewed.

All models

An out of balance wheel will produce a hammering effect through the steering at high speed. Spin the wheel several times. A well balanced wheel will come to rest in any position. One that comes to rest in the same position will have its heaviest part downward and weights must be added to a point diametrically opposite until balance is achieved. Where the tyre has a balance mark on its sidewall (usually a coloured spot), check it is in line with the valve.

To check the wheel bearings, grasp each wheel firmly at the top and bottom and attempt to rock it from side to side; any free play indicates worn bearings which must be renewed as described in Chapter 5. Make a careful check of the tyres, looking for signs of damage to the tread or sidewalls and removing any embedded stones etc. Renew the tyre if the tread is excessively worn or if it is damaged in any way.

9 General lubrication

At regular intervals the footrests, stand, the rear brake pedal and the throttle twistgrip must be dismantled so that all traces of corrosion and dirt can be removed, and the various components greased. This operation must be carried out to prevent excessive wear and to ensure that the various components can be operated smoothly and easily, in the interests of safety. The opportunity should be taken to examine closely each component, renewing any that show signs of excessive wear or of any damage.

The twistgrip is removed by unscrewing the screws which fasten both halves of the assembly. The throttle cable upper end nipple can be detached from the twistgrip with a suitable pair of pliers and the twistgrip slid off the handlebar end. Carefully clean and examine the handlebar end, the internal surface of the twistgrip, and the two halves of the switch cluster. Remove any rough burrs with a fine file, and apply a coating of grease to all the bearing surfaces. Slide the twistgrip back over the handlebar end, insert the throttle cable end nipple into the twistgrip flange, and reassemble the switch cluster. Check that the twistgrip rotates easily and that the throttle snaps shut as soon as it is released. Tighten the retaining screws securely, but do not overtighten them.

Although the regular checks will ensure that the control cables are lubricated and maintained in good order, it is recommended that a positive check is made on each cable at this mileage/time interval to ensure that any faults will not develop unnoticed to the point where smooth and safe control operation is impaired. If any doubt exists about the condition of any of the cables, the component in question should be removed from the machine for close examination.

To remove the front brake or clutch cable first slacken the adjuster locknut and screw in the adjuster to gain the maximum cable free play. Align the slot in the handlebar adjuster with that in the clamp and withdraw the cable upper end from the lever. In the case of the brake cable unscrew the adjusting nut and withdraw the cable, releasing it from any retaining clamps or ties. In the case of the clutch cable slide up the rubber carburettor cover, remove the metal carburettor cover and straighten the tang which secures the cable lower end nipple in the clutch release lever. Where necessary, unscrew fully the adjuster and disconnect the side stand raising cable from its linkage.

To remove the choke cable unscrew the lever pivot bolt, and carefully withdraw the lever until the cable end nipple can be disconnected from it. Slide up the black rubber carburettor cover, unscrew the plunger assembly from the carburettor, compress the spring and slide the nipple out of the plunger.

The throttle/oil pump cable can be removed only as a single assembly; disconnect the cable from the twistgrip as described above, then remove the oil pump cover and gently bend out the tang which retains the cable end nipple to the pump control lever. Remove the carburettor as described in Chapter 2, unscrew the carburettor top and disconnect the cable.

Check the outer cables for signs of damage, then examine the exposed portions of the inner cables. Any signs of kinking or fraying will indicate that renewal is required. To obtain maximum life and reliability from the cables they should be thoroughly lubricated using light machine oil. To do the job properly and quickly, use one of the hydraulic cable oilers available from most motorcycle shops. Free one end of the cable and assemble the cable oiler as described by the manufacturer's instructions. Operate the oiler until oil emerges from the lower end, indicating that the cable is lubricated throughout its length. This process will expel any dirt or moisture and will prevent its subsequent ingress.

If a cable oiler is not available, an alternative is to remove the cable from the machine. Hang the cable upright and make up a small funnel arrangement using plasticine or by taping a plastic bag around the upper end. Fill the funnel with oil and leave it overnight to drain through. Note that where nylon-lined cables are fitted, they should be used dry or lubricated with a silicone-based lubricant suitable for this application. On no account use ordinary engine oil because this will cause the liner to swell, pinching the cable. The throttle/oil pump cable junction box should be dismantled, cleaned and reassembled using a 'dry' graphite-based lubricant or any lightweight grease or aerosol spray lubricant. Do not use ordinary grease as this will at best make the throttle action heavy; at worst it could cause an accident by preventing the smooth return of the throttle.

Check all pivots and control levers, cleaning and lubricating them to prevent wear or corrosion. Where necessary, dismantle and clean any moving part which may have become stiff in operation.

When refitting the cables onto the machine, ensure that they are routed in easy curves and that full use is made of any guide or clamps that have been provided to secure the cable out of harm's way. Adjustment of the individual cables is described under Routine Maintenance tasks.

Be very careful to ensure that all controls are correctly adjusted and are functioning correctly before taking the machine out on the road.

To check the speedometer drive cable, remove the cable by unscrewing the knurled nut which secures each end with a pair of pliers, and withdraw the cable from the machine. The rev-counter fitted to KH100 G models is driven by a similar type of cable, but which in this case is retained in a socket in the crankcase by a screw that doubles as a cotter pin. Remove the screw completely, pull the cable out of its crankcase socket and withdraw it through the tunnel in the air filter casing.

Where possible, pull the inner out of the bottom end of the cable; if this is not possible the cable must be renewed if faulty. Check the cable, looking for cracks, kinks or other damage in the outer and for frayed strands or wear of the inner. Renew the cable if any damage is found; a jerky or sluggish movement at the instrument head is usually due to a cable fault.

Before greasing, clean the inner cable with a petrol soaked rag and examine the cable for broken strands or other damage. Do not check the cable for broken strands by passing it through the fingers or palm of the hand, this may well cause a painful injury if a broken strand snags the skin. It is best to wrap a piece of rag around the cable and pull the cable through it, any broken strands will snag the rag.

Regrease the cable with high melting-point grease, taking care not to grease the last six inches closest to the instrument head. If this precaution is not observed, grease will work into the instrument and immobilise the sensitive movement.

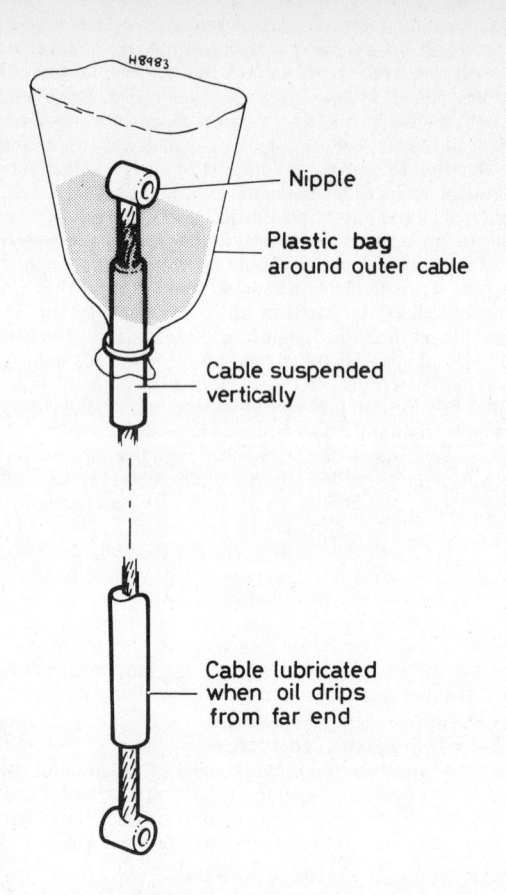

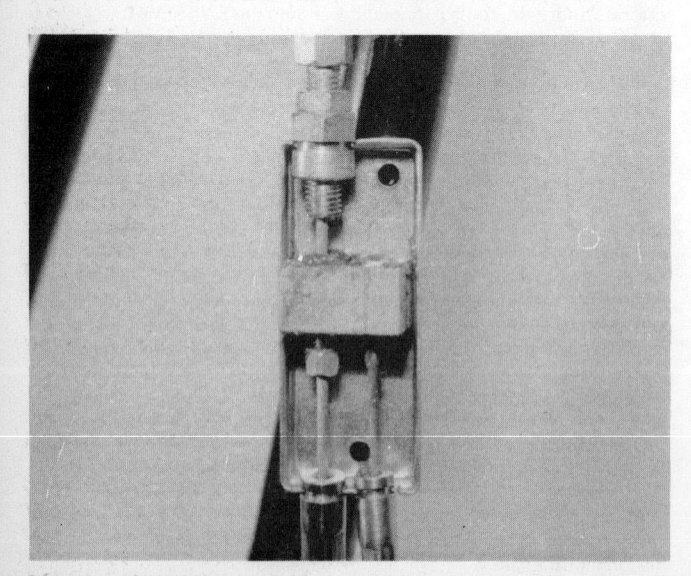

Oiling a control cable

Control cable junction boxes (where fitted) should not be omitted when lubricating cables

Unscrew knurled retaining ring to release instrument drive cable

Routine maintenance

When refitting a drive cable, always ensure that it has a smooth easy run to minimise wear, and check that it is secured where necessary by any clamps or ties provided for the purpose of keeping it away from any hot or moving parts. Note that refitting is aided if the wheel (or engine) is rotated slowly to help the various components engage each other correctly.

Annually, or every 5000 miles (8000 km)

Repeat all previous maintenance tasks, then carry out the following:

1 Change the transmission oil

Oil changes are much quicker and more efficient if the machine is taken for a journey long enough to warm the oil up to normal operating temperature so that it is thin and is holding any impurities in suspension. Position a container of at least 700cc (1.2 Imp pint, 1.75 US qt) capacity beneath the engine. Remove the drain plug from the underside of the engine unit. Whilst waiting for the oil to drain, examine the plug sealing washer and renew if damaged.

On completion of draining, refit the plug and tighten it securely, to the recommended torque setting where specified. Remove the filler plug adjacent to the kickstart lever and replenish the gearbox with 600cc (1.06 Imp pint, 0.63 US qt) of the specified oil. Check the oil level as described under the daily heading and refit the filler plug.

Remove drain plug to drain transmission oil

2 Change the brake fluid

If the brake fluid is not completely changed during the course of routine maintenance, it should be changed at least once annually. Brake fluid is hygroscopic, which means that it absorbs moisture from the air. Although the system is sealed, the fluid will gradually deteriorate and must be renewed before contamination lowers its boiling point to an unsafe level.

Before starting work, obtain a full can of new DOT 3 or SAE J1703 hydraulic fluid and read Chapter 5, Section 10. Prepare the clear plastic tube and glass jar in the same way as for bleeding the hydraulic system, open the bleed nipple by unscrewing it $1/4 - 1/2$ a turn with a spanner and apply the front brake lever gently and repeatedly. This will pump out the old fluid. Keep the master cylinder reservoir topped up at all times, otherwise air may enter the system and greatly lengthen the operation. The old brake fluid is invariably much darker in colour than the new, making it much easier to see when the old fluid has been pumped out and the new fluid has completely replaced it.

When the new fluid appears in the clear plastic tubing with no traces of old fluid contaminating it, close the bleed nipple, remove the plastic tubing and refit the rubber dust cap on the nipple. Top the master cylinder reservoir up to the upper mark. Carefully dry the diaphragm with a clean lint-free cloth, fold it into its compressed state, and refit the diaphragm and the reservoir cover, tightening the retaining screws securely. Wash off any surplus fluid with fresh water and check for any fluid leaks which may subsequently appear. Remember to check that full brake pressure is restored and that the front brake is working properly before taking the machine out on the road.

3 Change the fork oil

On machines not fitted with front fork drain plugs, each fork leg must be removed from the yokes and inverted to pump out the old oil. Refer to Chapter 4. On machines with drain plugs, place a sheet of cardboard against the wheel to keep oil off the brake and tyre, place a suitable container under the fork leg and remove the drain plug. Pump the forks several times to expel as much oil as possible, then repeat the process on the remaining leg. Leave the machine for a few minutes to allow any residual oil to drain to the bottom, then pump the forks again to remove it.

Renewing their sealing washers if worn or damaged, refit and tighten the drain plugs securely, then remove the fork leg top bolts. Add exactly the correct amount of the specified grade of fork oil (see Specifications) to each leg and use a length of welding rod or similar as a dipstick to check that the level of oil is correct and exactly the same in each leg. Add or remove oil to adjust the level. Tighten the top bolts securely, to the specified torque setting (where given) and check that the forks move smoothly and correctly throughout their full travel.

Note that the fork action can be varied to suit the rider's needs by using different grades of oil; the lighter the oil, the lighter the damping effect. Proprietary fork oils are available in several grades from most good motorcycle dealers.

4 Renew the air filter element

Irrespective of its apparent condition, the element must be renewed after it has been cleaned five times or at 5000 mile intervals. To keep check, mark the element with a permanent marker at each cleaning interval. When the machine is used in dusty conditions, the cleaning and renewal intervals should be reduced to compensate.

5 Clean the fuel system

Except for later KH100 G models, on which the fuel tap must be removed from the tank for cleaning, all models are fitted with separate filter bowls in the base of their fuel taps. These must be checked at regular intervals for the presence of dirt or water in the fuel.

Switch the tap to the 'Off' position and unscrew the filter bowl from the tap base. Check the condition of the sealing O-ring and renew it if it is seriously compressed, distorted or damaged. Thoroughly clean the filter bowl and, where fitted, the filter gauze; if signs of dirt or water are found in the petrol, the tap should be removed from the tank as described in Chapter 2 so that the tap filter can be cleaned and the tank flushed out.

On reassembly, tighten the bowl by just enough to nip up the O-ring; do not overtighten it as this will distort the O-ring and promote the risk of fuel leakage. If leaks are found they can be cured only by the renewal of the defective seal.

6 Check all nuts, bolts and fasteners

Work methodically round the machine checking that all nuts, bolts, screws and other fasteners are properly secured. Use the correct torque wrench settings, where given, to ensure that components are tightened correctly and not over-stressed. Pay particular attention to components which are subjected to a lot of vibration or to extremes of heat and renew any fasteners or locking devices (such as spring washers) which are found to be no longer serviceable.

Working evenly and in a diagonal sequence, check that the cylinder head nuts are securely fastened; again use the specified torque wrench setting where applicable.

Check the stand and lever pivots for security and lubricate them with light machine oil or engine oil. Make sure that the stand springs are in good condition.

It is advisable to lubricate the handlebar switches and stop lamp switches with WD40 or a similar water dispersant lubricant. This will keep the switches working properly and prolong their life, especially if the machine is used in adverse weather conditions. Check that all lights, turn signals, the horn and speedometer are working properly and that their mountings and connections are securely fastened.

Every two years, or 10 000 miles (16 000 km)

Complete the tasks listed under the previous headings, then carry out the following:

1 Grease the brake camshaft(s), wheel bearings and speedometer drive.

Referring to Chapter 5, remove each wheel in turn so that the above components can be removed, cleaned, checked for wear and renewed if necessary. On reassembly, pack the wheel bearings and speedometer drive with grease and apply a thin smear to the bearing surfaces of the brake camshaft(s).

2 Grease the steering head bearings

Referring to the relevant Sections of Chapter 4, dismantle the front forks and steering head. Clean all components and check them for wear, renewing any worn items. Reassemble, packing the steering head bearings with fresh grease.

3 Renew the front brake cable – except KH100 G

As a safety precaution, Kawasaki recommend that the brake cable be renewed at this interval to prevent the risk of sudden failure. Remove and fit the cable and adjust the brake as described under the *General lubrication* heading.

4 Renew the hydraulic seals – KH100 G

The manufacturer recommends that the seals in the master cylinder and front brake caliper be renewed as a safety precaution against their sudden failure. This task should be carried out, as described in Chapter 5, every two years.

Every four years, or 20 000 miles (32 000 km)

1 Renew the brake hose – KH100 G

Although the hose is strongly constructed it is under considerable strain and can deteriorate with age. For this reason the manufacturer recommends that the hose be renewed every four years, regardless of its apparent condition. This does not mean that the hose can be forgotten until its time is up; it must be inspected at regular intervals and renewed at the slightest sign of damage or deterioration.

2 Renew the fuel pipe

This must be renewed to prevent the risk of fuel spillage due to an age-hardened hose splitting or cracking. Refer to Chapter 2.

Additional routine maintenance

Certain tasks do not fall under the previous mileage/time headings as they concern items which deteriorate with age, whether the machine is used a great deal, or hardly at all. Two such items are given below:

1 Cleaning the machine

Keeping the motorcycle clean should be considered as an important part of the routine maintenance, to be carried out whenever the need arises. A machine cleaned regularly will not only succumb less speedily to the inevitable corrosion of external surfaces, and hence maintain its market value, but will be far more approachable when the time comes for maintenance or service work. Furthermore, loose or failing components are more readily spotted when not partially obscured by a mantle of road grime and oil.

Surface dirt should be removed using a sponge and warm, soapy water; the latter being applied copiously to remove the particles of grit which might otherwise cause damage to the paintwork and polished surfaces.

Oil and grease are removed most easily by the application of a cleaning solvent such as 'Gunk' or 'Jizer'. The solvent should be applied when the parts are still dry and worked in with a stiff brush. Large quantities of water should be used when rinsing off, taking care that water does not enter the carburettor, air cleaner or electrics.

Application of a wax polish to the cycle parts and a good chrome cleaner to the chrome parts will also give a good finish. Always wipe the machine down if used in the wet, and make sure the chain is well oiled. There is less chance of water getting into control cables if they are regularly lubricated, which will prevent stiffness of action.

2 Decarbonisation

The oily nature of any two-stroke engine's exhaust leads to layers of carbon being deposited in the combustion chamber and exhaust system. If not removed at regular intervals these deposits will build up to the point where the machine's performance and economy are significantly reduced.

It is very difficult to give a precise interval for decarbonisation as so many different factors have to be taken into account. For example if maintenance is neglected so that deposits build up at a faster rate through inefficient combustion, or if the wrong type of engine oil is used. Furthermore, the rider's driving style must be taken into account; any machine that is used principally for fast riding or for long journeys on open roads will not require decarbonisation as often as a machine which is used for short, low-speed commuting trips. Some machines will require decarbonisation, therefore, once a year while others will run satisfactorily for more than twice as long before attention is required. As an initial starting point it is suggested that this task be carried out once a year until sufficient experience is gained for the interval to be shortened or extended as necessary. Note that it may not be necessary to decarbonise the exhaust as frequently as the engine, or vice versa, so the two can be treated as separate items, with their own schedules.

Cylinder head and barrel

Refer to Chapter 1 and remove the cylinder head and barrel. It is necessary to remove all carbon from the head, barrel and piston crown whilst avoiding removal of the metal surfaces on which it is deposited. Take care when dealing with the soft alloy head and piston. Never use a steel scraper or screwdriver. A hardwood, brass or aluminium scraper is ideal as these are harder than the carbon but not harder than the underlying metal. With the bulk of carbon removed, use a brass wire brush. Finish the head and piston with metal polish; a polished surface will slow the subsequent build-up of carbon. Clean out the barrel ports to prevent the restriction of gas flow. Remove all debris by washing each component in paraffin whilst observing the necessary fire precautions. Renew the piston rings, if necessary, on reassembly.

Exhaust system

Instructions on the removal and refitting of the exhaust system are given in Chapter 2.

The exhaust system fitted to these machines is rather easier to clean than most modern machines since it incorporates a baffle tube that can be removed for cleaning purposes, and on KC and KH100 models, a detachable exhaust pipe which can be cleaned using ordinary scraping tools. To extract the baffle tube first remove the retaining bolt and withdraw the rear cover (KH100 models only) then, on all models, remove the retaining screw (KC100) or bolts (KE and KH100) and withdraw the tube. If it has seized in place attempt to work it free by rotating it. If this does not work some method will have to be devised of gripping the baffle hard enough to draw it out but without distorting it and wedging it even harder in the silencer. This is to a large extent up to the owner and the facilities available; do not resort to violence but take the complete exhaust system to an authorized Kawasaki dealer if all else fails.

When the baffle has been removed, strip off the wadding wrapped around it; this should be renewed if it is excessively fouled with carbon and oil deposits. On US KE100 A models remove its retaining bolt and remove the spark arrestor. If the build-up of carbon and oil on the baffle is not too great, wash it clean with petrol whilst taking the necessary fire precautions. Heavy deposits on the baffle may well indicate similar contamination within the silencer and pipe, in which case the system should be removed. **Ensuring that no petrol or vapour is present** so that the risk of fire is avoided, clean the baffle by running a blowlamp along its length to burn off the deposits, waiting for it to cool and tapping it sharply with a length of hardwood to dislodge any

Routine maintenance

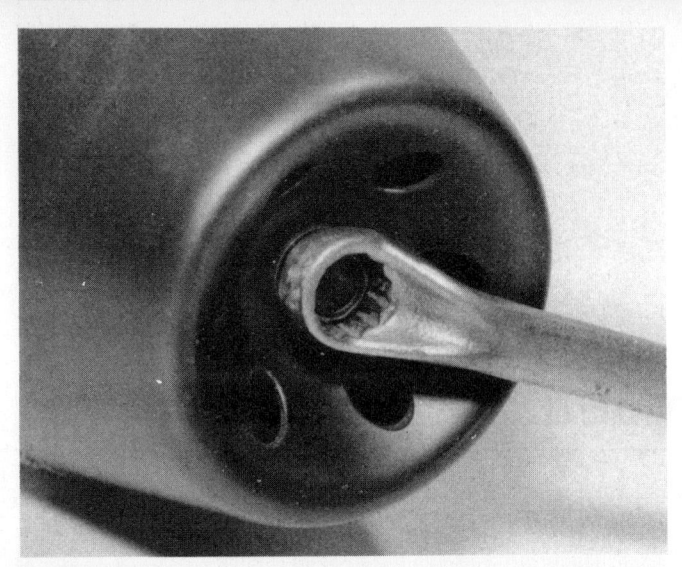

Removing exhaust baffle tube – KH100 – remove bolt to release rear cover ...

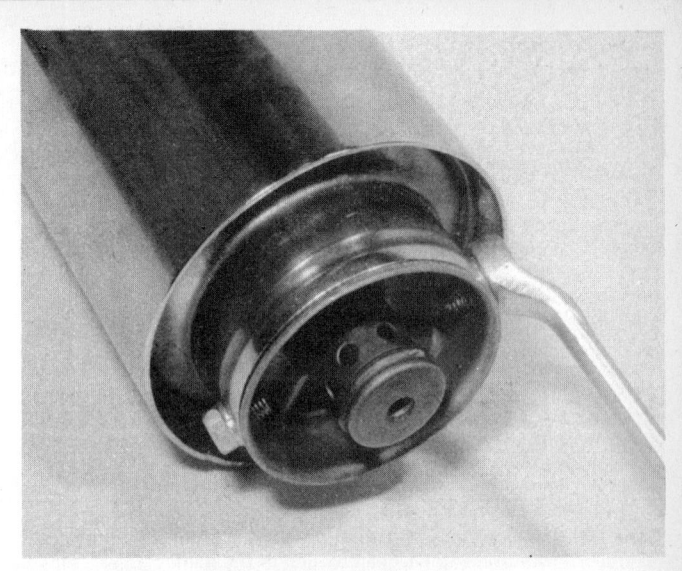

... and unscrew two bolts to release baffle tube

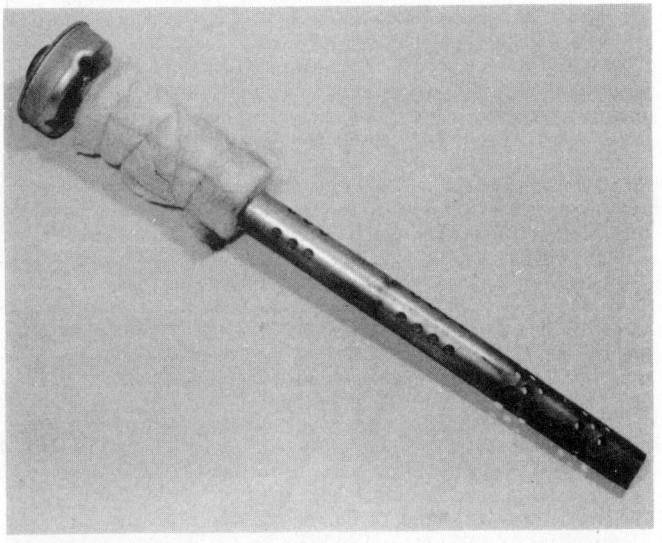

Clean all carbon deposits from baffle tube and renew wadding if heavily fouled

remaining deposits. Finish with a wire brush and check the baffle holes are clear.

Suspend the silencer from its rearmost end. Block the lower end with a cork or wooden bung ensuring enough protrudes for removal purposes. Mix up a caustic soda solution (3 lb to over a gallon of fresh water), adding the soda to the water gradually, whilst stirring. **Do not** pour water into a container of soda, this will cause a violent reaction to take place. Wear proper eye and skin protection; caustic soda is very dangerous. Eyes and skin contaminated by soda must be immediately flushed with fresh water and examined by a doctor. The solution will react violently with aluminium alloy, causing severe damage to any components.

Fill the system with solution, leaving the upper end open. Leave the solution overnight to allow its dissolving action to take place, but be careful that the deposits do not solidify on cooling to block the silencer again. If necessary, repeat the operation until the complete system is clean. Ventilate the area to prevent the build-up of noxious fumes. On completion, carefully pour out the solution and flush the system through with clean, fresh water.

Refit the system. Where applicable, renew the gasket sealing the joint between the two parts of the system.

Do not modify the baffle or run the machine with it removed. This will result in less performance and affect the carburation.

Chapter 1 Engine, clutch and gearbox

Contents

General description	1
Operations with the engine/gearbox unit in the frame	2
Operations with the engine/gearbox unit removed from the frame	3
Methods of preventing crankshaft rotation during overhaul	4
Removing the engine/gearbox unit from the frame	5
Dismantling the engine/gearbox unit: preliminaries	6
Dismantling the engine/gearbox unit: removing the cylinder head, barrel and piston	7
Dismantling the engine/gearbox unit: removing the crankcase right-hand cover	8
Dismantling the engine/gearbox unit: removing the clutch and primary drive pinion	9
Dismantling the engine/gearbox unit: removing the rotary disc valve	10
Dismantling the engine/gearbox unit: removing the gear selector external components	11
Dismantling the engine/gearbox unit: removing the flywheel generator	12
Dismantling the engine/gearbox unit: separating the crankcase halves	13
Dismantling the engine/gearbox unit: removing the crankshaft and gearbox components	14
Dismantling the engine/gearbox unit: removing oil seals and bearings	15
Examination and renovation: general	16
Examination and renovation: engine cases and covers	17
Examination and renovation: bearings and oil seals	18
Examination and renovation: cylinder head	19
Examination and renovation: cylinder barrel	20
Examination and renovation: piston and piston rings	21
Examination and renovation: crankshaft assembly	22
Examination and renovation: primary drive	23
Examination and renovation: clutch	24
Examination and renovation: kickstart mechanism	25
Examination and renovation: gearbox components	26
Gearbox shafts: dismantling and reassembly	27
Engine reassembly: general	28
Reassembling the engine/gearbox unit: preparing the crankcases	29
Reassembling the engine/gearbox unit: refitting the crankshaft and gearbox components	30
Reassembling the engine/gearbox unit: joining the crankcase halves	31
Reassembling the engine/gearbox unit: refitting the flywheel generator	32
Reassembling the engine/gearbox unit: refitting the gear selector external components	33
Reassembling the engine/gearbox unit: refitting the rotary disc valve	34
Reassembling the engine/gearbox unit: refitting the clutch and primary drive	35
Reassembling the engine/gearbox unit: refitting the right-hand crankcase cover	36
Reassembling the engine/gearbox unit: refitting the piston, cylinder barrel and head	37
Refitting the engine/gearbox unit into the frame	38
Starting and running the rebuilt engine	39
Taking the rebuilt machine on the road	40

Specifications

Engine

Bore	49.5 mm (1.95 in)
Stroke	51.8 mm (2.04 in)
Capacity	99 cc (6.04 cu in)
Compression ratio:	
KE100 A8, A9, A10 (US), KE100 B (US)	6.9 : 1
All other models	7.0 : 1
Compression pressure – @ 5 kicks, engine fully warmed up:	
KC100, KH100	7.8 – 12.1 kg/cm² (111 – 172 psi)
KE100	7.4 – 11.5 kg/cm² (105 – 164 psi)

Cylinder head – KE100, KH100

Gasket face maximum warpage	0.05 mm (0.0020 in)

Cylinder barrel – KE100, KH100

	Standard	Service limit
Bore ID:		
KE100 A8, A9, A10 (US), KE100 B (US)	49.499 – 49.514 mm (1.9488 – 1.949 in)	49.580 mm (1.9520 in)
All other KE100 models, KH100	49.508 – 59.523 mm (1.9491 – 1.9497 in)	49.600 mm (1.9528 in)
Taper and ovality	0 – 0.01 mm (0 – 0.0004 in)	0.05 mm (0.0020 in)

Chapter 1 Engine, clutch and gearbox

Piston – KE100, KH100

	Standard	Service limit
Outside diameter:		
KE100 A8, A9, A10 (US), KE100 B (US)	49.477 – 49.492 mm (1.9479 – 1.9485 in)	49.370 mm (1.9437 in)
All other KE100 A models	49.425 – 49.440 mm (1.9459 – 1.9465 in)	49.320 mm (1.9417 in)
KE100 B (UK)	49.380 – 49.395 mm (1.9441 – 1.9447 in)	49.250 mm (1.9390 in)
KH100	49.471 – 49.486 mm (1.9477 – 1.9483 in)	49.330 mm (1.9421 in)
Piston/cylinder clearance:		
KE100 A8, A9, A10 (US) KE100 B (US)	0.017 – 0.027 mm (0.0007 – 0.0011 in)	
All other KE100 A models	0.026 – 0.036 mm (0.0010 – 0.0014 in)	
KE100 B, KH100	0.032 – 0.042 mm (0.0013 – 0.0016 in)	
Gudgeon pin bore in piston:		
Standard	13.998 – 14.005 mm (0.5511 – 0.5514 in)	
Service limit – KE100	14.100 mm (0.5551 in)	
Service limit – KH100	14.070 mm (0.5539 in)	
Gudgeon pin OD:		
Standard – KE100, KH100	13.994 – 14.000 mm (0.5509 – 0.5512 in)	
Service limit – KE100, KH100	13.960 mm (0.5496 in)	
Gudgeon pin/piston standard clearance	0.002 – 0.011 mm (0.0001 – 0.0004 in)	

Piston rings

	Standard	Service limit
Installed end gap:		
KE100 A8, A9, A10 (US), KE100 B (US)	0.15 – 0.35 mm (0.0059 – 0.0138 in)	0.70 mm (0.0276 in)
All other KE100 models, KH100	0.15 – 0.35 mm (0.0059 – 0.0138 in)	0.65 mm (0.0256 in)
Free end gap – approx:		
KE100 B (UK)	6.0 mm (0.2362 in)	5.4 mm (0.2126 in)
All other KE100 models, KH100	7.0 mm (0.2756 in)	6.3 mm (0.2480 in)
Second piston ring thickness – KE100 B (UK)	1.470 – 1.490 mm (0.0579 – 0.0587 in)	1.400 mm (0.0551 in)
Second piston ring groove width – KE100 B (UK)	1.540 – 1.560 mm (0.0606 – 0.0614 in)	1.620 mm (0.0638 in)
Second piston ring/groove clearance – KE100 B (UK)	0.050 – 0.090 mm (0.0020 – 0.0035 in)	0.190 mm (0.0075 in)

Crankshaft – KE100, KH100

	Standard	Service limit
Connecting rod small-end ID	18.003 – 18.014 mm (0.7088 – 0.7092 in)	18.050 mm (0.7106 in)
Big-end bearing radial clearance:		
KE100	0.010 – 0.039 mm (0.0004 – 0.0015 in)	0.060 mm (0.0024 in)
KH100	0.010 – 0.022 mm (0.0004 – 0.0009 in)	0.060 mm (0.0024 in)
Big-end bearing side clearance:		
KE100	0.35 – 0.45 mm (0.0138 – 0.0177 in)	0.60 mm (0.0236 in)
KH100	0.35 – 0.45 mm (0.0138 – 0.0177 in)	0.65 mm (0.0256 in)
Crankshaft runout	0 – 0.03 mm (0 – 0.0012 in)	0.10 mm (0.0039 in)

Primary drive

Reduction ratio	3.524 : 1 (74/21T)	
Maximum permissible backlash – KE100, KH100	0.15 mm (0.0059 in)	

Clutch

	Standard	Service limit
Friction plate thickness	3.1 – 3.3 mm (0.1221 – 0.1299 in)	2.8 mm (0.1102 in)
Friction plate warpage – KE100	0 – 0.30 mm (0 – 0.0118 in)	0.40 mm (0.0158 in)
Friction plate warpage – KH100	0 – 0.30 mm (0 – 0.0118 in)	0.45 mm (0.0177 in)
Plain plate warpage	0.15 mm (0.0059 in)	0.30 mm (0.0118 in)
Friction plate tongue/outer drum clearance – KE100, KH100	0.15 – 0.40 mm (0.0059 – 0.0158 in)	0.50 mm (0.0197 in)
Outer drum centre bush/input shaft clearance:		
KE100 A, KE100 B (US)	0.028 – 0.062 mm (0.0011 – 0.0024 in)	0.162 mm (0.0064 in)
KE100 B (UK)	0.002 – 0.041 mm (0.0001 – 0.0016 in)	0.080 mm (0.0032 in)
KH100	0.028 – 0.058 mm (0.0011 – 0.0023 in)	0.158 mm (0.0062 in)
Spring free length:		
KE100 B (UK)	27.5 – 28.7 mm	26.7 mm (1.0512 in)
All other models	21.6 mm (0.8504 in)	20.0 mm (0.7874 in)

Gearbox

Reduction ratios:	KH100 A2, A3, KE100 A5 to A8	KE100 B (UK)	All other models
1st	2.917 : 1 (35/12T)	2.917 : 1 (35/12T)	2.917 : 1 (35/12T)
2nd	1.765 : 1 (30/17T)	1.733 : 1 (26/15T)	1.765 : 1 (30/17T)
3rd	1.300 : 1 (26/20T)	1.300 : 1 (26/20T)	1.300 : 1 (26/20T)
4th	1.091 : 1 (24/22T)	1.091 : 1 (24/22T)	1.091 : 1 (24/22T)
5th	0.958 : 1 (23/24T)	0.929 : 1 (26/28T)	0.929 : 1 (26/28T)

	Standard	Service limit
Gearchange shaft claw arm spring free length – KE100, KH100	29.4 mm (1.1575 in)	31.0 mm (1.2205 in)
Selector fork claw end thickness – KE100, KH100	4.90 – 5.00 mm (0.1929 – 0.1969 in)	4.80 mm (0.1890 in)
Gear pinion groove width – KE100, KH100	5.05 – 5.15 mm (0.1988 – 0.2028 in)	5.25 mm (0.2067 in)
Selector fork guide pin OD:		
KE100 A5, A6, A7, A8	7.985 – 8.000 mm (0.3144 – 0.3150 in)	7.940 mm (0.3126 in)
KH100 A2, A3	8.000 – 8.015 mm (0.3150 – 0.3156 in)	7.950 mm (0.3130 in)
All other KE100 models	5.900 – 6.000 mm (0.2323 – 0.2362 in)	5.850 mm (0.2303 in)
All other KH100 models	5.900 – 6.000 mm (0.2323 – 0.2362 in)	5.800 mm (0.2284 in)
Selector drum groove width:		
KE100 A5, A6, A7, A8	8.000 – 8.015 mm (0.3150 – 0.3156 in)	8.200 mm (0.3228 in)
KH100 A2, A3	8.050 – 8.200 mm (0.3169 – 0.3228 in)	8.250 mm (0.3248 in)
All other KE100 models	6.050 – 6.200 mm (0.2382 – 0.2441 in)	6.250 mm (0.2461 in)
All other KH100 models	6.050 – 6.200 mm (0.2382 – 0.2441 in)	6.300 mm (0.2480 in)
Kickstart gear ID – KE100, KH100	20.000 – 20.021 mm (0.7874 – 0.7882 in)	20.070 mm (0.7902 in)
Kickstart shaft OD (at gear) – KE100, KH100	19.959 – 19.980 mm (0.7858 – 0.7866 in)	19.920 mm (0.7843 in)
Gear pinion/shaft clearance – KE100, KH100:		
Input 5th gear, output 1st gear	0.016 – 0.045 mm (0.0006 – 0.0018 in)	0.145 mm (0.0057 in)
Input 4th gear, output 2nd, 3rd gear	0.016 – 0.052 mm (0.0006 – 0.0021 in)	0.152 mm (0.0060 in)
Kickstart idler gear/input shaft – KE100 A	0.040 – 0.074 mm (0.0016 – 0.0029 in)	0.174 mm (0.0069 in)
Kickstart idler gear/input shaft – KE100 B (UK)	0.028 – 0.058 mm (0.0011 – 0.0023 in)	0.100 mm (0.0039 in)
Kickstart idler gear/input shaft – all other models	0.028 – 0.058 mm (0.0011 – 0.0023 in)	0.158 mm (0.0062 in)
Kickstart idler gear/output shaft	0.032 – 0.061 mm (0.0013 – 0.0024 in)	0.161 mm (0.0063 in)

Final drive

Reduction ratio:	
KC100, KH100 G5, G6, G7, G8	2.643 : 1 (37/14T)
KE100 A	2.800 : 1 (42/15T)
KE100 B1 to B9 (UK), B1 to B12 (US)	2.667 : 1 (40/15T)
KE100 B10, B11 (UK)	2.600 : 1 (39/15T)
KH100 A, KH100 G2, G3, G4	2.786 : 1 (39/14T)
Chain size	428 (1/2" x 5/16")
Number of links:	
KC100	104
KE100 A	110
KE100 B	112
KH100	106
Length of 20 links – chain pulled taut, 1-21 pins:	
Standard	254.0 mm (10.0000 in)
Service limit	259.0 mm (10.1968 in)

Torque wrench settings – KE100, KH100

Component	kgf m	lbf ft
Spark plug	2.5 – 3.0	18 – 22
Cylinder head nuts	2.2	16
Neutral detent plunger bolt	2.8	20
Generator rotor nut:		
KE100 – maximum	5.0	36
KH100	4.2	30
Gearbox sprocket bolt	2.2 – 2.5	16 – 18

Chapter 1 Engine, clutch and gearbox

Component	kgf m	lbf ft
Primary drive pinion nut:		
KE100	7.0 – 7.5	50.5 – 54
KH100	4.8	34.5
Clutch spring bolts	0.4 – 0.5	3 – 3.5
Engine mounting bolt nuts:		
KE100	2.6 – 3.5	19 – 25
KH100	3.5	25
Kickstart pedal pinch bolt – KH100 A4, KH100 G	2.0	14.5
Transmission oil drain plug:		
KE100 A, KH100 G	0.7 – 1.0	5 – 7
KE100 B, KH100 A	2.0	14.5

1 General description

The crankshaft is a built-up assembly which rotates on two ball journal main bearings and is fitted with needle roller bearings at the connecting rod big- and small-ends. A flywheel generator is fitted on its left-hand end, a gear on its right-hand end providing the drive for the clutch; the disc valve assembly is fitted on the crankshaft, inboard of the primary drive gear. The clutch is a wet multi-plate type mounted on the right-hand end of the gearbox input shaft, the gearbox itself being of the constant-mesh type, built in unit with the engine. All engine and gearbox components are housed in the vertically split, aluminium alloy engine unit castings.

The engine/gearbox unit is simple in design and construction, requiring the bare minimum of special tools during dismantling and overhaul.

2 Operations with the engine/gearbox unit in the frame

The following items can be removed with the engine/gearbox unit in the frame:

  a) Cylinder head, barrel and piston, carburettor and disc valve
  b) Oil pump
  c) Clutch assembly and primary drive gear pinion
  d) Gear selector mechanism external components
  e) Kickstart spring
  f) Flywheel generator assembly
  g) Gearbox sprocket and neutral switch

3 Operations with the engine/gearbox unit removed from the frame

It will be necessary to remove the complete engine/gearbox unit from the frame and to separate the crankcase halves to gain access to the following components:

  a) Crankshaft, main bearings and oil seals
  b) Gearbox shafts and bearings, gear selector drum and forks
  c) Kickstart shaft and pawl assembly

Note that while it is possible to remove and refit the crankshaft main bearing oil seals without separating the crankcase halves, this requires a great deal of care and is not recommended for the reasons given in Section 15.

4 Methods of preventing crankshaft rotation during overhaul

When removing the flywheel generator and primary drive gear during an overhaul it will be necessary to prevent the crankshaft from turning as the retaining nut is slackened. Kawasaki produce some excellent service tools for this purpose, but these may prove expensive for the owner, who will require them infrequently. As an alternative, the approaches described below can be tried, choosing whichever method is the more convenient.

  a) Use a chain or strap wrench around the edge of the generator rotor. This method can be used for rotor nut removal, but is not suitable for slackening the primary drive pinion nut.

  b) If the engine is in the frame and the clutch is in place, select **top** gear and apply the rear brake to lock the entire drive train.

  c) If the engine is on the workbench, make up a simple holding tool using the gearbox sprocket (see photograph). Again, select **top** gear to minimise leverage effects.

  d) If the cylinder head, barrel and piston have been removed, place a hardwood block on each side of the crankcase mouth, then pass a **smooth** steel bar through the small end eye. The bar will rest against the blocks, preventing further rotation.

  e) If a tightly-wadded piece of rag is wedged in the angle between the teeth of the clutch outer drum and the primary drive pinion, this will save time if just the primary drive is to be dismantled.

5 Removing the engine/gearbox unit from the frame

1 If the machine is dirty, it is advisable to wash it thoroughly before starting any major dismantling work. This will make work much easier and will prevent the risk of disturbed lumps of caked-on dirt falling into some vital component.
2 Drain the transmission oil as described in Routine Maintenance. While the oil is draining remove the fuel tank and exhaust system as described in Chapter 2, Sections 2 and 11 respectively. Remove also the side panel(s).
3 Note that whenever any component is removed, all mounting nuts, bolts or screws should be refitted in their original locations with their respective washers and mounting rubbers (where fitted).
4 Work is made much easier if the machine is lifted to a convenient height on a purpose-built ramp or a platform constructed of planks and concrete blocks. Ensure that the wheels are chocked with wooden blocks so that the machine cannot move and that it is securely tied down so that it cannot fall, also that the stand is supporting it correctly.
5 Disconnect the battery (negative lead first), to prevent any risk of short circuits. If the machine is to be out of service for some time, remove the battery and give it regular refresher charges as described in Chapter 6.
6 Marking its shaft so that it can be refitted in the same position, fully remove the kickstart lever pinch bolt and pull the lever off the shaft splines. Remove its mounting screws and withdraw the oil pump cover from the rear of the crankcase right-hand cover. Remove the oil pump and the carburettor, described in Sections 12 and 14 of Chapter 2. Remove the oil and fuel feed pipes completely from the crankcase right-hand cover.
7 Referring to Routine Maintenance, disconnect the clutch cable and, on KH100 G models only, the rev-counter cable.
8 Slacken the clamp securing the air filter duct to the crankcase top surface and remove the single air filter mounting bolt. Withdraw the filter assembly.
9 Marking its shaft so that it can be refitted in the same position, fully remove the gearchange pedal pinch bolt and pull the lever off the shaft splines. Remove its mounting screws and withdraw the crankcase left-hand cover. On KC100 models remove the six mounting bolts and the clamp bolt at the rear end, then remove the two halves of the final drive chain enclosure.
10 Flatten back the raised portion of its locking washer, then remove the gearbox sprocket retaining bolt, applying the rear brake hard to prevent rotation. Remove the sprocket from the shaft, disengage it from the chain and hang the chain over the swinging arm pivot. Note that it may be necessary to slacken the chain adjustment to permit sprocket removal. On KC100 models, remove the two mounting screws and withdraw the chain enclosure inner plate.

Chapter 1 Engine, clutch and gearbox

11 Tracing the generator lead from the crankcase top surface up to the connectors joining it to the main loom, disconnect all electrical wires and release the lead from any clamps or ties securing it to the frame. Pull the spark plug cap off the plug. The engine/gearbox unit should now be retained only by its four mounting bolts; check carefully that all components have been removed or disconnected so that nothing will prevent or hinder the removal of the unit.

12 Remove their retaining nuts and tap out the four mounting bolts, then withdraw the engine/gearbox unit. If any bolt is locked in place with corrosion or dirt, apply a liberal quantity of penetrating fluid, allow time for it to work, and then release the bolt by rotating it with a spanner before tapping it out with a hammer and drift.

6 Dismantling the engine/gearbox unit: preliminaries

1 Before any dismantling work is undertaken, the unit should be thoroughly cleaned. This will prevent the contamination of the engine internals, and will also make working a lot easier and cleaner. A high flash-point solvent, such as paraffin can be used or a proprietary engine degreaser such as Gunk. Use old paintbrushes and toothbrushes to work the solvent into the various recesses. Take care to exclude solvent or water from the electrical components and inlet and exhaust ports. The use of petrol should be avoided because of the fire risk.

2 When clean and dry, arrange the unit on the workbench, leaving a suitable clear area for working. Gather a selection of small containers and plastic bags so that parts can be grouped together in an easily identifiable manner. Some paper and a pen should be on hand to permit notes to be made and labels attached where necessary. A supply of clean rag is also required.

3 Before commencing work, read through the appropriate section so that some idea of the necessary procedure can be gained. When removing the various engine components it should be noted that great force is seldom required, unless specified. In many cases, a component's reluctance to be removed is indicative of an incorrect approach or removal method. If in any doubt, re-check with the text.

5.6 Mark kickstart shaft prior to removal so that lever can be correctly refitted

5.7a Disconnecting rev-counter cable – KH100 G models – remove retaining screw ...

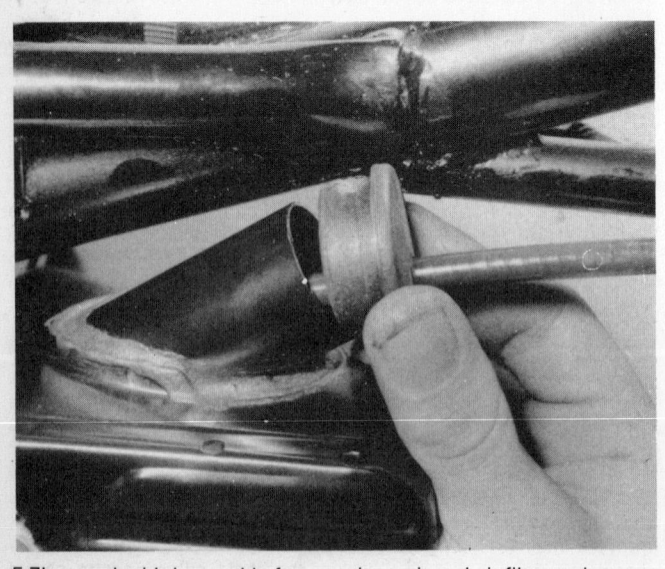

5.7b ... and withdraw cable from engine unit and air filter casing

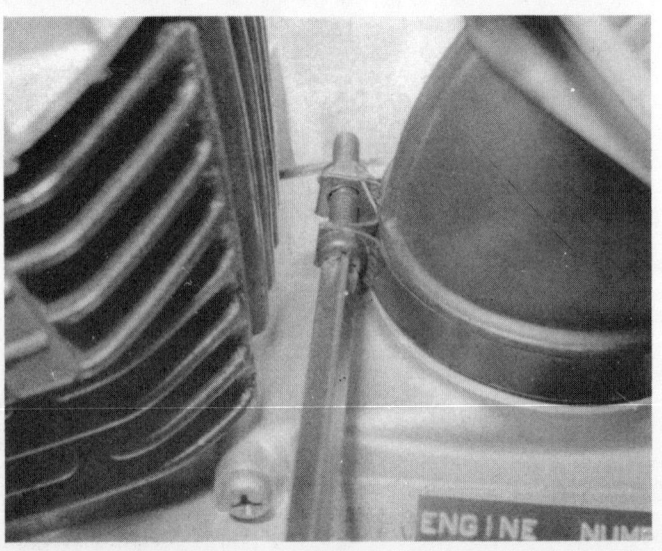

5.8 Slacken clamp to release air filter duct from crankcase top surface

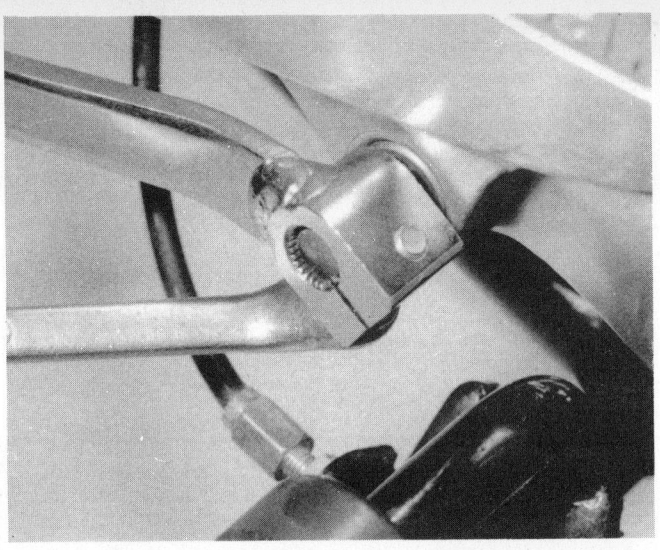

5.9 Mark gearchange lever shaft so that lever can be refitted in same position

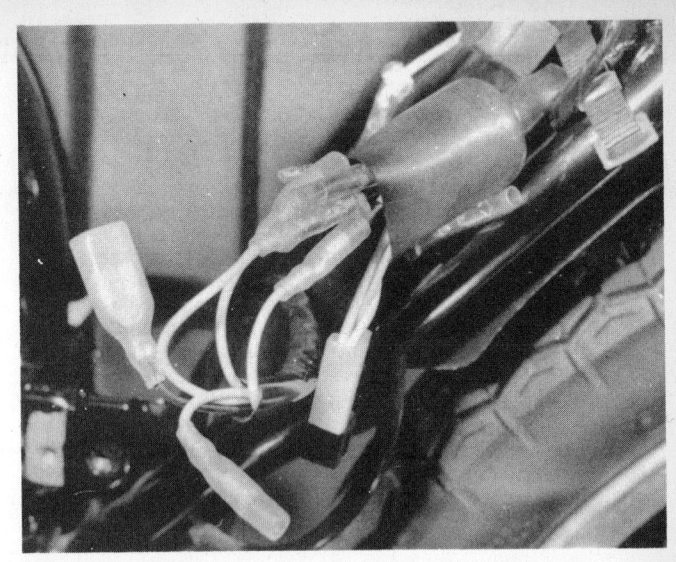

5.11 Trace generator lead up to connectors and disconnect from wiring loom

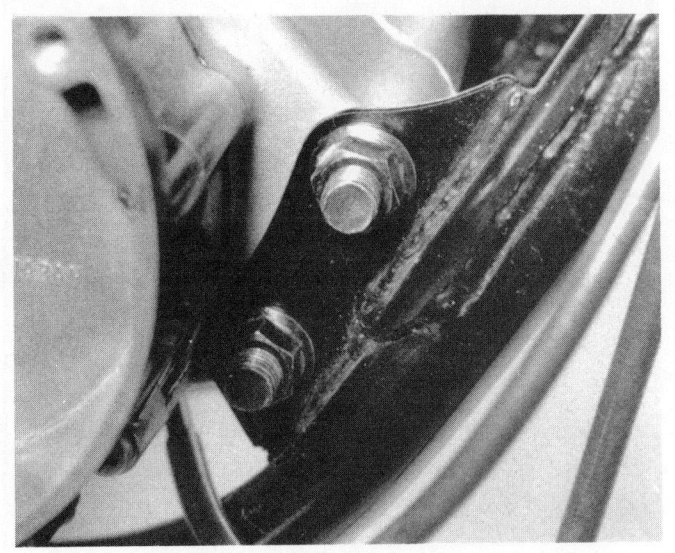

5.12a Remove their retaining nuts and withdraw the two engine front mounting bolts ...

5.12b ... the upper rear mounting bolt ...

5.12c ... and the lower rear mounting bolt

7 Dismantling the engine/gearbox unit: removing the cylinder head, barrel and piston

1 These components can be removed with the engine/gearbox unit in or out of the frame, but in the former case the HT lead must be disconnected from the spark plug, and the exhaust system must be removed. Disconnect the spark plug cap and remove the spark plug.

2 Working in a diagonal sequence and by one turn at a time to avoid the risk of distortion, remove the four cylinder head retaining nuts (and lock washers, where fitted), then lift away the head and its gasket.

3 Bring the piston to the top of its stroke, then lift the barrel just enough to expose the bottom of the piston skirt. It may be necessary to use a soft-faced mallet to tap gently around the barrel to break the seal of the base gasket; take great care not to damage the fins. Pack a wad of clean rag in the crankcase mouth to prevent dirt or debris from dropping in, then lift the barrel off the piston.

4 Use a sharp-pointed instrument or a pair of needle-nose pliers to remove one of the gudgeon pin retaining circlips, press out the gudgeon pin far enough to clear the connecting rod and withdraw the piston. Push out the small-end bearing. Discard the used circlips and obtain new ones for reassembly.

5 If the gudgeon pin is a tight fit in the piston, soak a rag in boiling water, wring it out and wrap it around the piston; the heat will expand the piston sufficiently to release its grip on the pin. If necessary, the pin may be tapped out using a hammer and drift, but take care to support the piston and connecting rod firmly while this is done.

6 The piston rings can be removed by holding the piston in both

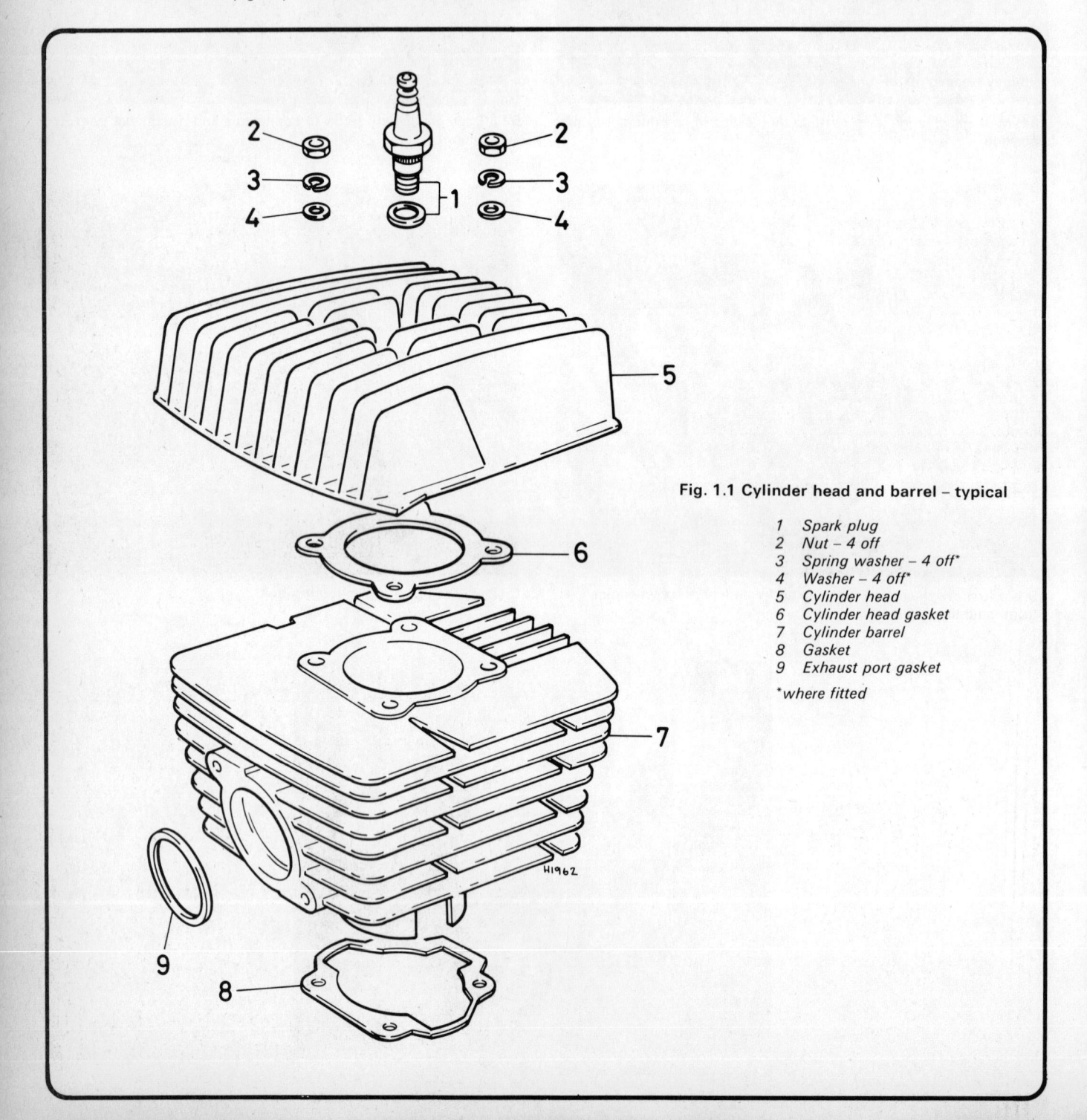

Fig. 1.1 Cylinder head and barrel – typical

1 Spark plug
2 Nut – 4 off
3 Spring washer – 4 off*
4 Washer – 4 off*
5 Cylinder head
6 Cylinder head gasket
7 Cylinder barrel
8 Gasket
9 Exhaust port gasket

*where fitted

Chapter 1 Engine, clutch and gearbox

hands and prising the ring ends apart gently with the thumbnails until the rings can be lifted out of their grooves and on to the piston lands, one side at a time. The rings can then be slipped off the piston and put to one side for cleaning and examination. If the rings are stuck in their grooved by excessive carbon deposits, use three strips of thin metal sheet to remove them, as shown in the accompanying illustration. Be careful, as the rings are brittle and will break easily if overstressed.

8 Dismantling the engine/gearbox unit: removing the crankcase right-hand cover

1 This may be done with the engine/gearbox unit in or out of the frame, but in the former case the transmission oil must be drained, the kickstart lever and carburettor removed, the oil pump cable and tank/pump feed pipe disconnected and, with the fuel feed pipe withdrawn clear of the crankcase cover. The clutch cable must be disconnected. It may be necessary to remove the right-hand footrest assembly (three bolts) on early KE100 A models, and the exhaust system on KH100 models. Refer to Section 5. While removal of the oil pump is not strictly necessary, it is advisable. Refer to Chapter 2, Section 12.
2 Working progressively and in a diagonal sequence from the outside in, slacken the cover retaining screws and withdraw them. Store them in a cardboard template of the cover to provde a guide to their correct positions on reassembly.
3 Pull the cover away, tapping gently with a soft-faced mallet to break the seal. Be careful that there are no components from the clutch release mechanism sticking to the cover and ensure that the oil transfer pipe remains in the valve cover. Peel off the gasket. Unless the locating dowel is firmly fixed in the crankcase, it should be removed and stored with the cover. Similarly, the oil transfer pipe should be stored in a safe place to prevent its loss.

9 Dismantling the engine/gearbox unit: removing the clutch and primary drive pinion

1 These components can be removed after the crankcase right-hand cover has been withdrawn as described in the previous Section.
2 Remove the components of the release mechanism from the clutch.

Note: while it is apparently possible to speed up work by merely removing the retaining circlip at this stage, this is inadvisable due to the possible effect of clutch spring pressure and the difficulty of aligning all components on reassembly. If the primary drive gear is to be removed, note the comments made in Section 4 before removing the clutch.
2 Avoid distortion of the clutch spring plate by slackening its retaining bolts evenly and in a diagonal sequence. Remove the bolts and springs, followed by the spring plate. Remove the retaining circlip and the thrust washer. Lift away the clutch centre and plates as a single assembly, followed by the single thrust washer.
3 On UK KE100 B models remove the second (wire type) retaining circlip. On all models remove the clutch outer drum. On UK KE100 B models this is followed by a wave washer and the Woodruff key from its keyway in the kickstart idler sleeve gear; on all other models the outer drum is followed by a coil spring and the second thrust washer.
4 Flatten back the raised portion of the primary drive gear retaining nut lock washer. Lock the crankshaft by one of the methods described in Section 4 of this Chapter and unscrew the nut. Remove the lock washer and withdraw the primary drive gear. Displace the Woodruff key from its crankshaft keyway.

7.4 Removing one of the gudgeon pin retaining circlips – always renew disturbed circlips

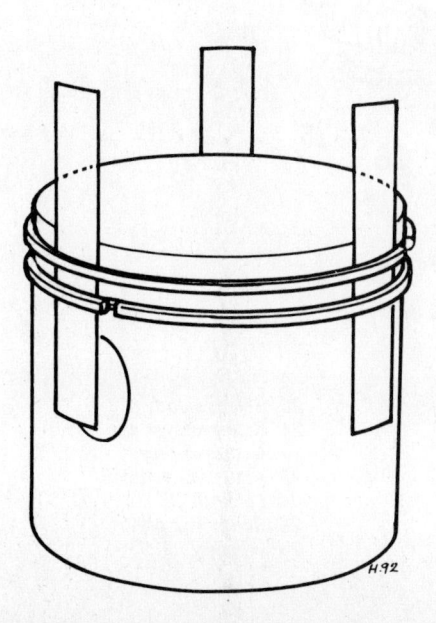

Fig. 1.2 Method of removing gummed piston rings

9.2 Before dismantling clutch check for any balancing marks which may require alignment on reassembly

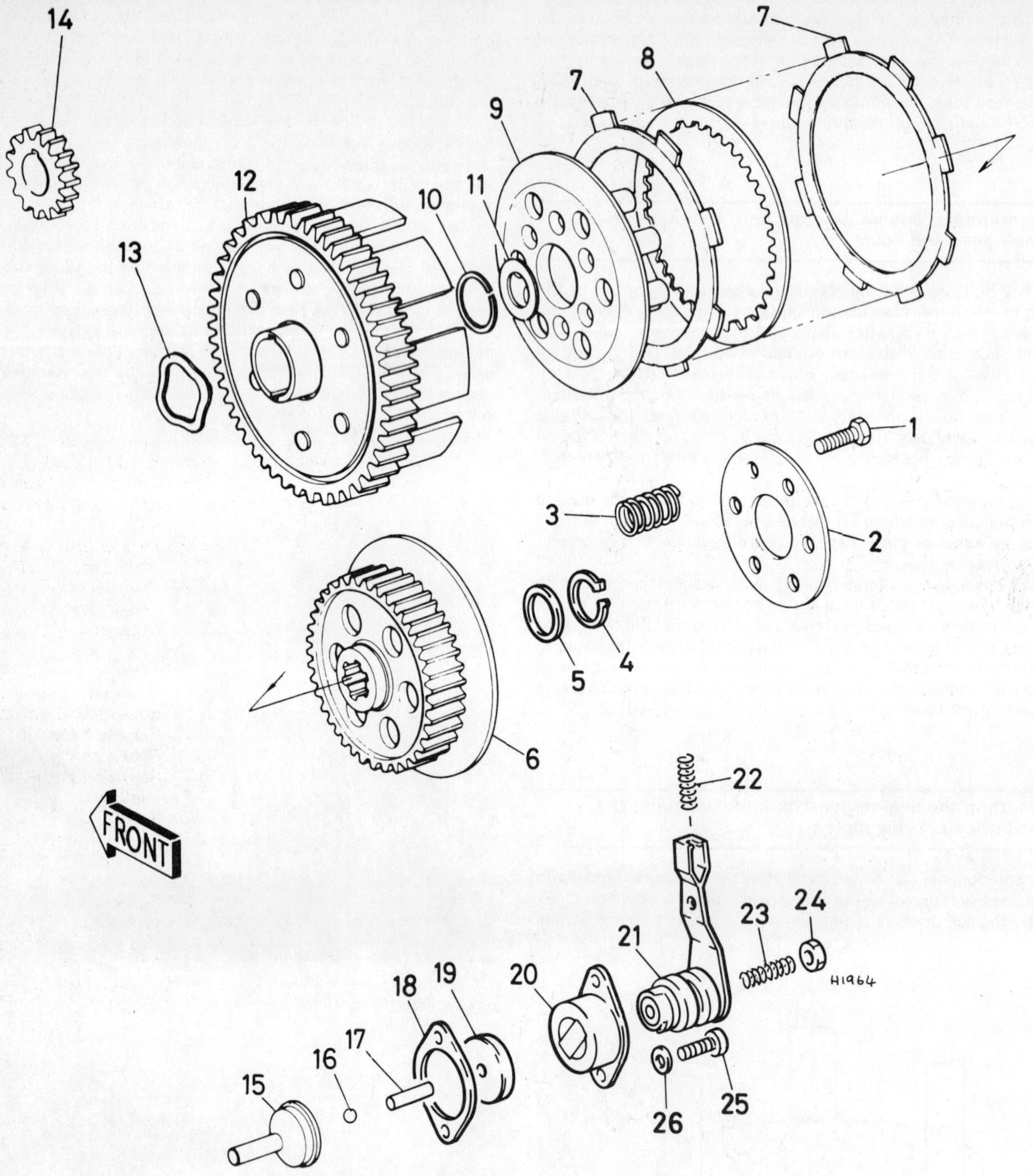

Fig. 1.3 Clutch – KE100 B UK model

1 Bolt – 6 off	10 Circlip	19 Oil seal
2 Spring plate	11 Thrust washer	20 Release body
3 Spring – 6 off	12 Outer drum	21 Release lever assembly
4 Circlip	13 Wave washer	22 Return spring
5 Thrust washer	14 Primary drive gear	23 Adjusting screw
6 Clutch centre	15 Lifter piece	24 Locknut
7 Friction plate – 4 off	16 Ball	25 Screw – 2 off
8 Plain plate – 3 off	17 Pushrod	26 Washer – 2 off
9 Pressure plate	18 Gasket	

Chapter 1 Engine, clutch and gearbox 55

1 Bolt – 6-off
2 Spring plate
3 Spring – 6 off
4 Circlip
5 Thrust washer
6 Clutch centre
7 Clutch centre – KH100 A2, KE100 A5, A6, A7
8 Clutch outer plate – KH100 A2, KE100 A5, A6, A7
9 Friction plate – 4 off
10 Plain plate – 3 off
11 Pressure plate
12 Thrust washer
13 Outer drum
14 Coil spring
15 Thrust washer
16 Primary drive gear
17 Release lever assembly
18 Release body
19 Return spring
20 Adjusting screw
21 Locknut
22 Screw – 2 off
23 Washer – 2 off
24 Oil seal
25 Gasket
26 Pushrod
27 Lifter piece
28 Ball*
29 Lifter piece*

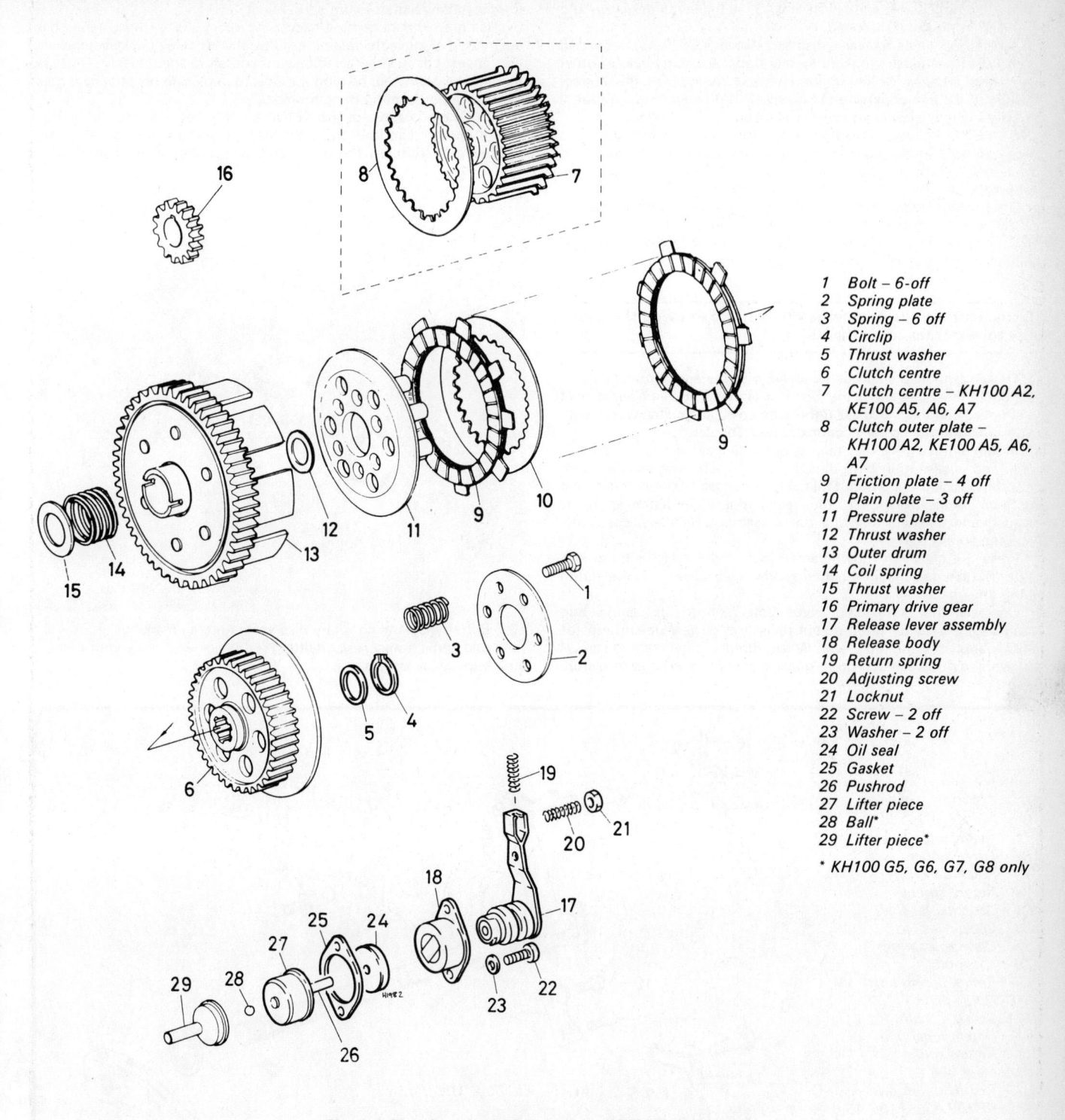

* KH100 G5, G6, G7, G8 only

Fig. 1.4 Clutch – all models except KE100 B UK model

10 Dismantling the engine/gearbox unit: removing the rotary disc valve

1 The valve assembly cannot be removed until the crankcase right-hand cover has been withdrawn and the clutch and primary drive have been dismantled, as described in the previous Sections of this Chapter.
2 **Note:** *before disturbing the valve, have ready a spirit-based felt marker or a similar means of marking delicate metal objects without* harming them. Mark the outer faces of all valve components as they are removed and take notes, if required. The valve timing marks are vague and not always clearly marked, so **confusion is easy on reassembly if suitable precautions are not taken**.
3 Using an impact driver if necessary and working evenly and in a diagonal sequence to avoid the risk of distortion, slacken and remove the valve cover retaining screws. Carefully withdraw the cover, noting the presence of the large sealing O-ring and the two locating dowels; **do not** disturb the valve disc yet.

Chapter 1 Engine, clutch and gearbox

4 On all models except the KE100 A5, withdraw the spacer and O-ring from the crankshaft end.
5 Carefully examine the valve disc and sleeve (KE100 A5) or splined collar (all other models) looking for any signs of timing marks or other identifying features which might be used to indicate the correct position of the components on reassembly. If no marks can be found, or if the existing ones are unclear, make your own as follows:
6 On KE100 A5 models mark the valve outer face and withdraw it; it is located by a single pin and cannot, therefore, be fitted wrongly in this respect. Withdraw the inner sleeve and O-ring and pin from the crankshaft.
7 On all other models mark the valve disc outer face and also make a 'timing' mark from the disc to the splined collar to show their exact relationship. Remove the valve disc. Mark the outer face of the splined collar and remove this, followed by its locating pin.

11 Dismantling the engine/gearbox unit: removing the gear selector external components

1 These components can be removed with the engine/gearbox unit in or out of the frame, but in the former case the gearchange pedal must be removed and the crankcase right-hand cover and clutch withdrawn, as described in the earlier Sections of this Chapter.
2 Before removing the mechanism, press the gearchange pedal up or down and release it; in both directions the mechanism should return smoothly, quickly and without free play to the rest position. If not, or if any free play is discernible, the gearchange shaft return spring is fatigued and must be renewed on reassembly. Similarly, check the claw arm spring.
3 Press the claw arm downwards clear of the selector drum and withdraw the gearchange shaft from the crankcase with the return spring and spacer.
4 Unscrew the shouldered pivot bolt, unhook its spring and withdraw the selector drum detent roller arm or gear detent arm (as appropriate) from the drum end. Where fitted, remove its two screws and withdraw the selector drum guide plate. There is need to disturb the gearchange shaft return spring post.
5 Using an impact driver if necessary where a cross-head screw has been fitted, remove the retaining screw and lift away the selector drum end cover. Lift out the selector pins and store them safely. On later models the drum end, holding the selector pins, can be detached; note the dowel pin locating the pin holder.
6 On all models except the KE100 A5 and A6, unscrew the neutral detent plunger cap from the crankcase top surface, noting its sealing washer, and withdraw the spring and the plunger or ball beneath it.

10.2 Before dismantling rotary disc valve note all markings (arrowed) which may assist refitting – if necessary, make your own timing marks as shown

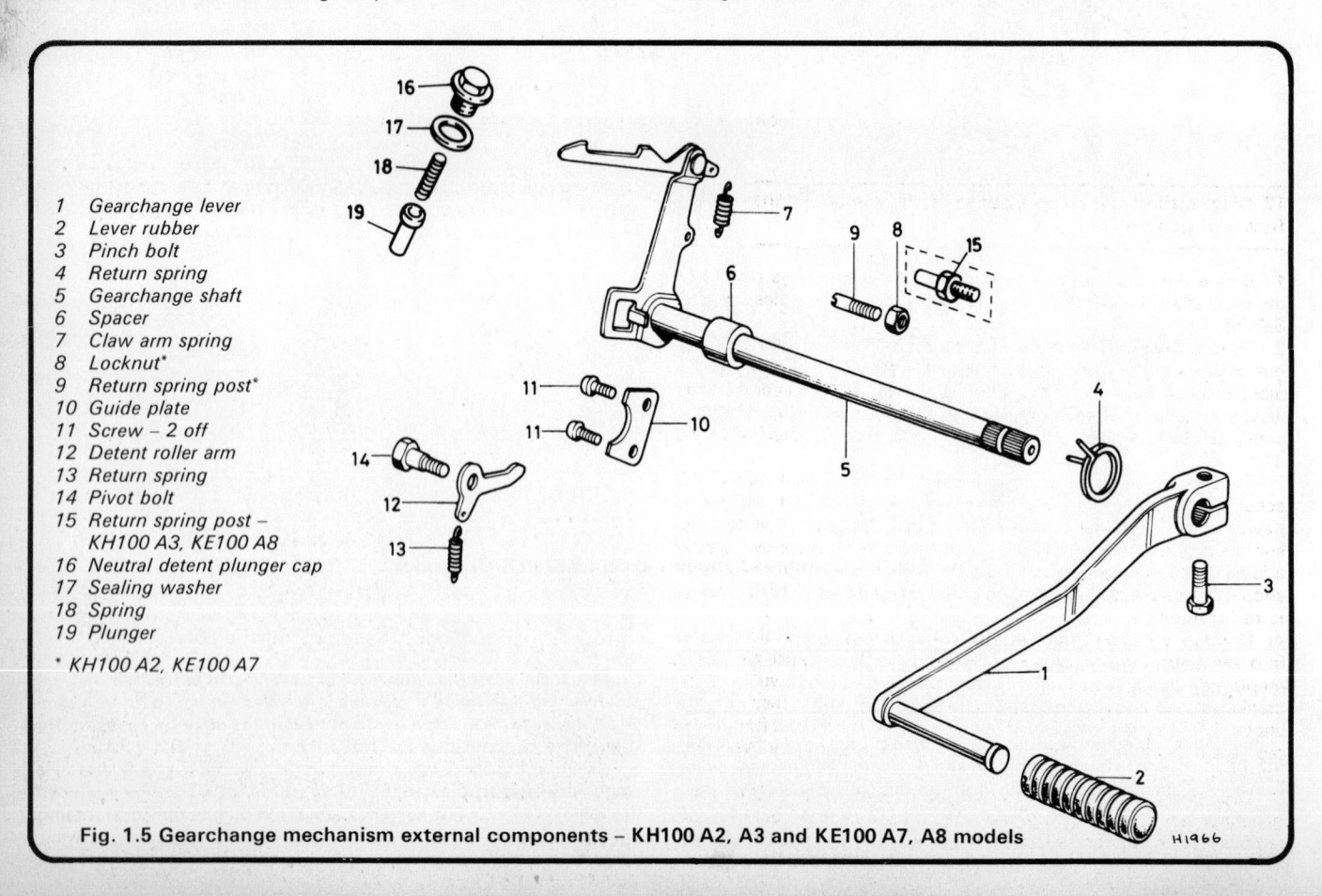

1 Gearchange lever
2 Lever rubber
3 Pinch bolt
4 Return spring
5 Gearchange shaft
6 Spacer
7 Claw arm spring
8 Locknut*
9 Return spring post*
10 Guide plate
11 Screw – 2 off
12 Detent roller arm
13 Return spring
14 Pivot bolt
15 Return spring post – KH100 A3, KE100 A8
16 Neutral detent plunger cap
17 Sealing washer
18 Spring
19 Plunger

* KH100 A2, KE100 A7

Fig. 1.5 Gearchange mechanism external components – KH100 A2, A3 and KE100 A7, A8 models

Chapter 1 Engine, clutch and gearbox 57

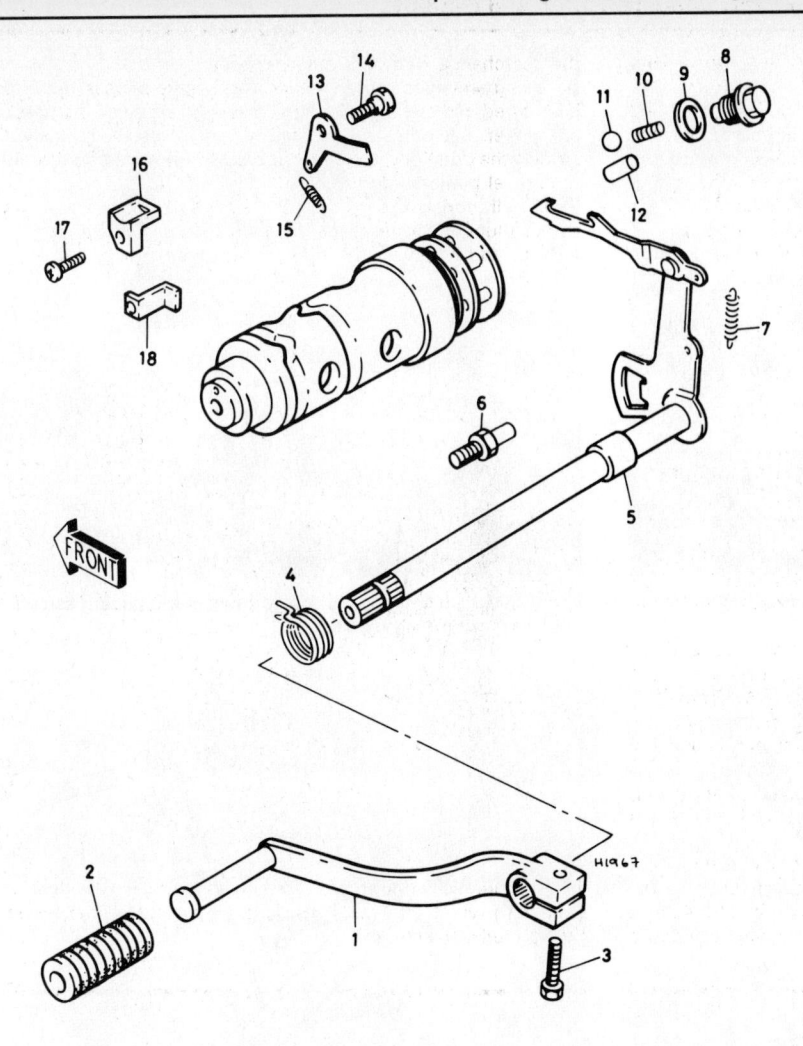

1 Gearchange lever
2 Lever rubber
3 Pinch bolt
4 Return spring
5 Gearchange shaft
6 Return spring post
7 Claw arm spring
8 Neutral detent plunger cap
9 Sealing washer
10 Spring
11 Detent ball – KC100 C3, C4, KE100 B, KH100 G
12 Detent plunger – all other models
13 Detent roller arm
14 Pivot bolt
15 Return spring
16 Cast alloy stop – KE100 B, KH100 G5, G6, G7, G8
17 Screw
18 Bent metal stop – all other models

Fig. 1.6 Gearchange mechanism external components – all other models

12 Dismantling the engine/gerbox unit: removing the flywheel generator

1 Before the generator can be removed, the gearchange pedal and the crankshaft left-hand cover must be withdrawn, as described in Section 5.
2 To lock the crankshaft to permit the removal of the rotor retaining nut use one of the methods outlined in Section 4 of this Chapter. A more positive form of holding tool is shown in the accompanying illustration and will prove generally useful once made up. With the crankshaft locked, remove the retaining nut and the washer(s) behind it.
3 Remove the rotor using only a centre bolt flywheel puller (see accompanying photograph). These are available from authorized Kawasaki dealers under various part numbers, but pattern versions to suit the majority of small-capacity Japanese machines are available at a lower price from most good motorcycle dealers. One of these tools should be considered essential; **no other method of rotor removal is recommended.**
4 Unscrew the tool centre bolt and screw the tool body anticlockwise into the thread in the centre of the rotor (a **left-hand thread** is employed) until it seats firmly. Tighten the centre bolt down on to the crankshaft end and tap smartly on the bolt head with a hammer. The shock should jar the rotor free, if not, tighten the centre bolt further and tap again. Withdraw the rotor and its Woodruff key from the crankshaft.
5 Disconnect the lead from the neutral indicator switch. Unclip the generator lead from the crankcase, note the stator fitted position and remove its two countersunk screws; lift away the stator plate.

6 If required, the neutral indicator switch can be removed from the crankcase by removing its two retaining screws. The switch contact is secured by a single screw to the selector drum end.

12.3 Generator rotor must be removed using only special tool, as shown

58   Chapter 1 Engine, clutch and gearbox

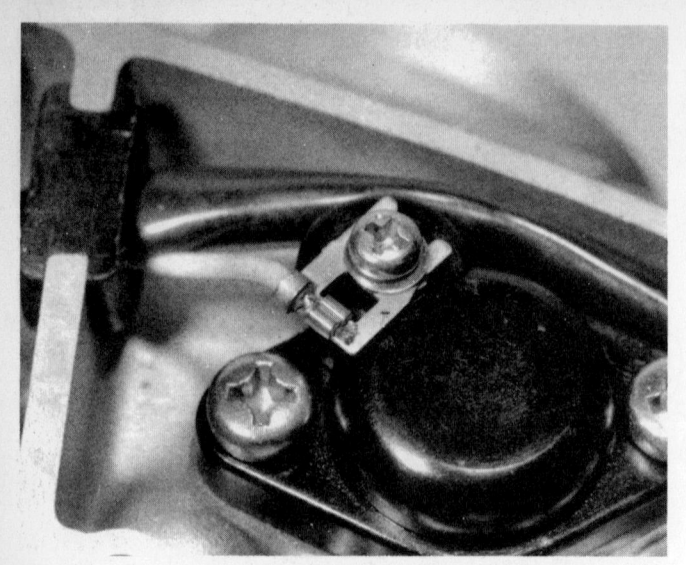

12.5a Neutral indicator switch lead is secured by a single screw

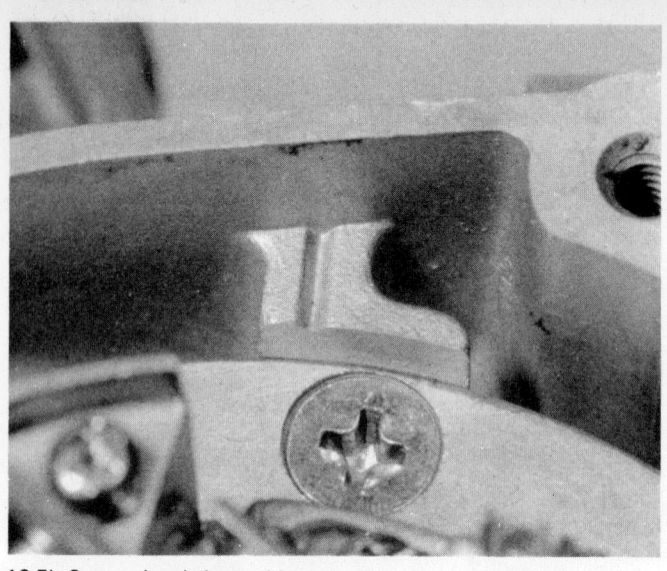

12.5b Stator plate is located by countersunk retaining screws – no need to make timing marks

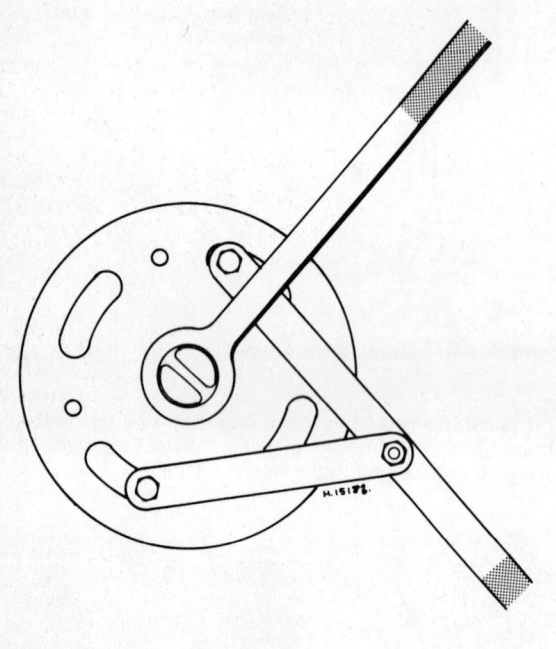

Method of use

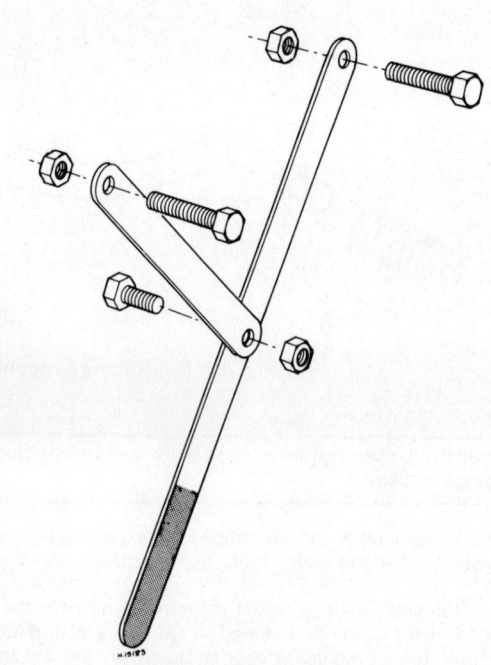

Construction of tool

Fig. 1.7 Fabricated rotor holding tool

13 Dismantling the engine/gearbox unit: separating the crankcase halves

1 The crankcases can be separated only after the engine/gearbox unit has been removed from the frame and all preliminary dismantling operations described in Sections 6 – 12 of this Chapter have been carried out.
2 Withdraw the guide from the kickstart return spring and grasp the spring right-hand end with a strong pair of pliers. Release the spring from the shaft and allow it to unwind slowly to the relaxed position, then withdraw it.

3 On all models except those fitted with rev-counters, withdraw the oil pump drive components from the crankcase. On models equipped with rev-counters the drive gear is on the inside of the crankcase so the shaft cannot be extracted until the crankcases have been separated. It is, however, possible to remove the retaining circlip and to release the drive gear, Woodruff key and thrust washers before the crankcases are separated.
4 Withdraw the spacer from the output shaft left-hand end, noting the O-ring behind it.
5 Make a final check that all components have been removed which might hinder crankcase separation. Make a cardboard template, marking their respective positions, and remove all crankcase fastening

Chapter 1 Engine, clutch and gearbox

screws, slackening them progressively and in a diagonal sequence from the outside inwards.

6   The crankcase left-hand half should be lifted away, leaving the crankshaft and gearbox components in the right-hand half.

7   Using only a soft-faced mallet, tap gently on the exposed ends of the crankshaft and gearbox shafts and all around the joint area of the two crankcase halves until initial separation is achieved. Lift off the half to be removed, ensuring it remains absolutely square so that the bearings do not stick on their respective shafts; tap gently on the shaft ends to assist removal. **Do not** use excessive force and **never** attempt to lever the cases apart. If undue difficulty is encountered, tap the cases back together and start again. If all else fails, take the assembly to an authorised Kawasaki dealer for the cases to be separated using special tool Part number 57001-153.

8   If difficulty is encountered in achieving initial separation, it may be because corrosion has formed on the locating dowel pins. Apply a quantity of penetrating fluid to the joint area and inside the various mounting bolt passages, allow time for it to work and start again.

9   When the casing is removed, check that there are no loose components such as dowels or thrust washers which might drop clear and be lost. Any such components should be refitted in their correct locations.

13.3a Where rev-counter is fitted, remove circlip to release oil pump drive pinion ...

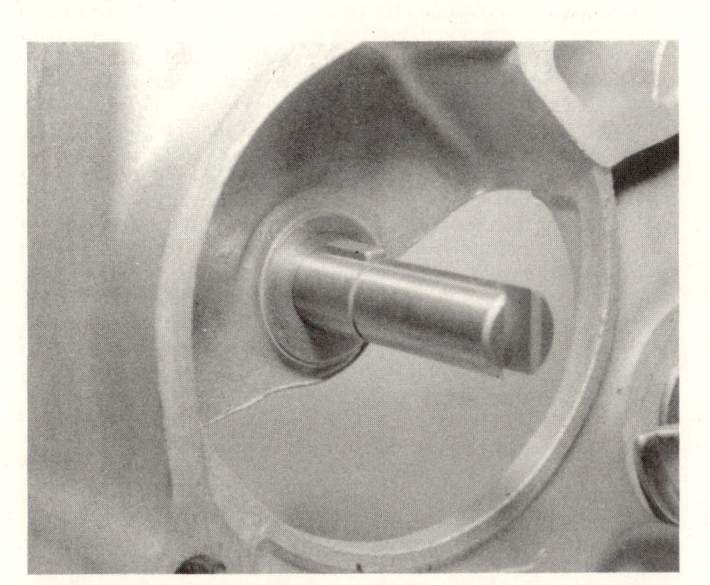

13.3b ... withdraw Woodruff key ...

13.3c ... and thrust washer

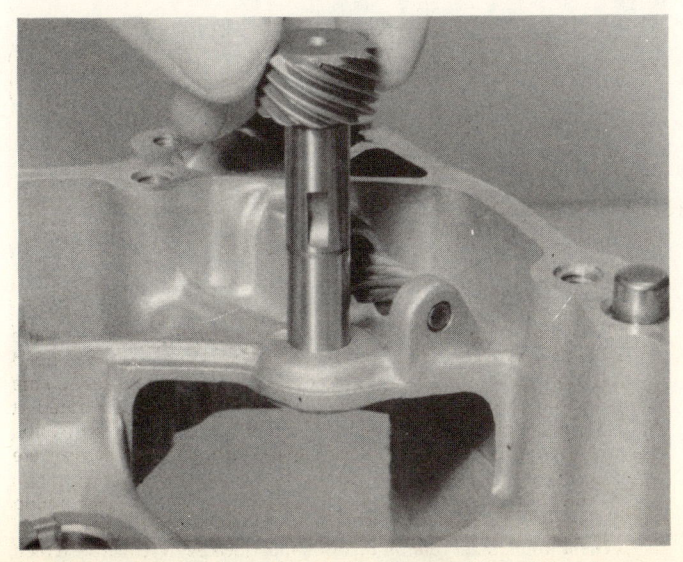

13.3d Driveshaft can be removed only after crankcase halves have been separated

13.4 Do not forget to remove output shaft spacer and neutral indicator switch before separating crankcases

14 Dismantling the engine/gearbox unit: removing the crankshaft and gearbox components

1 On all later models it is possible to remove the selector fork shafts so that the forks can be withdrawn, followed by the selector drum and then the two gearbox shafts. If this course is followed, be very careful to make accurate notes of where and which way round each fork is fitted as it is removed.

2 For example, on the machine featured in the accompanying photographs, the front (input shaft) fork was found to have the figure 8 embossed in its right-hand face. Of the output shaft forks the left-hand fork was found to be marked in a similar fashion while the right-hand fork was unmarked; note that both output shaft forks are fitted with their webs facing towards each other.

3 On early models hold both gearbox clusters, the selector drum and forks as a single assembly in one hand and pull them out of the crankcase while tapping gently with the soft-faced mallet on the exposed end of the shaft. Be very careful not to lose any small thrust washers or similar components which might drop clear; then withdraw the kickstart shaft assembly and the oil pump drive components (where necessary).

4 Firmly support the crankcase half with the crankshaft on wooden blocks so that the crankshaft end is clear of the work surface. Protect the threaded end of the shaft by refitting the primary drive gear retaining nut, place a copper drift against the shaft and strike it with a heavy hammer to drift the shaft down and clear of the crankcase. Do not use excessive force, a few firm blows should suffice. Be careful not to allow the crankshaft to drop clear; a thick layer of rag should be placed on the work surface underneath.

5 If all else fails, do not resort to excessive force, but take the assembly to an authorized Kawasaki dealer for the crankshaft to be pressed out.

15 Dismantling the engine/gearbox unit: removing oil seals and bearings

1 Before removing any oil seal or bearing, check that it is not secured by a retaining plate. If this is the case, use an impact driver or spanner, as appropriate, to release the securing screws or bolts and lift away the retaining plate.

2 Oil seals are easily damaged when disturbed and should be renewed as a matter of course during overhaul. Prise them out of position using the flat of a screwdriver and take care not to damage the alloy seal housings; note which way round the seals are fitted.

3 It is possible to remove the main bearing oil seals without separating the crankcases, either by screwing in two self-tapping screws so that the seal can be pulled out with two pairs of pliers, or by simply digging the seal out with a sharply-pointed instrument. This latter method is not recommended as it is difficult to carry out without scratching the seal housing or crankshaft, or without damaging one or both of the main bearings. Furthermore, it is almost impossible to fit a new seal without damaging it and nothing can be done to trace and rectify the fault which caused the seal to fail in the first place.

4 The crankshaft and gearbox bearings and bushes are a press fit in their respective crankcase locations. To remove a bearing, the crankcase casting must be heated so that it expands and releases its grip on the bearing, which can be drifted or pulled out.

5 To prevent casting distortion, it must be heated evenly to a temperature of about 100°C by placing it in an oven; if an oven is not available, place the casting in a suitable container and carefully pour boiling water over it until it is submerged.

6 Taking care to prevent personal injury when handling heated components, lay the casting on a clean surface and tap out the bearing using a hammer and a suitable drift. If the bearing is to be re-used, apply the drift only to the bearing outer race, where this is accessible, to avoid damaging the bearing. In some cases it will be necessary to apply pressure to the bearing inner race; in such cases closely inspect the bearing for signs of damage before using it again. When drifting a bearing from its housing it must be kept square to the housing to prevent tying with the resulting risk of damage. Where possible, use a tubular drift such as a socket spanner which bears only on the bearing outer race; if this is not possible, tap evenly around the outer race to achieve the same result. A similar approach can be used to tap out the kickstart idler sleeve gear, should this be necessary.

7 In some cases, bearings are pressed into blind holes in the castings. These bearings must be removed by heating the casting and tapping it face downwards on to a clean wooden surface to dislodge the bearing under its own weight. If this is not successful, the casting should be taken to a motorcycle service engineer who has the correct internally expanding bearing puller. If a bearing sticks to its shaft on removal, it can only be removed safely, using a knife-edged bearing puller.

14.1 Note position and markings of selector forks before removal, so that they can be correctly refitted

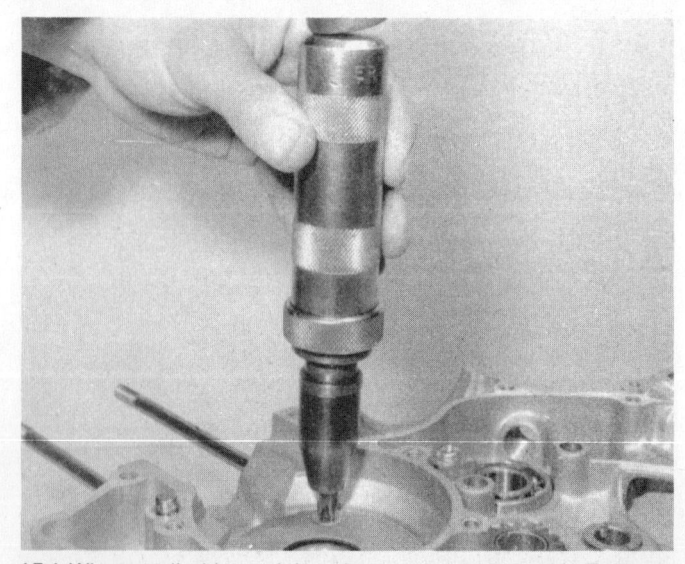

15.1 Where applicable, retaining plates must be removed before bearings or oil seals are withdrawn

Chapter 1 Engine, clutch and gearbox   61

15.2 Removing oil seals – note which way round seal is fitted before removing and take care not to damage seal housing

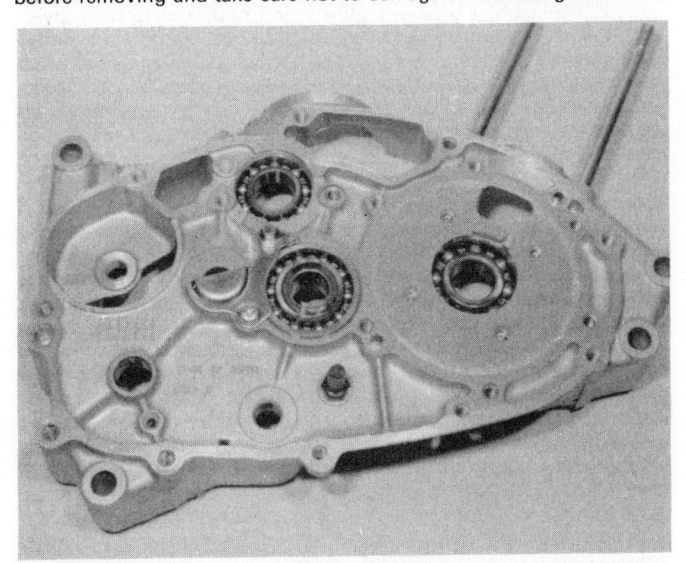

15.4 It may be necessary to heat crankcase casting to remove and refit bearings

15.6 Removing the kickstart idler sleeve gear

16 Examination and renovation: general

1 Before examining the parts of the engine unit they should be cleaned thoroughly. Use a petrol/paraffin mix or a high flash-point solvent to remove all traces of old oil and dirt. Where petrol is used, normal fire precautions should be taken and cleaning should be carried out in a well ventilated place.
2 Examine carefully each part to determine the extent of wear, checking with the tolerance figures listed in the Specifications section of this Chapter or in the main text. If there is any doubt about the condition of a particular component, play safe and renew.
3 Use a clean lint-free rag for cleaning and drying the various components. This will obviate the risk of small particles obstructing the internal oilways.
4 Various instruments for measuring wear are required, including a vernier gauge or external micrometer and a set of standard feeler gauges. Both an internal and external micrometer will be required to check wear limits. Additionally, although not absolutely necessary, a dial gauge and mounting bracket are invaluable for accurate measurement of endfloat, and play between components of very low diameter bores – where a micrometer cannot reach. After some experience has been gained, the state of wear of many components can be determined visually, or by feel, and a decision on their suitability for re-use can be made without resorting to direct measurement.

17 Examination and renovation: engine cases and covers

1 Small cracks or holes in aluminium castings may be repaired with an epoxy resin adhesive, such as Araldite, as a temporary measure. Permanent repairs can only be effected by argon-arc welding, and only a specialist in this process is in a position to advise on the economics or practicability of such a repair.
2 Damaged threads can be economically reclaimed by using a diamond section wire insert, of the Helicoil type, which is easily fitted after drilling and re-tapping the affected thread. Most motorcycle dealers and small engineering firms offer a service of this kind.
3 Sheared studs or screws can usually be removed with screw extractors, which consist of tapered, left-hand thread screws, of very hard steel. These are inserted by screwing anticlockwise into a pre-drilled hole in the stud, and usually succeed in dislodging the most stubborn stud or screw. If a problem arises which seems to be beyond your scope, it is worth consulting a professional engineering firm before condemning an otherwise sound casting. Many of these firms advertise regularly in the motorcycle papers.

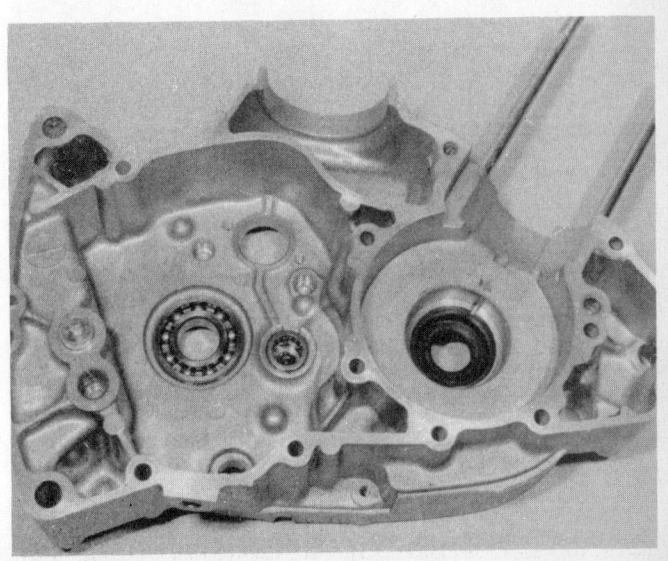

17.1 Check crankcase castings for cracks or other damage, especially at sealing surfaces

18 Examination and renovation: bearings and oil seals

1 The crankshaft and gearbox bearings can be examined while they are still in place in the crankcase castings. Wash them thoroughly to remove all traces of oil, then feel for free play by attempting to move the inner race up and down, then from side-to-side. Examine the bearing balls or rollers and the bearing tracks for pitting or other signs of wear, then spin the bearing hard. Any roughness caused by the defects in the bearing balls or rollers or in the bearing tracks will be felt and heard immediately.
2 If any sign of free play are discovered, or if the bearing is not free and smooth in rotation but runs roughly and slows down jerkily, it must be renewed. Bearing removal is described in Section 15 of this Chapter, and refitting in Section 29.
3 One or two bushes are used in the gearbox. While these should prove very long-lasting they will become worn in time. Examine each one for signs of wear such as scoring, chipping or excessive free play when the shaft is temporarily refitted. Any sign of wear or damage will mean that the bush must be renewed.
4 To prevent oil leaks occurring in the future, all oil seals and O-rings should be renewed whenever they are disturbed during the course of an overhaul, regardless of their apparent condition. This is particularly true of the main bearing oil seals, which are a weak point on any two-stroke.

18.1 Bearings can be checked while in place in casting – note output shaft bush

19 Examination and renovation: cylinder head

1 Check that the cylinder head fins are not clogged with oil or road dirt, otherwise the engine will overheat. If necessary, use a degreasing agent and brush to clean between the fins. Check that no cracks are evident, especially in the vicinity of the spark plug or stud holes.
2 Check the condition of the thread in the spark plug hole. If it is damaged an effective repair can be made using a Helicoil thread insert. This service is available from most motorcycle dealers. Always use the correct plug and do not overtighten, see Routine Maintenance.

3 Leakage between the head and barrel will indicate distortion. Check the head by placing a straight-edge across several places on its mating surface and attempting to insert a 0.05 mm (0.0020 in) feeler gauge between the two.
4 Remove excessive distortion by rubbing the head mating surface in a slow circular motion against emery paper placed on plate glass. Start with 200 grade paper and finish with 400 grade and oil. Do not remove an excessive amount of metal. If in doubt consult an authorised Kawasaki dealer.
5 Note that most cases of cylinder head distortion can be traced to unequal tensioning of the cylinder head securing nuts or to tightening them in the incorrect sequence.

20 Examination and renovation: cylinder barrel

1 The usual indication of a badly worn cylinder barrel and piston is piston-slap, a metallic rattle that occurs when there is little or no load on the engine.
2 Clean all dirt from between the cooling fins. Carefully remove the ring of carbon from the bore mouth so that bore wear can be accurately assessed and check the barrel/cylinder head mating surface as described in the previous Section. Clean all carbon from the exhaust port and all traces of old gasket from the cylinder base.
3 Examine the bore for scoring or other damage, particularly if broken rings are found. Damage will necessitate reboring or renewal and a new piston, regardless of the amount of wear. A satisfactory seal cannot be obtained if the bore is not perfectly finished.
4 There will probably be a lip at the uppermost end of the cylinder bore which marks the limit of travel of the top of the piston ring. The depth of the lip will give some indication of the amount of bore wear that has taken place even through the amount of wear is not evenly distributed.
5 The most accurate method of measuring bore wear is by the use of a cylinder bore DTI (Dial Test Indicator) or a bore micrometer. Measure the bore at the points indicated in the appropriate accompanying illustration. Note that in each case measurements should be taken both along the gudgeon pin axis and at right angles to it.
6 Late US KE100 models are fitted with barrels which cannot be honed or rebored as the bore is plated with an electrically-deposited coating to provide a hard-wearing but lightweight bearing surface. This means that if either the barrel or the piston are found to be worn to the specified service limit or beyond, they must be renewed.
7 For all other models, if any of the measurements obtained exceed the service limit for that size of bore (if oversized, the bore wear limit is found by adding 0.1 mm (0.004 in) to the diameter that the cylinder was bored out to; if this is unknown, measure the bore at the bottom of the barrel, where it will be relatively unworn) the barrel must be rebored and an oversized piston and rings must be fitted. If the rebore will enlarge the bore diameter to 50.5 mm (1.9882 in) or more, the cylinder must be renewed. Also if there is more than 0.05 mm (0.002 in) difference between any two measurements, the barrel must be rebored.
8 If a rebore is necessary, first obtain the oversize piston and measure its diameter carefully (see Section 21). Add to this figure the standard piston/cylinder clearance figure given in the Specifications Section of this Chapter; this is the diameter of the rebore. Note that oversize piston and rings are available in increments of 0.5 mm (0.020 in) and 1.0 mm (0.040 in). When finished and allowed to cool down, measure the rebored cylinder as described in paragraph 5 above; there should be no more than 0.01 mm (0.0004 in) difference between any of the measurements obtained.
9 After a rebore has been carried out, the reborer should hone the bore lightly to provide a fine cross-hatched, surface so that the new piston and rings can bed in correctly. Also the edges of the ports should be chamfered, first with a scraper, then with fine emery, to prevent the rings from catching on them and breaking.
10 If a new piston and/or rings are to be run in a part-worn bore the surface should be prepared first by glaze-busting. This involves the use of a special honing attachment with (usually) an electric drill to provide a surface similar to that described above. Most motorcycle dealers have such equipment and will be able to carry out the necessary work for a small charge.

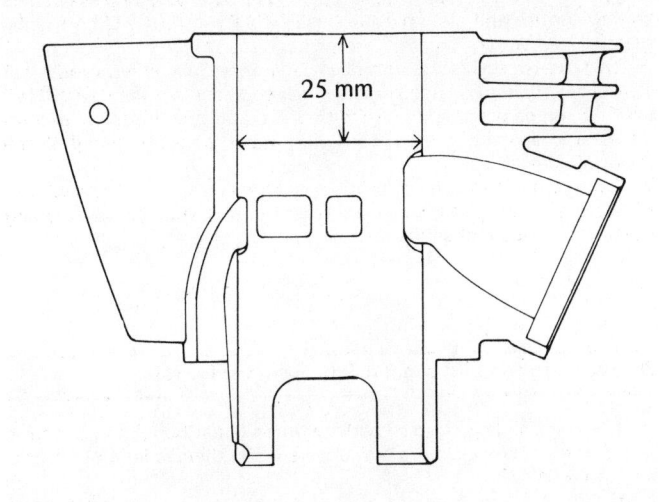

Fig. 1.8 Cylinder bore wear measurement point – KE100 A8, A9, A10 (US) and KE100 B (US) models

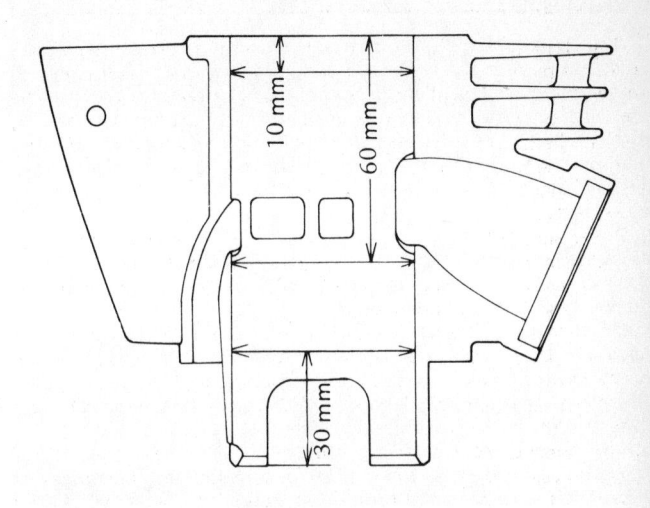

Fig. 1.9 Cylinder bore wear measurement points – all other models

21 Examination and renovation: piston and piston rings

1 Disregard the existing piston and rings if a rebore is necessary; they will be replaced with oversize items. Note also that it is considered a worthwhile expense by many mechanics to renew the piston rings as a matter of course, regardless of their apparent condition.
2 Measure the piston diameter at right angles to the gudgeon pin axis, at a point 5 mm/0.2 in above the base of the skirt. If the piston is worn to the specified service limit or beyond, it must be renewed. If the piston is oversize the wear limit is established by subtracting 0.15 mm (0.006 in) from the original oversize piston diameter.
3 Piston/cylinder clearance is calculated by subtracting the piston diameter from the smallest of the cylinder bore measurements (see Section 20.) The standard clearance figures must be observed whenever the cylinder is rebored or renewed, but may be exceeded by a very small amount if only the piston is renewed.
4 Piston wear usually occurs at the skirt, especially on the forward face, and takes the form of vertical score marks. Reject any piston which is badly scored or which has been badly blackened as the result of the blow-by of gas. Slight scoring of the piston can be removed by careful use of a fine swiss file. Use chalk to prevent clogging of the file teeth and the subsequent risk of scoring. If the ring locating pegs are loose or worn, renew the piston.
5 The gudgeon pin should be a firm press fit in the piston. Check for scoring on the bearing surfaces of each part and where damage or wear is found, renew the part affected. Note that the degree of wear can be assessed by direct measurement if the equipment is available. The pin circlip retaining grooves must be undamaged; renew the piston rather than risk damage to the bore through a circlip becoming detached. Discard the circlips themselves; these should **never** be re-used.
6 Any build-up of carbon in the ring grooves can be removed by using a section of broken piston ring, the end of which has been ground to a chisel edge. Using a feeler gauge, measure each ring to groove clearance. Renew the piston if the measurement obtained exceeds that given in the Specifications and if the rings themselves appear unworn.
7 The condition of the rings can be assessed by measuring their free end gaps; if this is found to be less than the service limit on either ring it will have lost its elasticity and must be renewed; always renew the rings as a set.
8 Measure ring wear by inserting each ring into part of the bore which is unworn and measuring the gap between the ring ends with a feeler gauge. If the measurement exceeds that given in the Specifications, renew the ring. Use the piston crown to locate the ring squarely in the bore approximately 20 mm (0.8 in) from the gasket surface.
9 Reject any rings which show discoloured patches on their mating surfaces; these should be brightly polished from firm contact with the cylinder bore.
10 Do not assume when fitting new rings that their end gaps will be correct; the installed end gap must be measured as described above to ensure that it is within the specified tolerances; if the gap is too wide another piston ring set must be obtained (having checked again that the bore is within specified wear limits), but if the gap is too narrow it must be widened by the careful use of a fine file.

21.2 Measuring the piston outside diameter

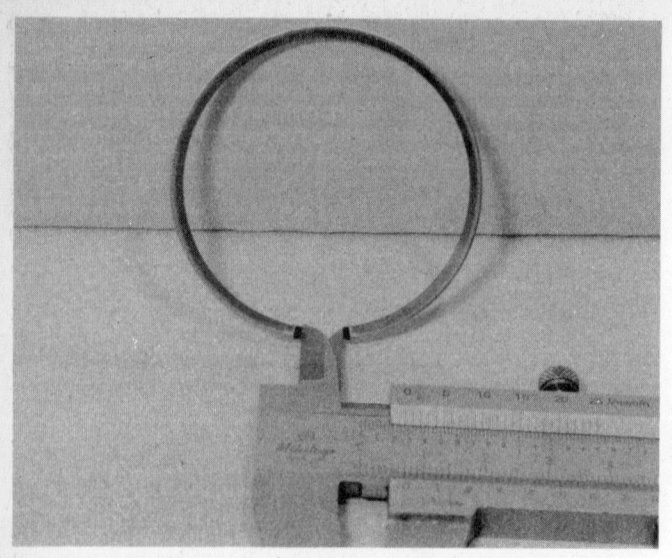

21.7 Measuring piston ring free end gap

21.8 Measuring piston ring installed end gap

22 Examination and renovation: crankshaft assembly

1 Big-end failure is characterised by a pronounced knock which will be most noticeable when the engine is worked hard. The usual causes of failure are normal wear, or failure of the lubrication supply. In the case of the latter, the noise will become apparent very suddenly, and will rapidly worsen.
2 Check for wear with the crankshaft set in the TDC (top dead centre) position, by pushing and pulling the connecting rod. No discernible movement will be evident in an unworn bearing, but care must be taken not to confuse endfloat, which is normal, and bearing wear.
3 If a dial gauge is available, set its pointer against the big-end eye. Measure the radial deflection of the connecting rod. Renew the big-end bearing if the measurement exceeds the specified service limit.
4 Push the connecting rod to one side and use a feeler gauge to measure the big-end side clearance. Renew the big-end bearing, the crank pin and its washers and the connecting rod if the clearance exceeds that specified.
5 Set the crankshaft on V-blocks positioned on a completely flat surface and measure the amount of deflection with the dial gauge pointer set against the inboard end of either mainshaft. Renew the crankshaft if the measurement exceeds 0.10 mm (0.004 in).
6 Push the small-end bearing into the connecting rod eye and push the gudgeon pin through the bearing. Hold the rod steady and feel for movement between it and the pin or check for wear by direct measurement. If movement is felt, renew the pin, bearing or connecting rod, as necessary, so no movement exists. Renew the bearing if its roller cage is cracked or worn.
7 Do not attempt to dismantle the crankshaft assembly, this is a specialist task. If a fault is found, return the assembly to an authorized Kawasaki dealer who will supply a new or service-exchange item.

23 Examination and renovation: primary drive

1 The primary drive consists of a crankshaft pinion which engages a large gear integral with the clutch outer drum.
2 If wear or damage is discovered it will be necessary to renew the component concerned. In the case of the large driven gear it will be necessary to purchase a complete clutch outer drum because the two items cannot be obtained separately. Note that the large driven gear/clutch outer drum assembly has a shock absorber fitted. Check for play in this by holding the clutch outer drum and attempting to twist or rotate the primary driven gear backwards and forwards. Unfortunately no figures are given with which to assess the state of the shock absorber unit, and it will therefore be a matter of experience to decide whether or not renewal is necessary. Seek an expert opinion if any doubt exists about the amount of play discovered.
3 If the necessary equipment is available, ie a dial gauge and stand, the degree of wear of the primary drive gear teeth can be assessed by measuring the amount of backlash present. Place the gauge so that its tip leans against one of the teeth of the clutch outer drum, then hold steady the primary drive pinion while rocking the outer drum backwards and forwards. The difference between the highest and lowest figures is the amount of backlash. If this exceeds the service limit specified, both components must be renewed.

24 Examination and renovation: clutch

1 Clean the clutch components in a petrol/paraffin mix whilst observing the necessary fire precautions.
2 Overworn friction plates will cause clutch slip. Renew each plate if its thickness is less than the specified limit. Check the plate tangs and drum slots for indentations caused by clutch chatter. If slight, damage can be removed with a fine file. If damage is excessive or the specified tang to slot clearance is exceeded, renewal is necessary.
3 Check all plates for distortion by laying each one on a sheet of plate glass and attempting to insert a feeler gauge beneath it. Refer to Specifications for maximum allowable warpage.
4 Examine each plain plate for scoring and signs of overheating in the form of blueing. Remove slight damage to the plate tangs and hub slots with a fine file, otherwise renewal is necessary. Relubricate all plates before assembly.
5 Examine the spring retaining plate and centre piece, the hub and the end plate for cracks, overheating and excessive distortion. Look for hairline cracks around the base of each end plate spring location and around the centre boss of the hub.
6 Any wear or scoring of the thrust washers and drum centre bush will necessitate their renewal. Renew the drum or bush if worn beyond the specified limits. Pay special attention to the slots in the drum centre which accommodate the drive dogs of the kickstart driven pinion.
7 The operating mechanism can be pulled from the right-hand crankcase cover. Check each part for wear and renew as necessary. Regrease the mechanism during assembly. Check the ball bearing is secure in the spring retaining plate centre piece.

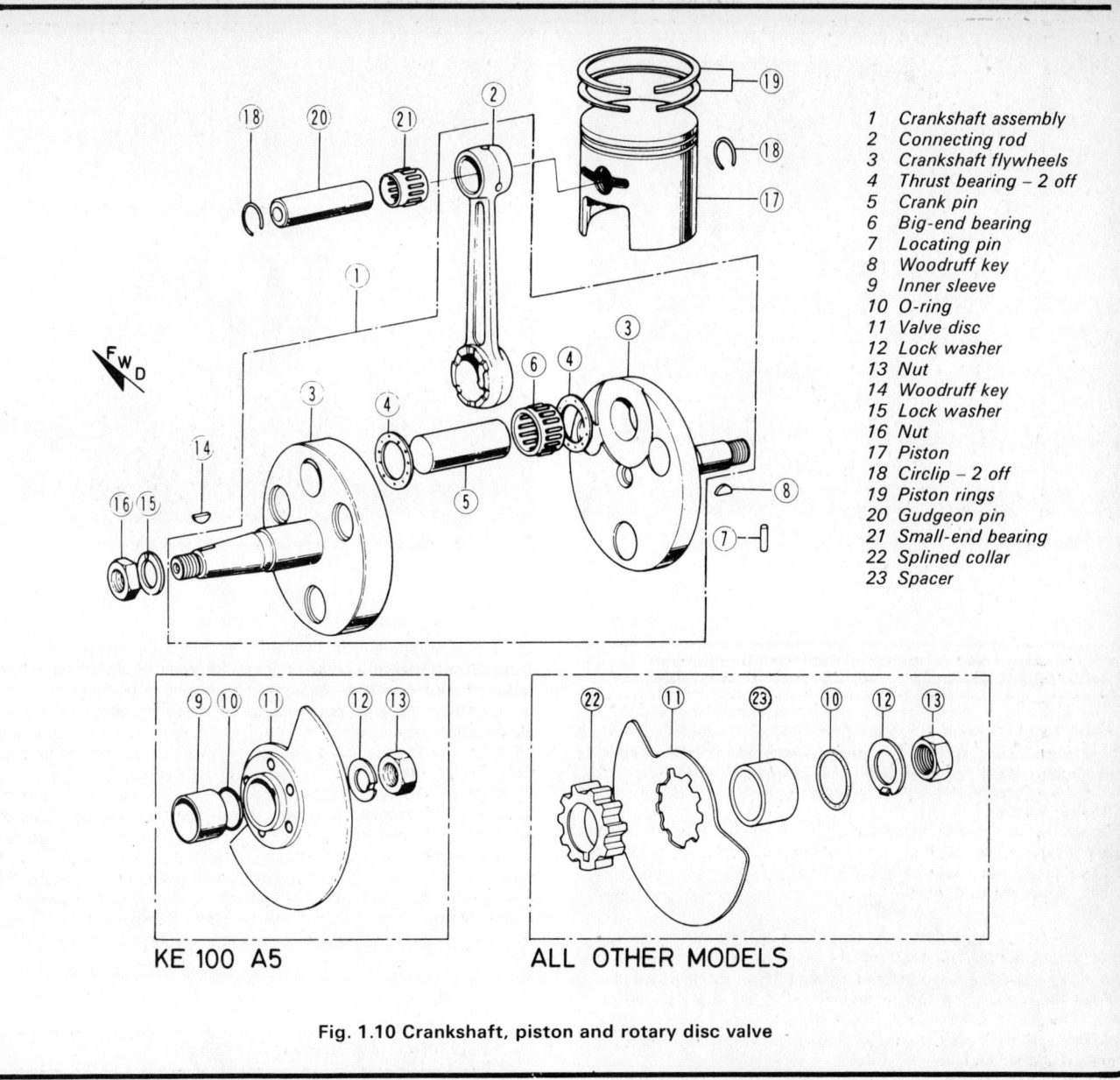

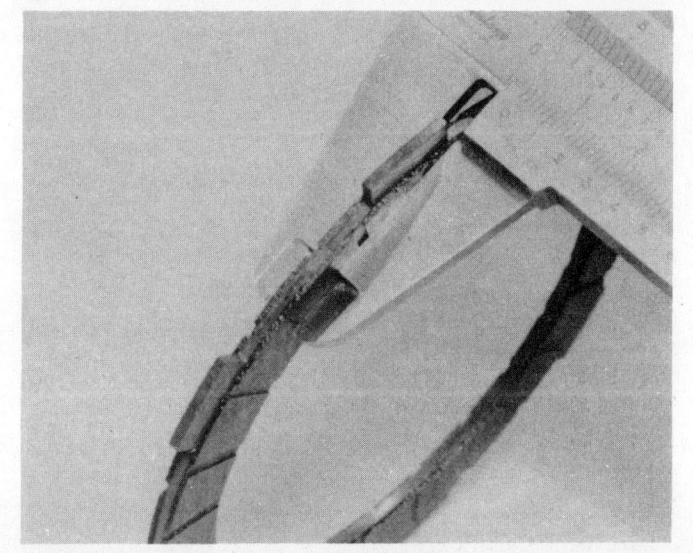

1. Crankshaft assembly
2. Connecting rod
3. Crankshaft flywheels
4. Thrust bearing – 2 off
5. Crank pin
6. Big-end bearing
7. Locating pin
8. Woodruff key
9. Inner sleeve
10. O-ring
11. Valve disc
12. Lock washer
13. Nut
14. Woodruff key
15. Lock washer
16. Nut
17. Piston
18. Circlip – 2 off
19. Piston rings
20. Gudgeon pin
21. Small-end bearing
22. Splined collar
23. Spacer

KE 100 A5

ALL OTHER MODELS

Fig. 1.10 Crankshaft, piston and rotary disc valve

24.2 Measuring clutch friction plate thickness

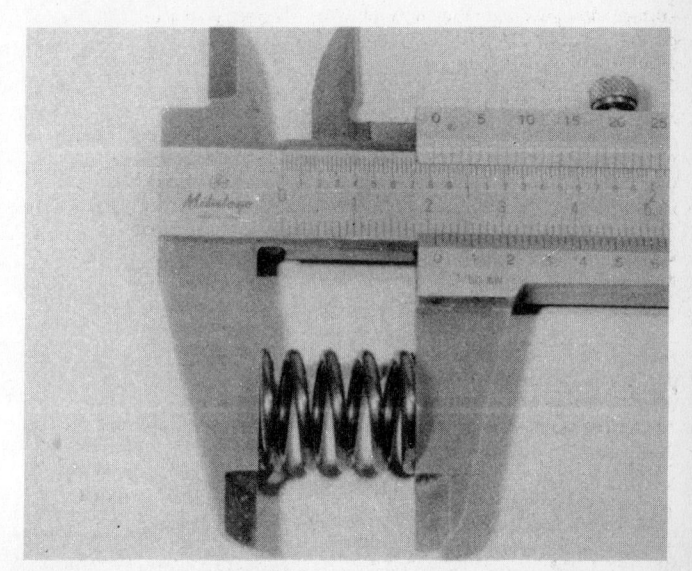

24.5 Check clutch springs for wear by measuring free length – renew any spring that is shorter than specified service limit

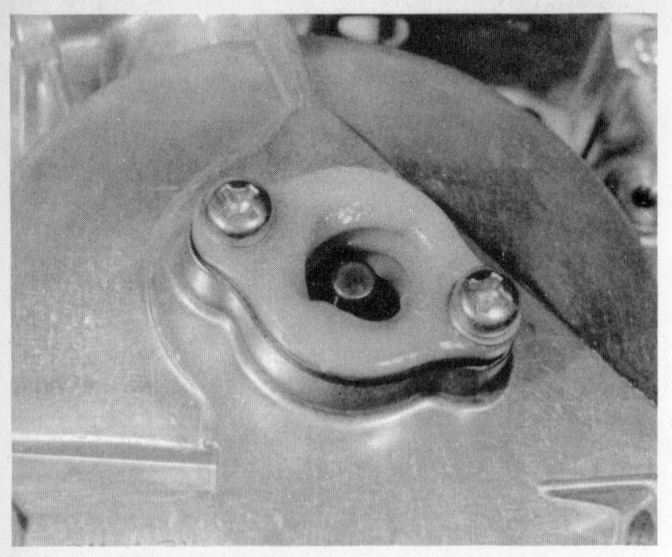

Chapter 1 Engine, clutch and gearbox

24.7a Check operating mechanism components for wear or damage ...

24.7b ... do not forget to renew oil seal if leaks are evident

25 Examination and renovation: kickstart mechanism

1 The kickstart mechanism is a robust assembly and should not normally require attention. Apart from obvious defects such as a broken return spring, the ratchet spring is the only component likely to cause problems if it becomes worn or weakened.

2 If the equipment is available, the various components of the kickstart mechanism can be checked for wear by direct measurement; any that are found to be excessively worn must be renewed.

3 Carefully check all components for signs of wear or damage and renew all defective items.

25.1 Kickstart assembly is robust – ratchet spring is only component likely to give trouble

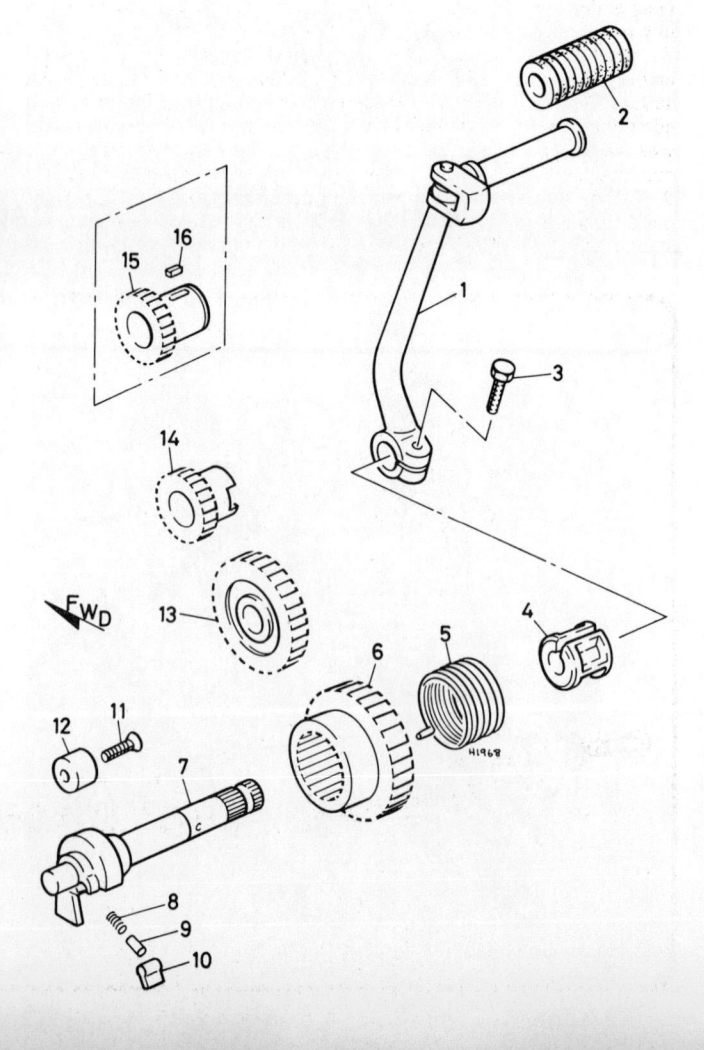

Fig. 1.11 Kickstart mechanism

1 Kickstart lever
2 Lever rubber
3 Pinch bolt
4 Spring guide
5 Return spring
6 Kickstart gear
7 Kickstart shaft
8 Spring
9 Plunger
10 Ratchet pawl
11 Screw
12 Kickstart stop
13 Idler gear
14 Idler sleeve gear
15 Idler sleeve gear*
16 Woodruff key*

* KE100 B (UK) models only

Chapter 1 Engine, clutch and gearbox

26 Examination and renovation: gearbox components

1 Give the gearbox components a close visual inspection for signs of wear or damage such as broken or chipped teeth, worn dogs, damaged or worn splines and bent selectors. Replace any parts found unserviceable because they cannot be reclaimed in a satisfactory manner.

2 The gearbox shafts are unlikely to sustain damage unless the lubricating oil has been run low or the engine has seized and placed an unusually high loading on the gearbox. Check the surfaces of the shafts, especially where a pinion turns on them, and renew the shafts if they are scored or have picked up. The shafts can be checked for trueness by setting them up in V-blocks and measuring any bending with a dial gauge. The procedure for dismantling and rebuilding the gearbox shaft assemblies is given in the following Section.

3 Examine the gear selector claw assembly, noting that worn or rounded ends on the claw can lead to imprecise gear selection. The springs in the selector mechanism and the detent or stopper arm should be unbroken and not distorted or bent in any way.

4 Examine the selector forks carefully, ensure that there is no sign of scoring on the bearing surface of either of their claw ends, their bores, or their selector drum guide pins. Check for any signs of cracking around the edges of the bores or at the base of the fork arms. Where information is given in the Specifications Section of this Chapter, check the various components for wear by direct measurement and renew any that are found to be excessively worn.

5 On later models, check each selector fork shaft for straightness by rolling it on a sheet of plate glass and checking for any clearance between the shaft and the glass with feeler gauges. A bent shaft will cause difficulty in selecting gears. There should be no sign of any scoring on the bearing surface of the shaft or any discernible play between each shaft and its selector fork(s).

6 The tracks in the selector drum should not show signs of undue wear or damage. Check also that the selector drum bearing surfaces are unworn; renew the drum if it is worn or damaged.

7 Finally, carefully inspect the splines of both shafts and pinions for any signs of wear, hairline cracks or breaking down of the hardened surface finish. If any one of these defects is apparent, the offending component must be renewed. It should be noted that damage and wear rarely occur in a gearbox which has been properly used and correctly lubricated, unless a very high mileage has been covered.

8 Clean the gearbox sprocket thoroughly and examine it closely, paying particular attention to the condition of the teeth. The sprocket should be renewed if the teeth are hooked, chipped, broken or badly worn. It is considered bad practice to renew one sprocket on its own; both sprockets should be renewed as a pair, preferably with a new final drive chain. Examine the splined centre of the sprocket for signs of wear. If any wear is found, renew the sprocket as slight wear between the sprocket and shaft will rapidly increase due to the torsional forces involved. Remember that as the output shaft will probably wear in unison with the sprocket it will be necessary to carry out a close inspection of the shaft splines.

9 Carefully examine the gear selector components. Any obvious signs of damage, such as cracks, will mean that the part concerned must be renewed. Check that the springs are not weak or damaged and examine all points of contact, eg gearchange shaft rollers and selector claw arm, for signs of excessive wear. If doubt arises about the condition of any component, it must be compared with a new part to assess the amount of wear that has taken place, and renewed if found to be damaged or excessively worn. Do not forget to check the pins set in the selector drum, these must be renewed if bent or worn.

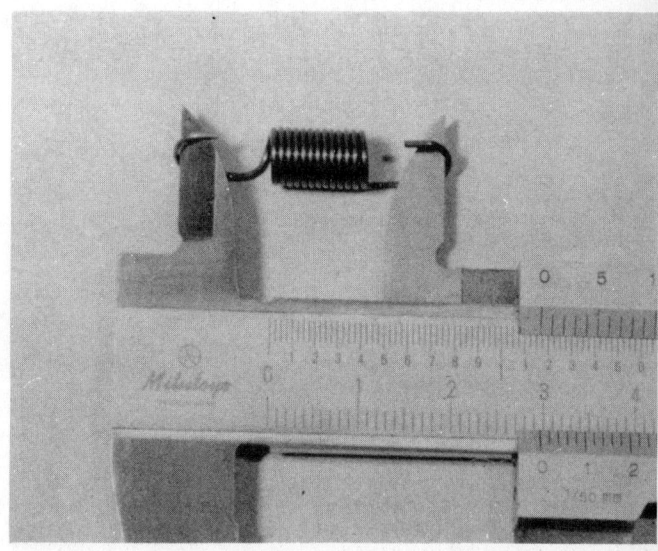

26.3 Measuring gearchange claw arm return spring free length

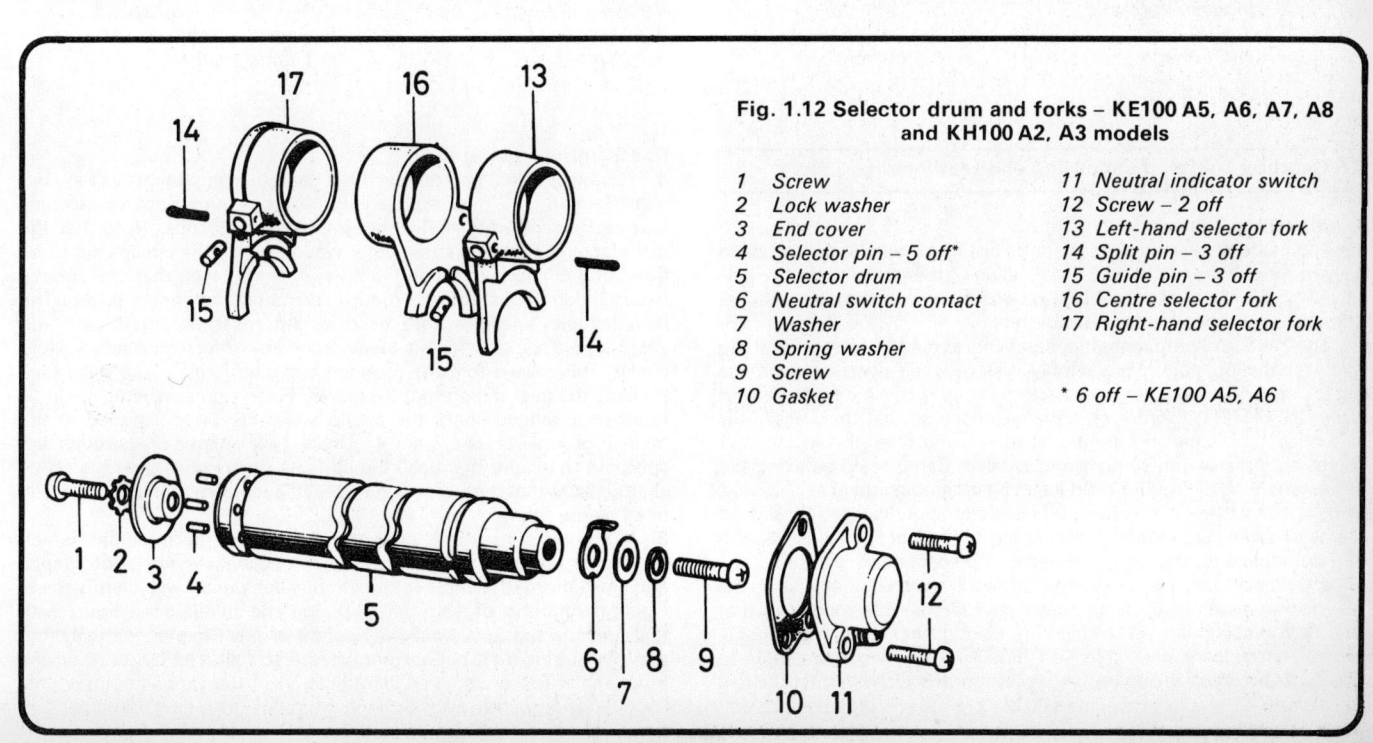

Fig. 1.12 Selector drum and forks – KE100 A5, A6, A7, A8 and KH100 A2, A3 models

1 Screw
2 Lock washer
3 End cover
4 Selector pin – 5 off*
5 Selector drum
6 Neutral switch contact
7 Washer
8 Spring washer
9 Screw
10 Gasket
11 Neutral indicator switch
12 Screw – 2 off
13 Left-hand selector fork
14 Split pin – 3 off
15 Guide pin – 3 off
16 Centre selector fork
17 Right-hand selector fork

* 6 off – KE100 A5, A6

Chapter 1 Engine, clutch and gearbox

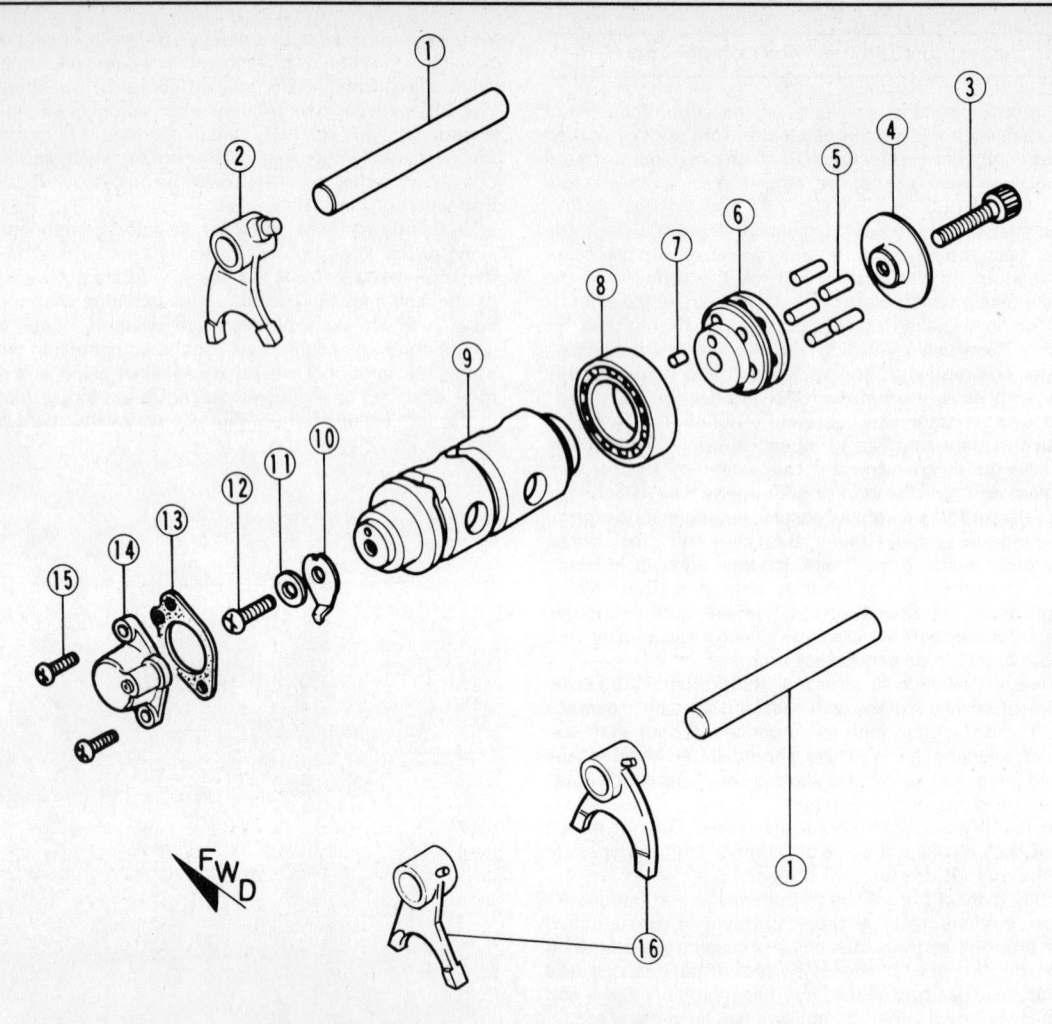

Fig. 1.13 Selector drum and forks – all other models

1 Selector fork shaft – 2 off
2 Input shaft fork
3 Screw
4 End cover
5 Selector pin – 5 off
6 Pin holder
7 Dowel pin
8 Bearing
9 Selector drum
10 Neutral switch contact
11 Washer
12 Screw
13 Gasket
14 Neutral indicator switch
15 Screw – 2 off
16 Output shaft fork

27 Gearbox shafts: dismantling and reassembly

Dismantling – general

1 The gearbox clusters should not be disturbed needlessly, and need only be stripped when careful examination of the whole assembly fails to reveal the cause of a problem, or where obvious damage, such as stripped or chipped teeth, is discovered.
2 The input and output shaft components should be kept separate to avoid confusion during reassembly. Using circlip pliers, remove the circlip and plain washer which retain each part. As each item is removed, place it in order on a clean surface so that the reassembly sequence is obvious and the risk of parts being fitted the wrong way round or in the wrong sequence is avoided. Care should be exercised when removing circlips to avoid straining or bending them excessively. The clips must be opened just sufficiently to allow them to be slid off the shaft. Note that a loose or distorted circlip might fail in service, and any dubious items must be renewed as a precautionary measure.
3 The accompanying illustration shows how both clusters of the gearbox are assembled on their respective shafts. It is imperative that the gear clusters are assembled in exactly the correct sequence, otherwise constant gear selection problems will occur. In order to eliminate the risk of misplacement, make rough sketches as the clusters are dismantled. Also strip and rebuild as soon as possible to reduce any confusion which might occur at a later date.

Reassembly – general

4 Having checked and renewed the gearbox components as required (see Section 26), reassemble each shaft, referring to the accompanying line drawing and photographs for guidance. Note that the manufacturer specifies a particular way in which the circlips are to be fitted. If a circlip is examined closely, it can be seen that one surface has rounded edges and the other has sharply-cut square edges. The manufacturer specifies that each circlip must be fitted with the sharp-edged surface facing away from any thrust received by that circlip. This means that the rounded surface of any circlip must face towards the gear pinion that it secures. Furthermore, when a circlip is fitted to a splined shaft, the circlip ears must be positioned in the middle of one of the splines. These two simple precautions are specified to ensure that each circlip is as secure as is possible on its shaft. Ensure that the bearing surfaces of each component are liberally oiled before fitting.
5 If problems arise in identifying the various gear pinions which cannot be resolved by reference to the accompanying photographs and illustrations, the number of teeth on each pinion will identify them. Count the number of teeth on the pinion and compare this figure with that given in the Specifications Section of this Chapter, remembering that the output shaft pinions are listed first, followed by those on the input shaft. The problem of identification of the various components should not arise, however, if the instructions given in paragraphs 2 of this Section are followed carefully.

27.4a Input shaft reassembly, KH100 G6 – 4th gear pinion is located by a thrust washer and secured by a circlip

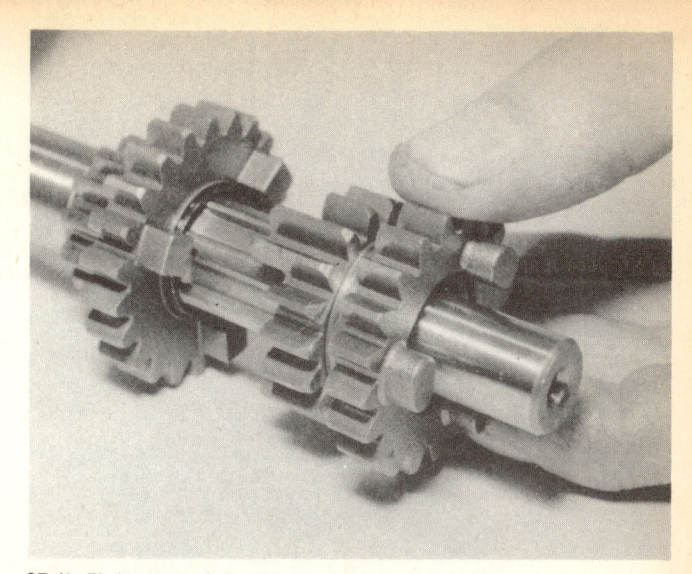

27.4b Fit input shaft 2nd/3rd gear pinion as shown ...

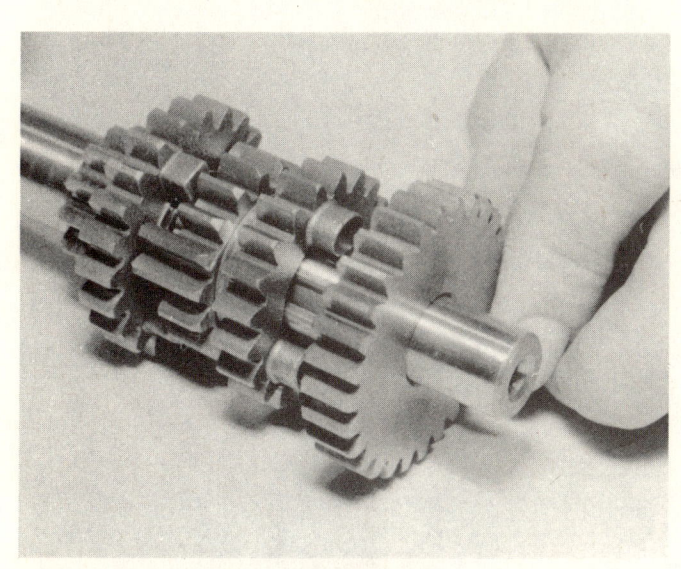

27.4c ... followed by the 5th gear pinion

27.4d Components fitted to input shaft left-hand end vary according to model – ensure that all are refitted in their correct positions

27.5a Take the bare output shaft and fit the 5th gear pinion as shown over its right-hand end, then fit a circlip to the groove shown ...

27.5b ... followed by a splined thrust washer ...

27.5c ... the 3rd gear pinion, recessed face to the left ...

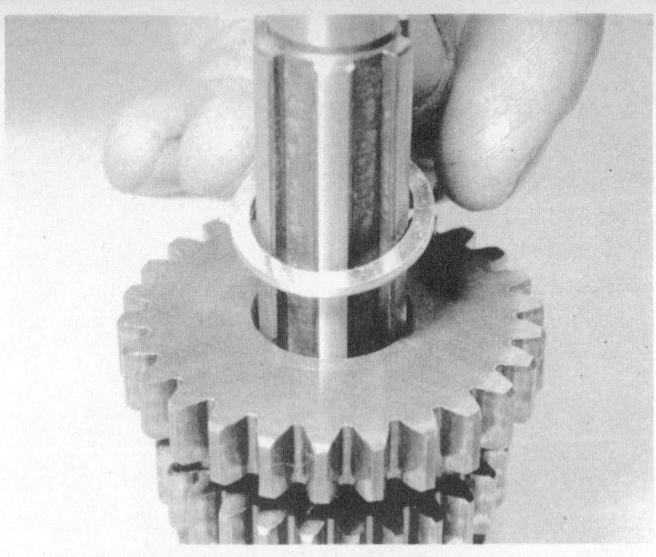

27.5d ... is followed by the plain thrust washer

27.5e The 2nd gear pinion, fitted with its recessed face to the right, ...

27.5f ... is located by a splined thrust washer and secured by a circlip

27.5g Fit the 4th gear pinion as shown ...

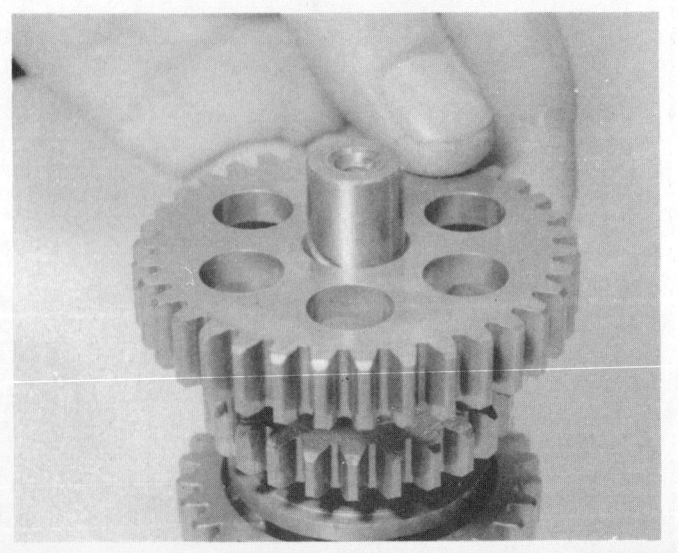

27.5h ... followed by the 1st gear pinion

27.5i Do not forget the kickstart idler gear

Fig. 1.14 Gearbox shafts

1 Input shaft
2 Input shaft 4th gear
3 Thrust washer
4 Circlip
5 Input shaft 2nd/3rd gear
6 Input shaft 5th gear
7 Copper thrust washer – KE100 B (UK), KH100 G5, G6, G7, G8
8 Steel thrust washer – KE100 B (UK), KH100 G5, G6, G7, G8
9 Roller – KE100 B (UK), KH100 G5, G6, G7, G8
10 Steel ball
11 Thrust washer – KC100, KE100 A, KE100 B (US), KH100 A, KH100 G2, G3, G4
12 Output shaft
13 Output shaft 5th gear
14 Circlip
15 Splined thrust washer
16 Output shaft 3rd gear
17 Thrust washer
18 Output shaft 2nd gear
19 Splined thrust washer
20 Circlip
21 Output shaft 4th gear
22 Output shaft 1st gear
23 O-ring
24 Spacer
25 Final drive sprocket
26 Washer
27 Lock washer
28 Bolt

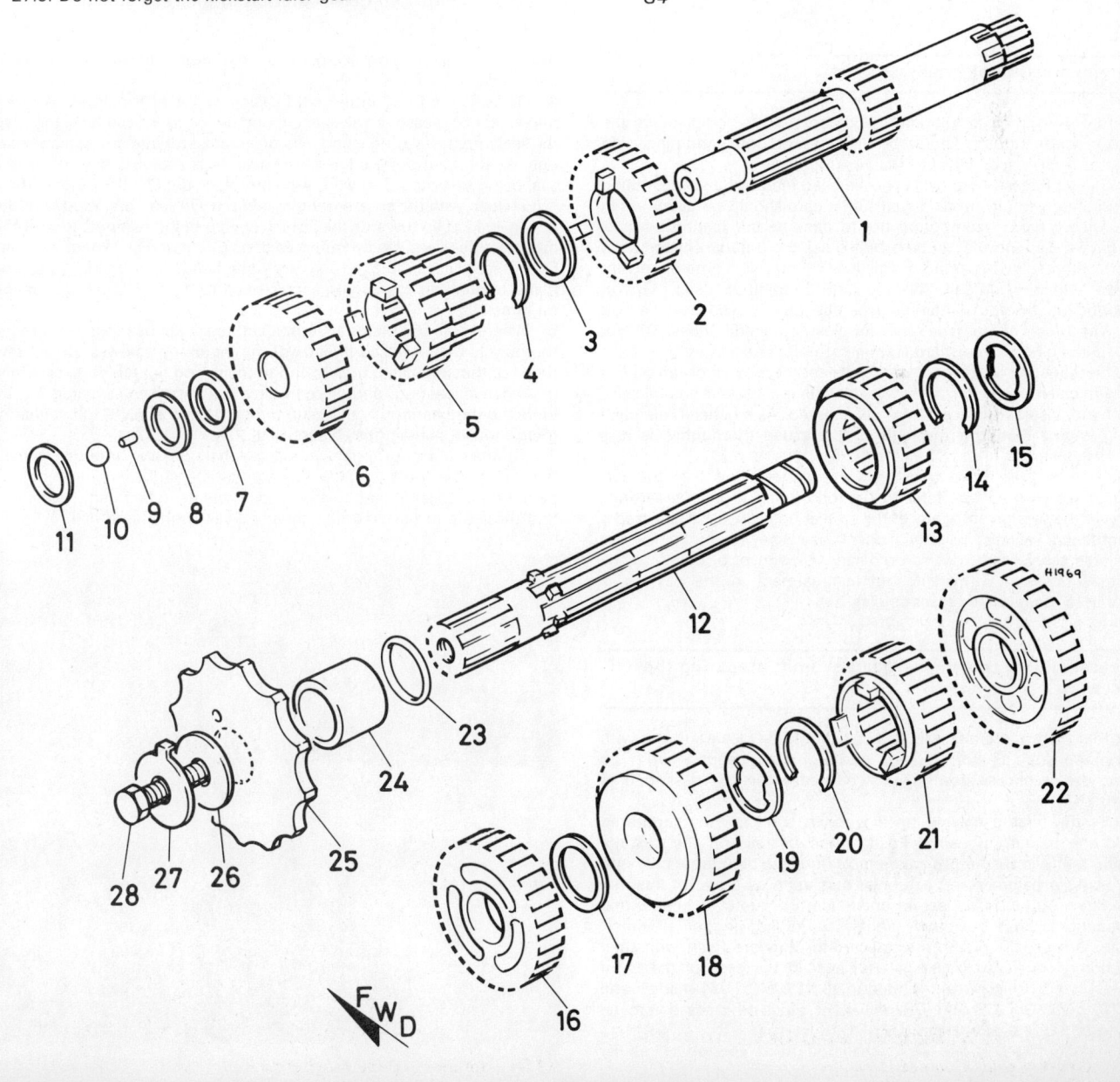

Fig. 1.15 Cross-section of gearbox shafts

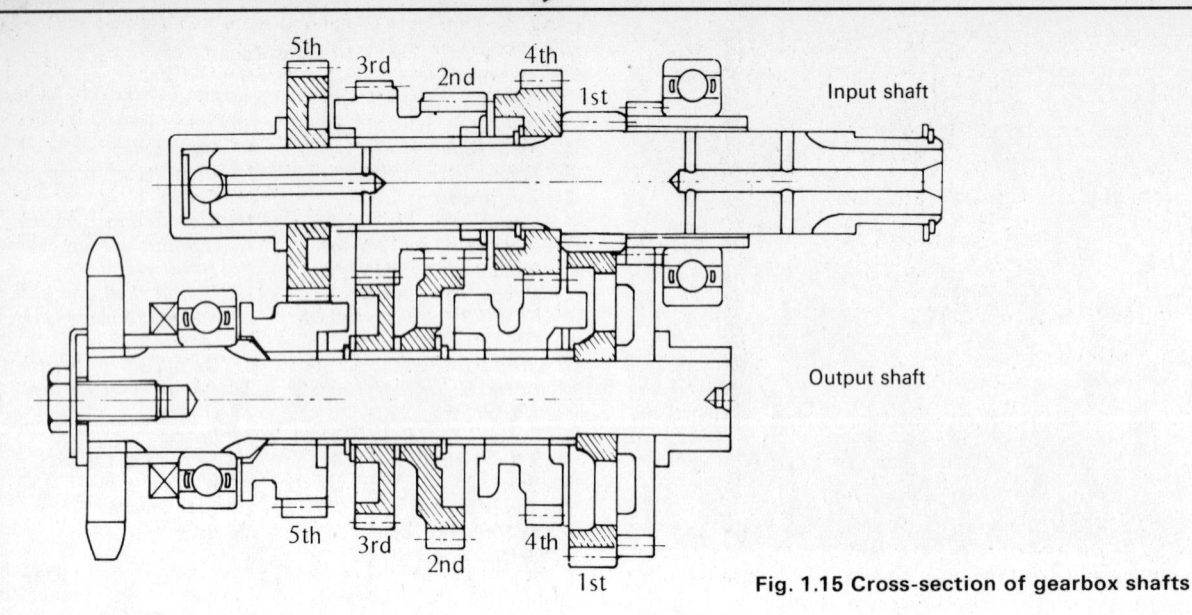

28 Engine reassembly: general

1   Before reassembly of the engine/gearbox unit is commenced, the various component parts should be cleaned thoroughly and placed on a sheet of clean paper, close to the working area.
2   Ensure all traces of old gaskets have been removed and the mating surfaces are clean and undamaged. Great care should be taken when removing old gasket compound not to damage the mating surface. Most gasket compounds can be softened using a suitable solvent such as methylated spirits, acetone or cellulose thinner. The type of solvent required will depend on the type of compound used. Gasket compound of the non-hardening type can be removed using a soft brass-wire brush of the type used for cleaning suede shoes. Do not resort to scraping with a sharp instrument unless necessary.
3   Gather together all the necessary tools and have available an oil can filled with clean engine oil. Make sure that all new gaskets and oil seals are to hand, also all replacement parts required. As a general rule each moving engine component should be lubricated thoroughly as it is fitted into position.
4   Make sure that the reassembly area is clean and that there is adequate working space. Refer to the torque and clearance settings whenever they are given. Many of the smaller bolts are easily sheared if overtightened. Always use the correct size screwdriver bit for the cross-head screws and never an ordinary screwdriver or punch. If the existing screws show evidence of maltreatment in the past, it is advisable to renew them as a complete set.

29 Reassembling the engine/gearbox unit: preparing the crankcases

1   At this stage the crankcase castings should be clean and dry with any damage, such as worn threads, repaired. If any bearings are to be refitted, the crankcase casting must be heated first as described in Section 15.
2   Place the heated casting on a wooden surface, fully supported around the bearing housing. Position the bearing on the casting, ensuring that it is absolutely square to its housing, then tap it fully into place using a hammer and a tubular drift such as a socket spanner which bears only on the bearing outer race. Be careful to ensure that the bearing is kept absolutely square to its housing at all times. Bearings are usually driven into place until they are flush with their surrounding crankcase housings. However in the case of the input shaft left-hand (roller) bearing fitted to all KE100 B (UK) models and to KH100 G5, G6, G7 and G8 models, it must be pressed into its housing to a depth of 1.0 mm (0.04 in).
3   The bush fitted to the input shaft left-hand end on all earlier models, and that fitted to the output shaft right-hand end on all models, must be driven into the heated crankcase so that the oil passage in each bush aligns with the notch in the surrounding crankcase.
4   Oil seals are fitted into a cold casing in a similar manner. Apply a thin smear of grease to the seal circumference to aid the task, then tap the seal into its housing using a hammer and a tubular drift which bears only on the hard outer edge of the seal, thus avoiding any risk of the seal being distorted. Tap each seal into place until its flat outer surface is just flush with the surrounding crankcase. Oil seals are fitted with the spring-loaded lip towards the liquid (or gas) being retained, ie with the manufacturer's marks or numbers facing outwards. Where double-lipped seals are employed, eg the right-hand main bearing, use the marks or numbers to position each seal correctly if no notes were made on removal.
5   Where retaining plates are employed to secure bearings or oil seals, thoroughly degrease the threads of the mounting screws, apply a few drops of thread locking compound to them, and tighten them securely.
6   When all bearings and oil seals have been fitted and secured, lightly lubricate the bearings with clean engine oil and apply a thin smear of grease to the sealing lips of each seal.
7   Support the appropriate crankcase half on two wooden blocks placed on the work surface, there must be sufficient clearance to permit the crankshaft and gearbox components to be fitted. Remember that these are to be fitted into the crankcase right-hand half.

29.2a Fitting bearings and oil seals

Chapter 1 Engine, clutch and gearbox

29.2b Note that certain bearings must be fitted to a precise depth – do not forget to refit retaining plates, where applicable

30 Reassembling the engine/gearbox unit: refitting the crankshaft and gearbox components

1 Refit the rotor retaining nut to protect the crankshaft threaded end and insert the crankshaft as far as possible into its main bearing, using a smear of oil to ease the task. Align the connecting rod with the crankcase mouth, check that the crankshaft is square to the crankcase and support the flywheels at a point opposite the crankpin to prevent distortion while the crankshaft is driven home with a few firm blows from a soft-faced mallet. Do not risk damaging the crankshaft by using excessive force; if undue difficulty is encountered, take the assembly to an authorised Kawasaki dealer for the crankshaft to be pressed into place.

2 When the crankshaft is fitted, remove the protecting nut and check that the crankshaft revolves easily with no traces of distortion.

3 Fit the kickstart shaft, with the ratchet pawl, spring and plunger into the kickstart gear and fit the assembly into the crankcase. Rotate it so that the stop lug is pointing towards the locating dowel, then fit the idler sleeve gear if this was removed.

4 On early KE100 A and KH100 A models the selector forks, if disturbed, must now be refitted to the drum. The left-hand fork, nearest the neutral indicator switch contact, will engage with the output shaft 5th gear pinion and is identified by having the side notched that is opposite the claw ends. The centre fork engages with the input shaft 2nd/3rd gear and is identified by the two projections, one on each side of its drum boss. The right-hand fork engages with the output shaft 4th gear pinion and is identified by the notch cut in its drum boss on the same side as the claw ends.

5 All three forks should be fitted over the selector drum with their extended bosses facing to the drum selector pin, or right-hand, side. As each fork is aligned with its drum groove insert the guide pin and fit a new split pin, passing it through from right to left. Trim the pin ends neatly and bend them flat against the fork so they cannot foul any other component.

6 Fit the gear cluster together, ensuring that all pinions are correctly meshed, then offer up the selector drum and forks. Fit the assembly as a single unit to the crankcase, pressing it carefully into place. A few taps with a soft-faced mallet may be necessary, but be very careful not to damage any component. Rotate the selection drum to the neutral position.

7 For all later models, mesh the two gear clusters together and fit them to the crankcase as described above. Fit the selector drum and rotate it to the neutral position. Refit the selector forks.

8 The input shaft fork is identified by its guide pin, which is slightly offset when compared with the other two, or by the notes made on dismantling. It engages the input shaft 2nd/3rd gear pinion and the selector drum centre track; its webbed face must be to the right. The right-hand (lower) output shaft fork engages the output shaft 4th gear and is fitted with its webbed face upwards, to the left, so that it engages with the drum right-hand track. The left-hand output shaft fork engages the output shaft 5th gear pinion and is fitted with its webbed face downwards to the right, so that it engages with the drum left-hand track. When all the forks are in position, oil them thoroughly and refit the fork shafts, pressing them firmly into their crankcase recesses.

9 On all models rotate the selector drum and check that all gears can be selected with reasonable ease. Return the drum to the neutral position and oil the bearing surfaces liberally.

10 On UK KE100 B models and KH100 G5, G6, G7, G8 models refit to the input shaft end the copper thrust washer followed by the steel thrust washer, then insert into the shaft end the roller and steel ball. Thoroughly grease all components to prevent them being dislodged as the crankcase halves are reassembled.

11 On all earlier models fit the steel ball and thrust washer to the shaft end and grease them thoroughly.

30.1 Take great care to prevent damage when refitting crankshaft – note nut protecting threads and flywheel support wedged opposite crankpin

30.3a Place kickstart gear on crankcase as shown ...

30.3b ... fit ratchet spring and plunger into kickstart shaft bore ...

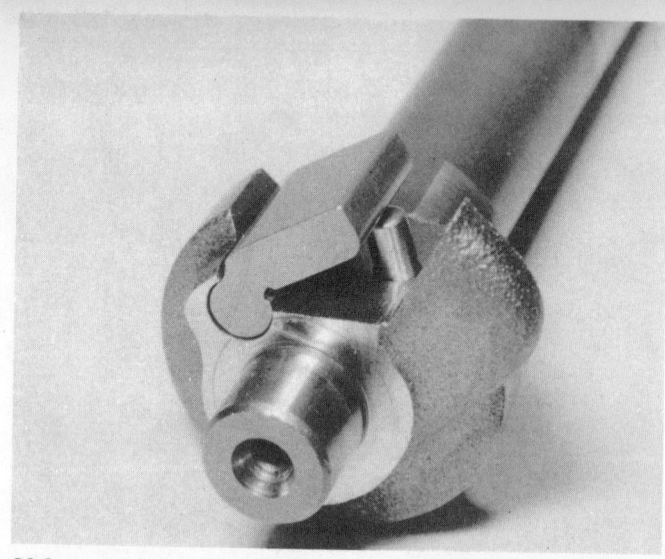

30.3c ... and fit ratchet as shown

30.3d Fit shaft assembly into gear and position stop lug as shown

30.7a Gearbox reassembly, later models – mesh pinions together and fit clusters as a single unit – note input shaft kickstart idler sleeve gear in position

30.7b Insert selector drum and rotate to neutral position

30.8a Input shaft selector fork is identified by offset guide pin

Chapter 1 Engine, clutch and gearbox　　75

30.8b Output shaft right-hand fork is positioned as shown

30.8c When forks are in position, oil fork shaft and refit

30.10 Use grease to secure components on input shaft left-hand end

31 Reassembling the engine/gearbox unit: joining the crankcase halves

1 Apply a thin film of sealing compound to the gasket surface of the right-hand crankcase half, then press the two locating dowels firmly into their recesses in the crankcase mating surface. Make a final check that all components such as the rev-counter/oil pump drive are in position and that all bearings and bearing surfaces are lubricated.
2 Lower the upper crankcase half into position, using firm hand pressure only to push it home. It may be necessary to give a few gentle taps with a soft-faced mallet to drive the casing fully into place. Do not use excessive force; instead check that all shafts and dowels are correctly fitted and accurately aligned, and that the crankcase halves are exactly square to each other. If necessary, pull away the upper crankcase half to rectify the problem before starting again.
3 When the two halves have joined correctly and without strain, refit the crankcase retaining screws, using the cardboard template to position each screw correctly. Working in a diagonal sequence from the centre outwards, progressively tighten the screws until all are securely and evenly fastened.

4 Wipe away any excess sealing compound from around the joint area, check the free running and operation of the crankshaft and gearbox components. If a particular shaft is stiff to rotate, a smart tap on each end using a soft-faced mallet will centralise the shaft in its bearing. If this does not work, or if any other problem is encountered, the crankcases must be separated again to find and rectify the fault. Pack clean rag into the crankcase mouth to prevent the entry of dirt, then refit the drain plug, if removed, tightening it to the specified torque setting.
5 Smear the shaft end and seal lips with a generous smear of grease and fit a new O-ring over the output shaft left-hand end. Refit the spacer, noting that one of its inner edges is recessed; this recessed end must face inwards so as to fit over the O-ring and protect it from distortion. Take care not to damage the output shaft seal lips as the spacer is pressed fully into place.
6 Assemble the oil pump drive components, if disturbed, and refit them to the crankcase.
7 Rotate the kickstart shaft as far as possible in a clockwise direction and fit the return spring inner end into the hole in the crankcase wall. Grasping the spring right-hand end with a strong pair of pliers bring it around anticlockwise until it can be hooked into the drilling in the kickstart shaft. Fit the spring guide, then temporarily refit the kickstart lever and operate it once or twice to check that it is correctly installed and functioning properly.
8 On KH100 G5, G6, G7, G8 models and all UK KE100B models if the crankcases have been renewed, the input shaft endfloat must be set. The shaft left-hand end is retained by a plug screwed into the crankcase. To adjust the plug remove it and clean its threads, then coat them with a non-permanent locking compound. Refit the plug, ensuring that the steel ball is correctly refitted and tighten it down until it seats lightly on the roller and ball to take up all endfloat. Be careful not to overtighten the plug so that it drags on the shaft. Refer to the accompanying illustration for details.

32 Reassembling the engine/gearbox unit: refitting the flywheel generator

1 Relocate the stator in its previously noted position, using the wiring as a guide, and tighten its retaining screws. Clip the electrical leads to the crankcase and reconnect the neutral indicator switch.
2 Degrease the rotor and crankshaft mating surfaces. Insert the Woodruff key into the crankshaft and push the rotor over it. Gently tap the rotor centre with a soft-faced hammer to seat it and fit the plain washer, spring washer and nut. Lock the crankshaft and tighten the nut to the specified setting (where applicable).

31.1a Do not forget to refit rev-counter (where fitted) and oil pump drive components before joining crankcase halves

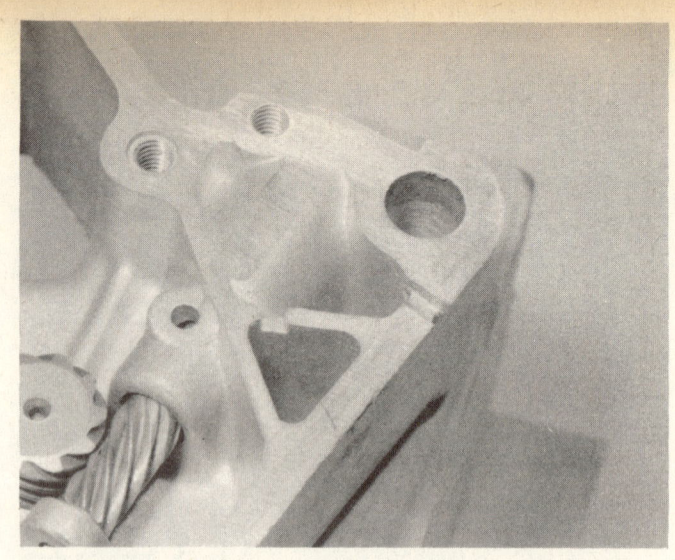

31.1b Note breather passage in gasket surface of crankcase right-hand half – ensure it is not blocked with jointing compound

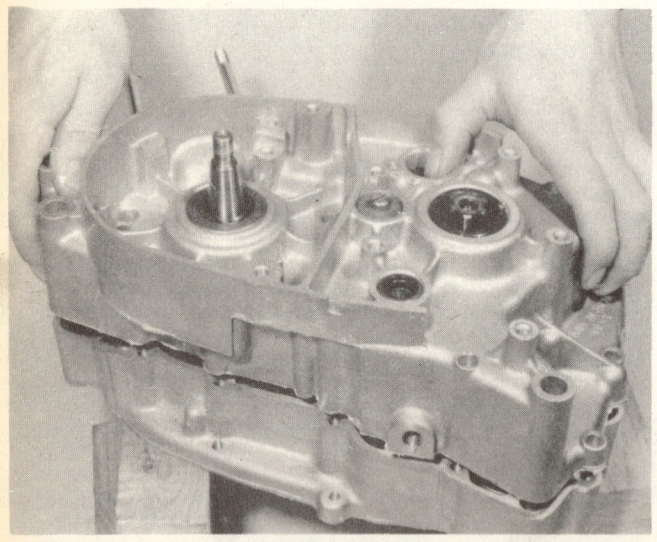

31.2 Check that all components are correctly refitted and lubricated before refitting crankcase left-hand half

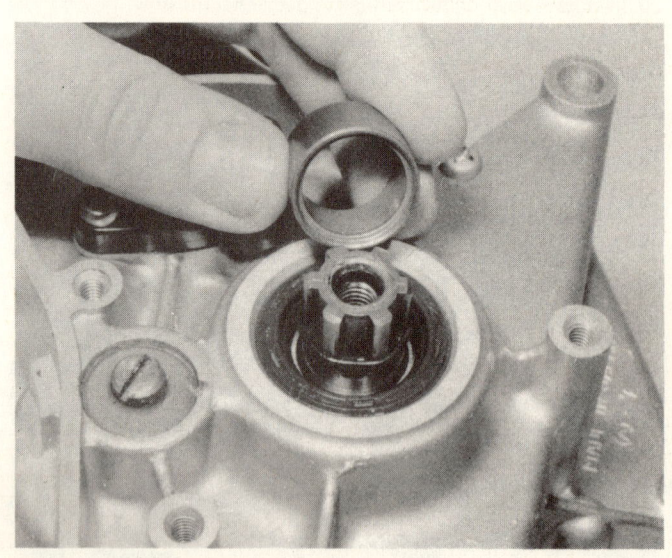

31.5 Output shaft spacer recessed end must fit over sealing O-ring, as shown

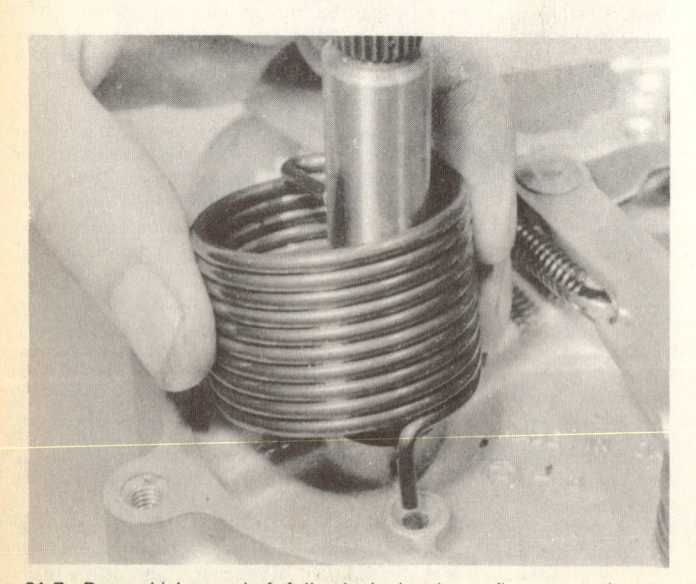

31.7a Rotate kickstart shaft fully clockwise then refit return spring ...

31.7b ... and guide. Check kickstart operation before proceeding with reassembly

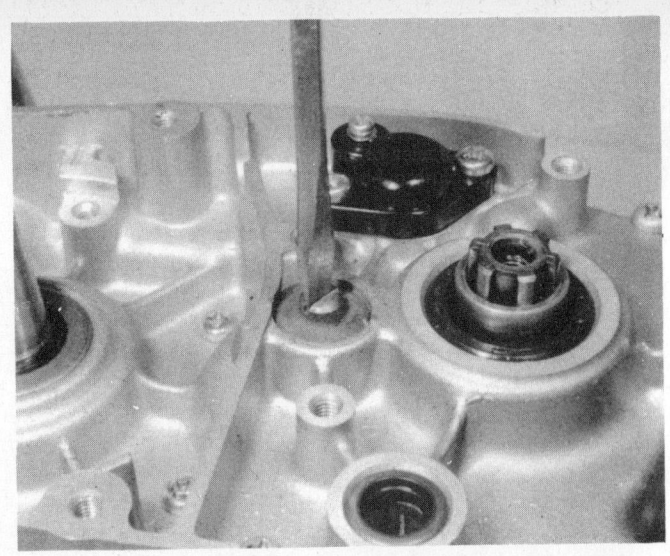

31.8 If any components have been renewed, input shaft endfloat must be reset on later models

Fig. 1.16 Adjusting the input shaft endfloat – KE100 B (UK) and KH100 G5, G6, G7, G8 models

Note: KE100 clutch shown

1. Adjuster
2. Steel ball
3. Bearing
4. Input shaft
5. Wave washer
6. Woodruff key
7. Clutch
8. Crankcase
9. Steel washer
10. Copper washer
11. 5th gear pinion
12. Kickstart idler gear
13. Bearing
14. Pushrod
15. Steel ball

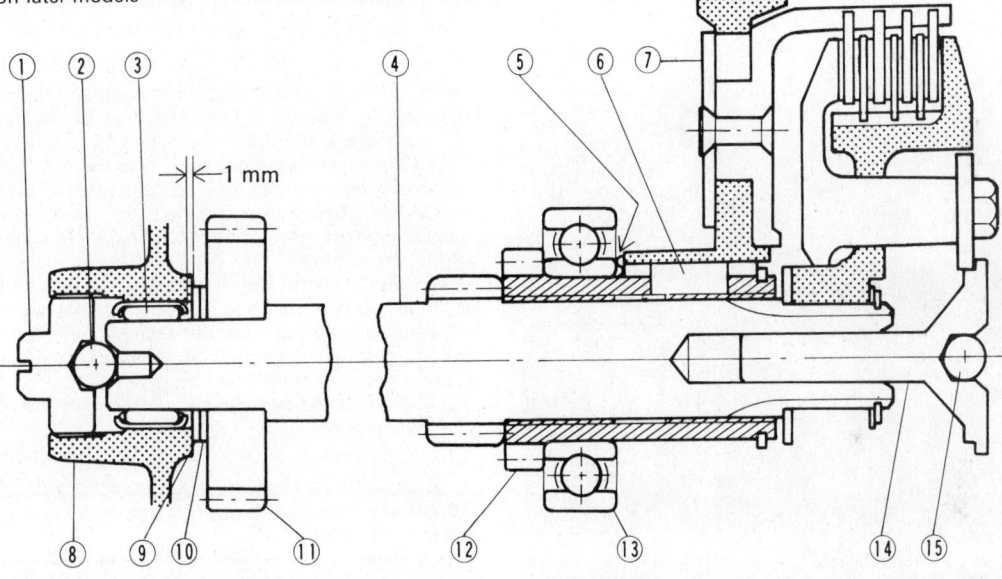

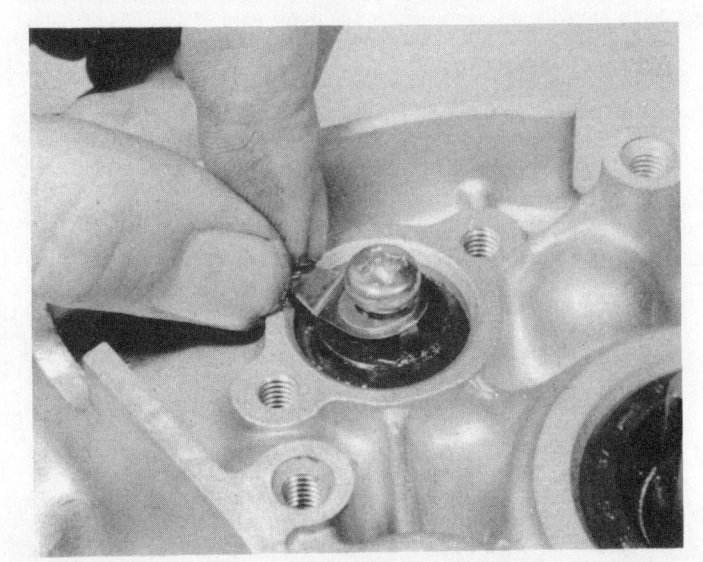

32.1a Fit neutral indicator switch contact to selector drum, ensuring that contact locating tang is fitted into hole in drum end

32.1b Do not overtighten switch retaining screws or it may be cracked

32.1c Position generator stator as shown and tighten retaining screws

32.2a Degrease rotor and crankshaft taper and refit Woodruff key

32.2b Note method used to lock crankshaft while tightening rotor nut

33 Reassembling the engine/gearbox unit: refitting the gear selector external components

1 On later models refit the selector pin holder to the end of the drum, noting that the locating dowel pin will fit in only one of the three holes in the drum.
2 On all models, insert the selector pins into the selector drum, apply thread locking compound to the threads of its retaining screw and refit the drum end cover, using an impact driver (if applicable) to tighten the screw securely.
3 Where fitted, refit the selector drum guide plate, apply thread locking compound to the threads of its screws and tighten them securely.
4 Hook its spring end on to the bearing retainer and refit the detent roller arm/gear detent arm to the end of the selector drum. Refit and tighten securely (applying thread locking compound to its threads) the shouldered pivot bolt. Before the bolt is finally tightened ensure that the arm is correctly engaged on its shoulder so that it is securely fastened but free to move against spring pressure.
5 Lubricate the ball or plunger and spring of the detent assembly, check that they slide easily in the plug, and refit the assembly with a new sealing washer. Tighten the plug to the torque setting recommended, where appropriate.
6 Fit the spacer and return spring to the gearchange shaft as shown in the accompanying photograph and liberally grease the shaft splines and seal lips before inserting the shaft into the crankcase. Engage the claw arm on the selector drum and the return spring ends on each side of the spring post.
7 On early KH and KE100 A models the post should have a separate locknut and a slotted end. This means that it is in fact an eccentric adjusting screw and must be rotated until the claw arm stops are exactly centred in relation to the selector drum pins. Tighten the locknut securely when adjustment is correct.
8 Check that all gears can be selected with relative ease and return to the neutral position.

34 Reassembling the engine/gearbox unit: refitting the rotary disc valve

1 Fit the locating pin to the crankshaft drilling and refit the inner sleeve (KE100 A5) or splined collar (all other models), aligning the sleeve or collar keyway with the pin. In the case of the splined collar, use the marks or notes made on dismantling to ensure that it is refitted exactly as it was found.
2 Check that the valve chamber surfaces are absolutely clean, apply a coat of clean 2-stroke engine oil to both faces of the disc and refit it. On KE100 A5 models align the keyway in the disc hub with the locating pin and ensure that the disc is refitted with its marked face outwards. On all other models ensure that the disc marked face is outwards, also that the 'timing' marks made on dismantling are aligned exactly as it is refitted, then refit the spacer.
3 On all models, fit a new O-ring around the crankshaft end and press it down against the inner sleeve/spacer. Smear a liberal coat of grease over the crankshaft end and sleeve/spacer and press the two locating dowels into place.
4 With a new seal fitted as previously described into its centre, place a new O-ring in the groove around the periphery of the valve cover inner face. Use a smear of grease to stick the new O-ring in place and apply a smear to the seal lips.
5 Carefully install the valve cover, taking care not to disturb the disc or the sealing O-ring and not to damage the seal lips as they pass over any sharp edges. Rotate the cover to align with the locating dowels when the carburettor stub is in the 12 o'clock position.
6 Tighten the cover retaining screws evenly and progressively until all are secure. Fit the oil transfer pipe into its recess in the cover; do not forget to fit new O-rings to the pipe and to the carburettor stub before refitting the crankcase right-hand cover.

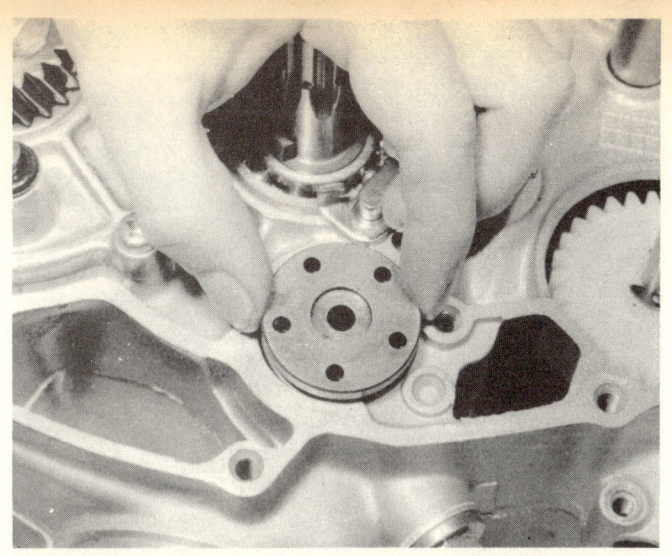

33.1a On later models, fit selector pin holder to drum end ...

33.1b ... noting that holder locating dowel pin will fit into only one of drum holes

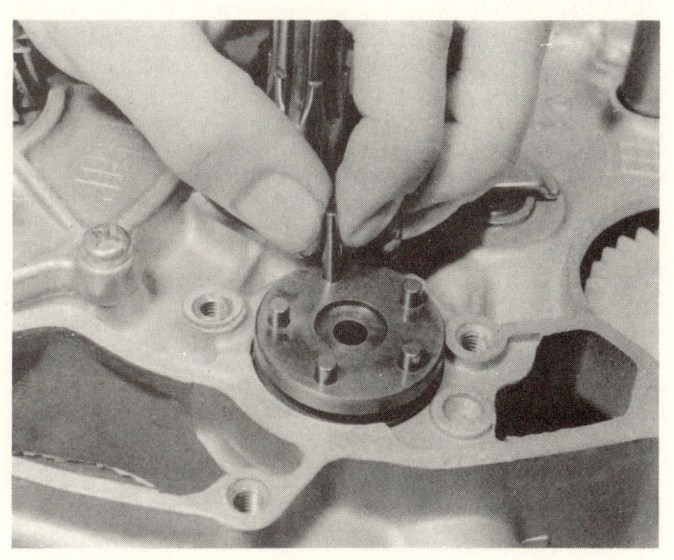

33.2a Insert selector pins into holder, followed by end cover

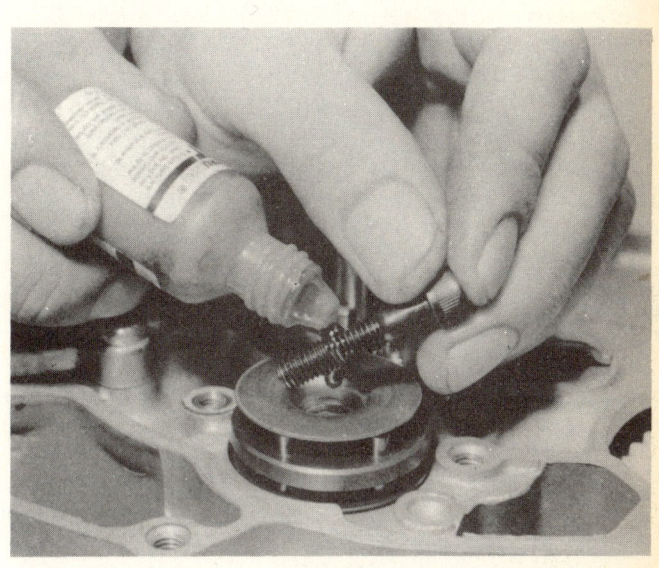

33.2b Apply thread locking compound to threads of retaining screw

33.4 Selector drum detent arm is fitted as shown – ensure it is free to move against spring pressure once shouldered pivot bolt has been tightened

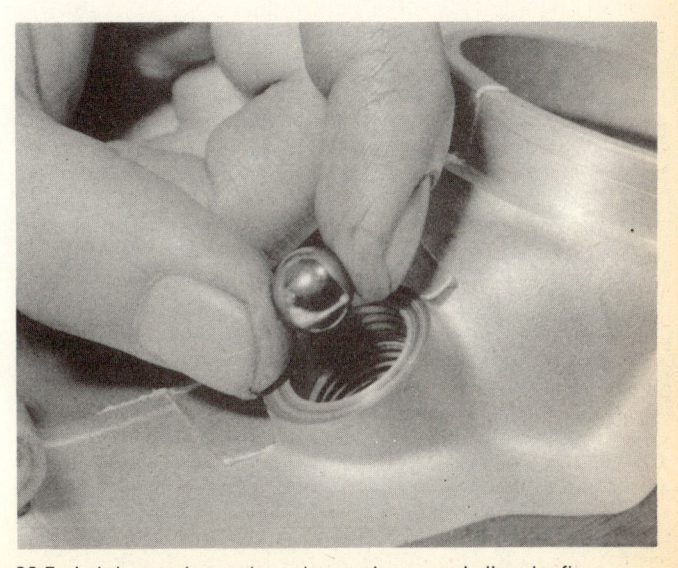

33.5a Lubricate selector drum detent plunger or ball and refit ...

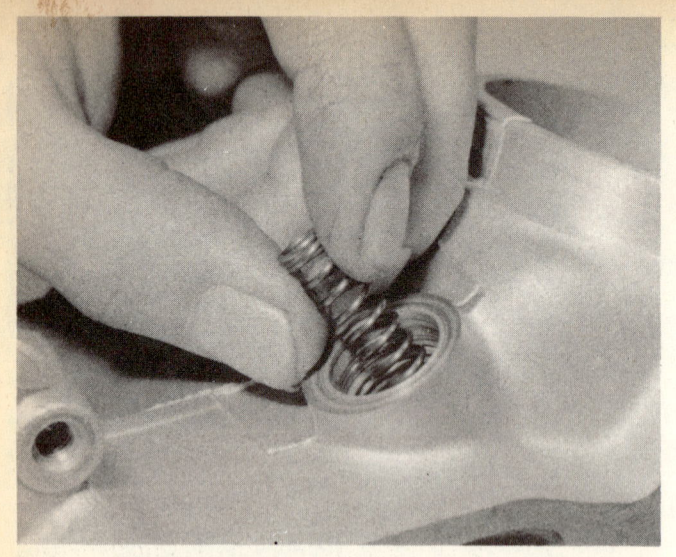

33.5b ... followed by detent spring

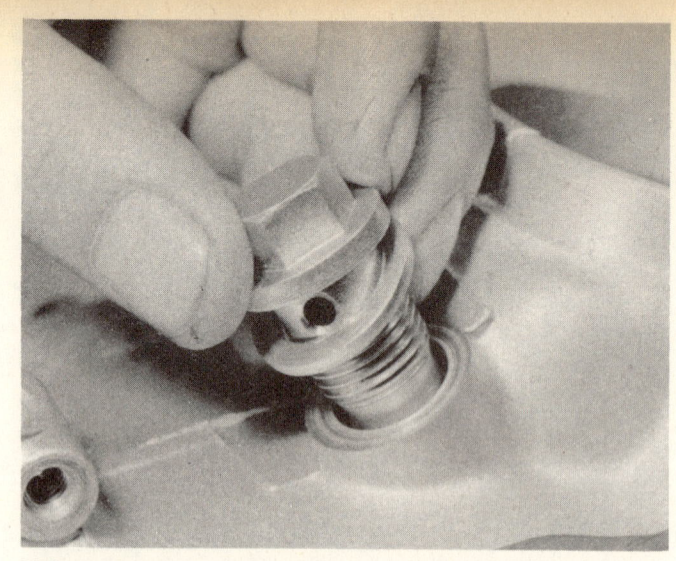

33.5c Renew sealing washer when refitting detent plug

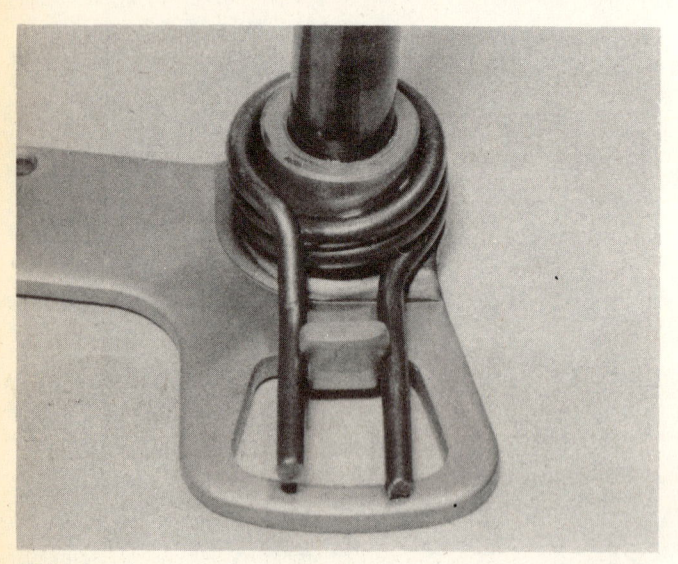

33.6a Gearchange shaft return spring and spacer should be fitted as shown

33.6b Engage gearchange shaft return spring and claw arm as shown

34.1a Fit locating pin into crankshaft drilling

34.1b Later models – ensure splined collar is refitted the correct way round, using marks or notes made on removal

34.2 Oil valve disc and align timing marks made on dismantling, using marks (arrowed) to ensure that it is refitted the correct way round

34.4 Always renew valve cover centre seal and O-ring to prevent leaks

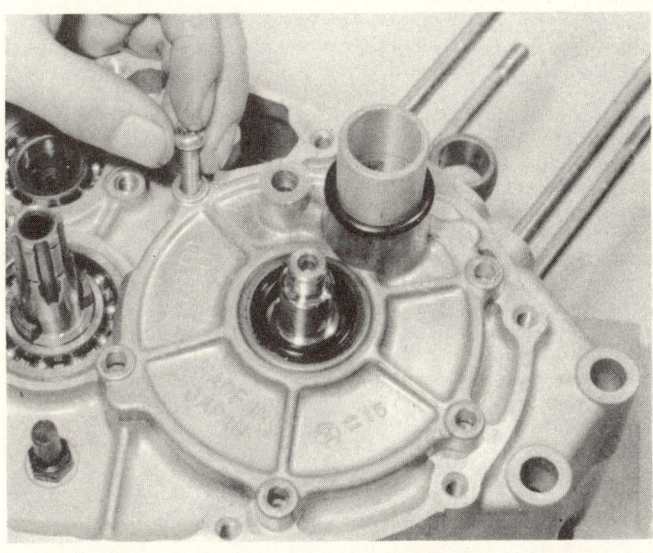

34.5a Fit valve cover over locating dowels and tighten retaining screws evenly

34.5b Later models – use grease to prevent damage to centre seal lips as spacer is refitted ...

34.5c ... do not forget to refit sealing O-ring against valve spacer/inner sleeve

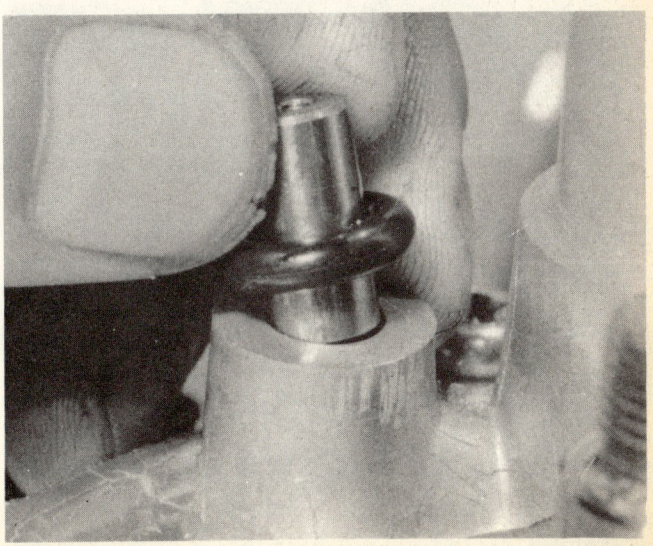

34.6a Refit oil transfer pipe to valve cover – always renew sealing O-ring

82 Chapter 1 Engine, clutch and gearbox

34.6b Also, always renew carburettor stub O-ring to prevent leaks

35 Reassembling the engine/gearbox unit: refitting the clutch and primary drive

1 Fit the Woodruff key to the end of the crankshaft. Place the primary drive pinion over the crankshaft with the hole in one face outwards and ensuring that it seats fully. If this proves difficult, check that the Woodruff key is parallel to the crankshaft. Fit the lock washer with its tang engaged in the pinion hole, then fit the retaining nut. Hold the crankshaft by whichever method was employed during removal, and tighten the nut securely. Where applicable, use the specified torque wrench setting. Secure the nut by bending up against one of its flats an unused portion of the lock washer.
2 On UK KE100 B models, fit the wave washer over the kickstart idler sleeve gear and press the Woodruff key firmly into the sleeve gear keyway. Tap the clutch outer drum onto the sleeve gear, being careful to align its keyway with the key. When the outer drum is fully in place, refit the wire circlip to retain it.
3 Place the clutch centre upside down on a clean work surface and refit the clutch plates, starting with a friction plate, then a plain plate and alternating the two until all are fitted; note that you should finish with a friction plate. Note also that all friction plates should be coated with a film of engine oil if they have been renewed. Where the friction plates have diagonal rather than straight grooves, they must be fitted with the grooves running towards the centre, in the direction of clutch rotation (anticlockwise, seen from the engine right-hand side). See Fig. 1.17, and remember that the direction of rotation is the reverse of that which may seem apparent as the clutch is being assembled on the bench. Finally, fit the pressure plate to the assembly.
4 Align the friction plate tongues, fit the thrust washer to the input shaft end and refit the clutch plate assembly, rotating the outer drum and the input shaft (the latter is achieved by selecting a gear and turning the output shaft back and forth) so that the splines of the input shaft align with those of the clutch centre while the friction plate tongues align with the slots in the outer drum.
5 Press the plate assembly into place and secure it by refitting the thrust washer and circlip. Note that the circlip must be aligned as shown in the accompanying illustration. Fit the clutch springs over their posts, then refit the spring plate and bolts. Tighten the bolts evenly and progressively to a torque setting of 0.4 – 0.5 kgf m (3 – 3.5 lbf ft). Install the clutch release centre piece with the steel ball at its centre.
6 On all other models, apply a coat of engine oil to the friction plates if they have been renewed. Place the clutch centre upside down on a clean work surface. On early KE and KH100 A models, fit first the (much thicker) steel outer plate. On all models start with a friction plate, followed by a plain plate and fit all the plates alternately to finish with a friction plate. Lastly fit the pressure plate to the rear of the assembly. Turn the plate assembly over and refit the springs, spring plate and bolts, but tighten the bolts only lightly at this stage. Place the smaller thrust washer in the centre of the clutch outer drum, then fit the plate assembly to the outer drum, aligning as necessary the friction plate tongues so that they engage with the slots in the outer drum.
7 Select a gear and fit the large thrust washer and coil spring over the kickstart idler sleeve gear. Offer up the complete clutch assembly, rotating the sleeve gear via the kickstart shaft so that the sleeve gear drive dogs engage with the slots cut in the rear face of the outer drum. It may also prove necessary to rotate the input shaft (by turning the output shaft back and forth) to ensure that its splines engage with those in the clutch centre. When the assembly is fully in position, refit the thrust washer and circlip to retain it. Note that the circlip must be aligned as shown in the accompanying illustration.
8 Progressively and evenly tighten the clutch spring bolts until all are secure; apply the torque wrench setting, where specified. Refit the clutch release centre piece.

35.1a Fit Woodruff key to crankshaft keyway ...

35.1b ... followed by primary drive pinion – lock washer locating tang must fit into pinion hole, as shown

35.1c Note method used to lock crankshaft while retaining nut is tightened

35.1d Bend up unused portion of lock washer as shown to secure nut

35.6a Place clutch centre upside down and refit clutch plates starting with a friction plate ...

35.6b ... followed by a plain plate ...

35.6c ... fit pressure plate last

35.6d Align balancing marks noted on dismantling when refitting plate assembly to outer drum

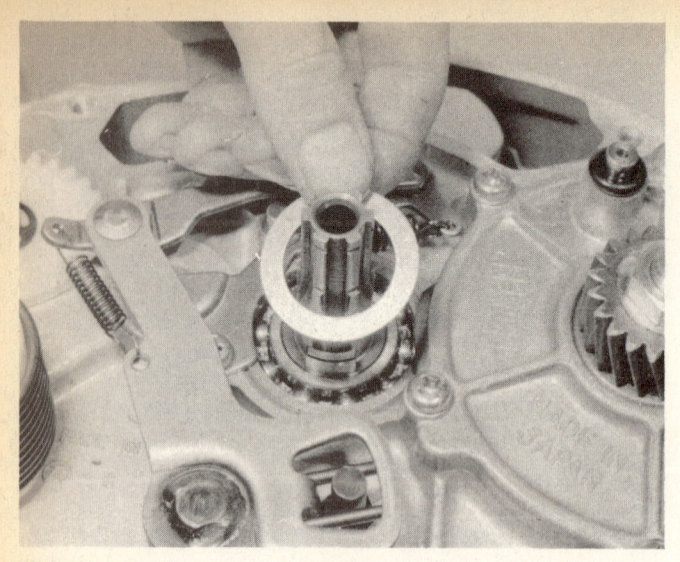

35.7a Fit large thrust washer over input shaft end ...

35.7b ... followed by coil spring

35.7c Outer drum must engage with kickstart sleeve gear dogs and must be pressed down against spring pressure

35.7d Smaller thrust washer is fitted between outer drum and plate assembly

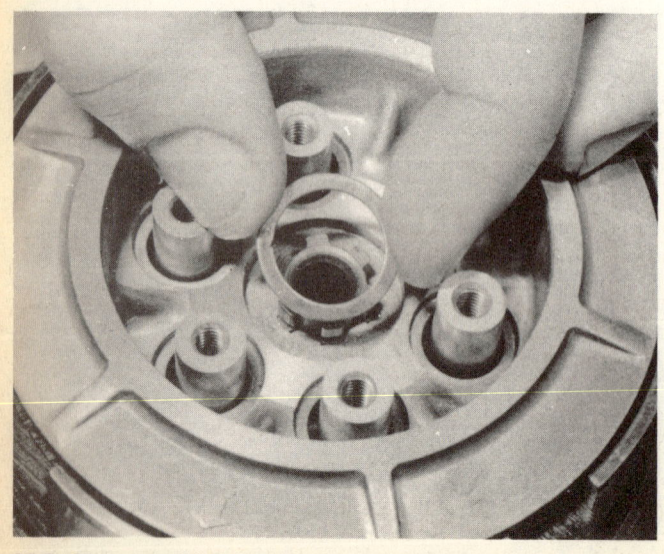

35.7e Plate assembly is located by a thrust washer ...

35.7f ... and retained by a circlip. Ensure circlip is correctly positioned

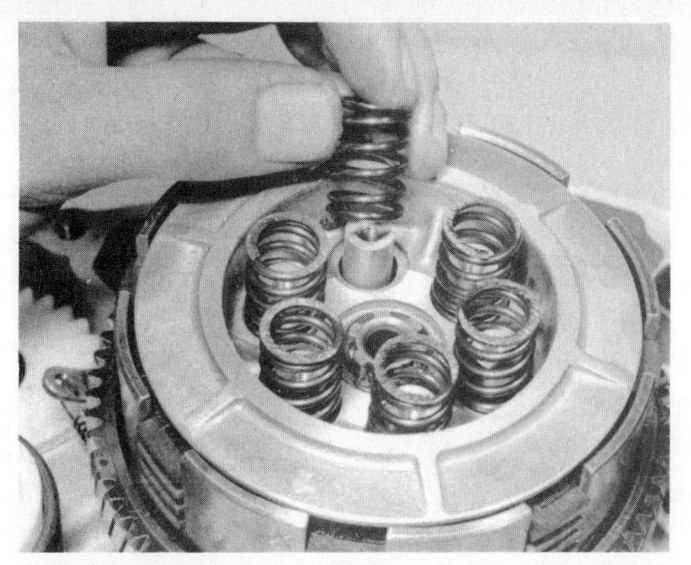

Chapter 1 Engine, clutch and gearbox

85

35.7g Refit clutch springs to pressure plate posts

35.8a Tighten securely clutch spring bolts ...

35.8b ... and refit release mechanism centre piece

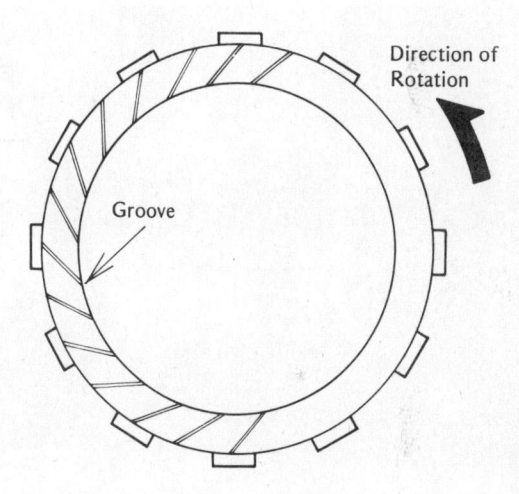

Fig. 1.17 Correct fitting of clutch plate – KE100 B (UK) model

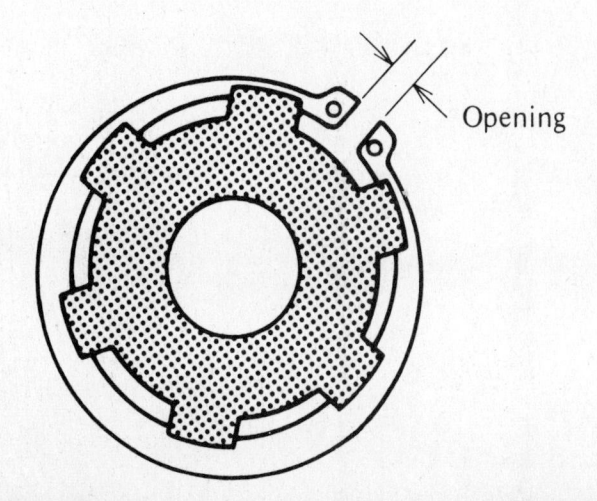

Fig. 1.18 Correct fitted position of clutch retaining circlip

36 Reassembling the engine/gearbox unit: refitting the right-hand crankcase cover

1 Check all components within the cover are properly fitted and lubricated. Grease the splined end of the kickstart shaft and the lip of the cover oil seal. Fit new sealing O-rings to the oil transfer pipe and carburettor stub protruding from the rotary valve cover.
2 Degrease the cover and crankcase mating surfaces, press the locating dowel into the crankcase and locate a new gasket over it. Push the cover over the dowel, tapping it lightly with a soft-faced mallet to seat it properly.
3 Fit each screw into its previously noted position. Tighten the screws evenly whilst working in a diagonal sequence from the centre outwards.

37 Reassembling the engine/gearbox unit: refitting the piston, cylinder barrel and head

1 The piston rings are refitted using the same technique as on removal. If neither surface is marked to show which is the top and the original rings are being refitted they must be installed the original way up, using the wear marks for identification. Keystone rings have a slight inwards taper on their top surfaces and are usually marked with an

identifying letter on that surface; check carefully before refitting to ensure that the tapered surface is upwards.

2 In all cases, rotate the rings so that their end gaps align with the locating peg in each ring groove to stop the rings rotating and catching in one of the ports. See the accompanying photograph.

3 Lubricate the big-end bearing and main bearings, bring the connecting rod to the top of its stroke, then pack the crankcase mouth with clean rag. Lubricate and fit the small-end bearing. With the arrow stamped in the piston crown facing forward, place the piston over the rod and fit the gudgeon pin. If necessary, warm the piston to aid fitting.

4 Retain the gudgeon pin with **new** circlips. Check each clip is correctly located; if allowed to work loose it will cause serious damage.

5 Check the piston rings are still correctly fitted. Clean the barrel and crankcase mating surfaces and fit a new base gasket. Lubricate the piston rings and cylinder bore. Lower the barrel over its retaining studs and ease the piston into the bore, carefully squeezing the ring ends together. Do not use excessive force.

6 Remove the rag from the crankcase mouth and push the barrel down onto the crankcase. Clean the barrel to head mating surfaces and fit a new head gasket. Fit the head and its retaining nuts with washers. Tighten the nuts evenly and in a diagonal sequence to the specified torque setting (where given).

7 Check its gap as described in Routine Maintenance and refit the spark plug, tightening it to the specified torque setting.

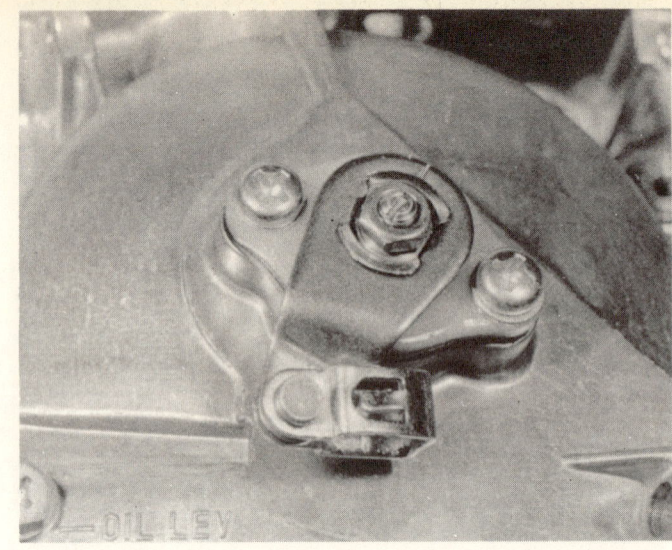

36.1 Check all components of clutch release mechanism are in place

36.2a Fit new gasket over locating dowel

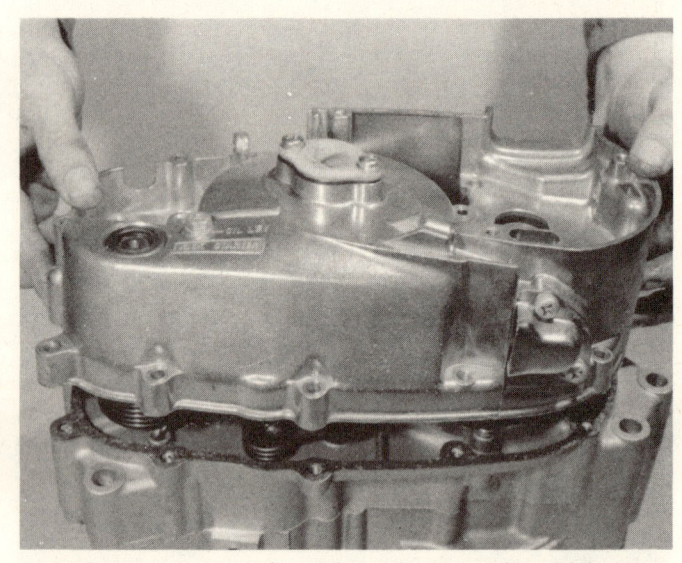

36.2b Grease kickstart shaft seal lips and press crankcase right-hand cover into place

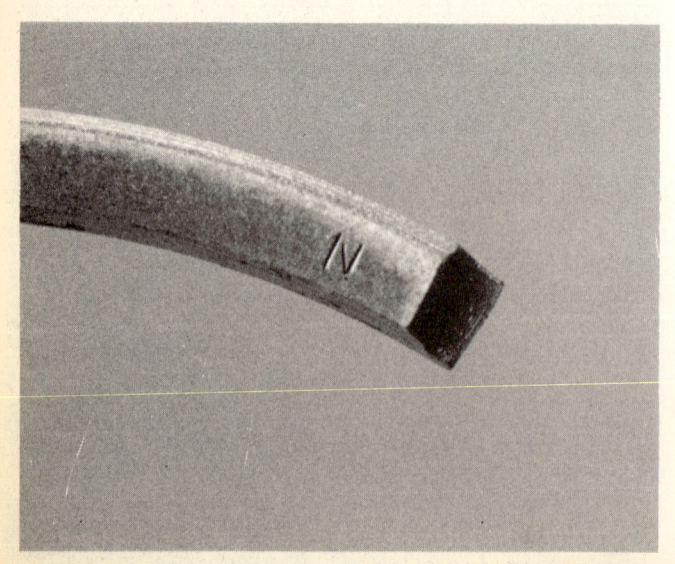

37.1 Keystone-type piston rings – top surface is usually marked with a stamped letter

37.2 Ensure piston ring end gaps are next to locating pegs as shown

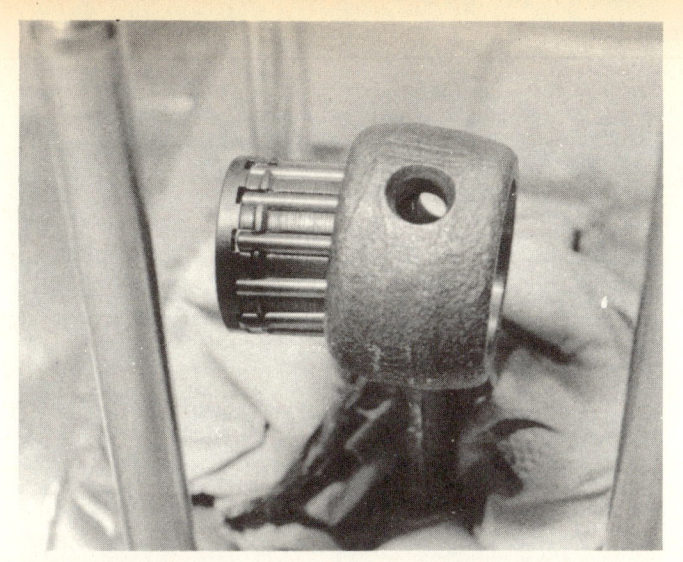

37.3a Oil all crankshaft bearings before reassembly – note rag packing crankcase mouth

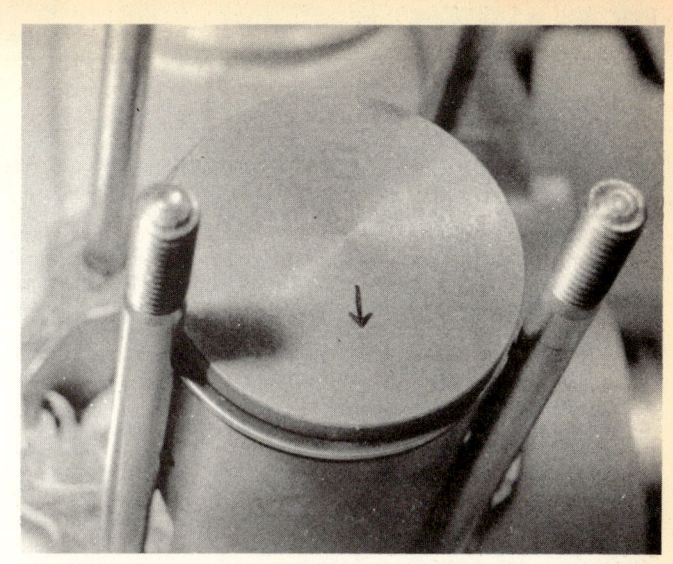

37.3b Refit piston so that arrow stamped in crown faces forwards, to exhaust port

37.4 Always use new circlips to secure gudgeon pin and ensure that each is correctly seated

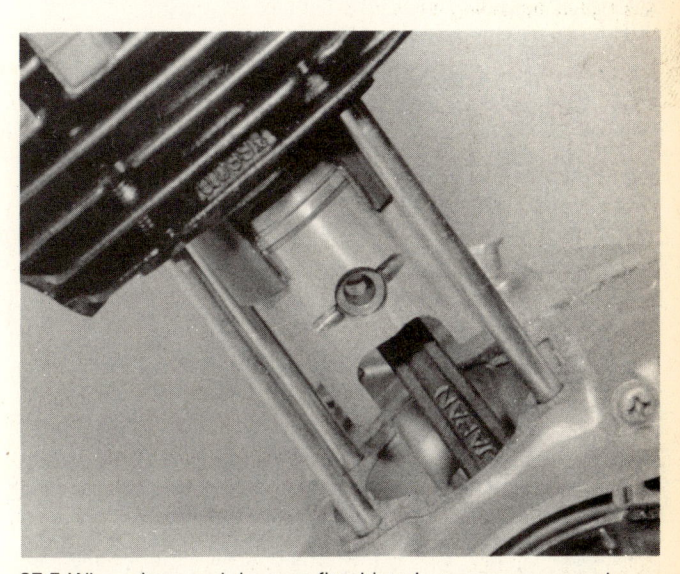

37.5 When piston and rings are fitted into bore, remove rag and press barrel down on to new base gasket

37.6a Always renew cylinder head gasket ...

37.6b ... and tighten head retaining nuts evenly to specified torque setting

Chapter 1 Engine, clutch and gearbox

38 Refitting the engine/gearbox unit into the frame

1 Check that no part has been omitted during reassembly. Prepare the machine and ease the engine into position. Align the engine and fit the mounting bolts and nuts; tighten these to the specified torque setting, where available.
2 Route the main generator lead from the crankcase top up to the frame, connect it again to the main loom and secure it with any clamps or ties provided. Connect the HT lead to the spark plug.
3 On KC100 models, refit the chain enclosure inner plate and tighten securely the two screws. On all models, engage the sprocket on the chain, fit the sprocket to the output shaft end and refit the toothed washer, flat washer and bolt. Apply the rear brake hard and tighten the bolt securely to the recommended torque setting, where specified. Secure the bolt by bending up against one of its flats an unused portion of the toothed washer.
4 Working as described in Routine Maintenance check, and adjust if necessary, the chain tension, rear brake adjustment and stop lamp rear switch setting. On KC100 models, refit the chain enclosure components.
5 Check, and reset if necessary, the ignition timing as described in Routine Maintenance, then refit the crankcase left-hand cover and the gearchange pedal.
6 Refit the air filter assembly, tightening securely its mounting bolt and clamp screw.
7 Refit the rev-counter cable (KH100 G models only). Refit the clutch cable and adjust the clutch release mechanism as described in Routine Maintenance.
8 Working as described in the relevant Sections of Chapter 2 and in Routine Maintenance, install the oil pump and bleed any air from the tank/pump feed pipe, refit the carburettor and the oil pump control cable and check the adjustment of the throttle and oil pump cables. If necessary check the carburettor settings as described in Section 8 of Chapter 2 before refitting the carburettor covers. Check that the oil feed pipes are correctly routed, then refit the oil pump cover.
9 Ensuring that it is installed in its original position, refit the kickstart lever and tighten securely the pinch bolt. Refit the exhaust system, as described in Section 11 of Chapter 2.
10 Reconnect the battery. Check the terminals are clean and prevent corrosion occurring by smearing them with petroleum jelly. Route the vent pipe clear of the lower frame.
11 Refit the fuel tank, checking there is no metal-to-metal contact which will split the tank. Reconnect the fuel feed pipe, turn on the tap and check for leaks, which must be cured immediately.
12 Refit the seat, the side panel(s) (as applicable), check that the transmission oil drain plug is refitted and tightened to the specified torque setting, then refill the gearbox with oil to the correct level as described in Routine Maintenance. Note, however, that slightly more oil may be required after a full engine rebuild.
13 Make a final check that all components have been refitted and that all are securely fastened and correctly adjusted.

39 Starting and running the rebuilt engine

1 Start the engine using the procedure for a cold engine. A certain amount of perseverance may prove necessary to coax the engine into activity even if new parts have not been fitted. Should the engine persist in not starting, check that the spark plug has not become fouled by the oil used during reassembly. Failing this, go through the fault-finding charts and work out what the problem is methodically.
2 When the engine does start, keep it running as slowly as possible to allow the oil to circulate. Open the choke as soon as the engine will run without it. During the initial running, a certain amount of smoke may be in evidence due to the oil used in the reassembly sequence being burnt away. The resulting smoke should gradually subside. As soon as the engine will run smoothly, carry out the procedure described in Chapter 2 to bleed air from the oil pump/engine feed pipe.
3 Check the engine for blowing gaskets and oil leaks. Before using the machine on the road, check that all the gears select properly, and that the controls function correctly.

38.3a Refit gearbox sprocket to output shaft followed by flat washer, fitted as shown ...

38.3b ... toothed washer locating tang must fit into plain washer cutout and into sprocket hole, as shown

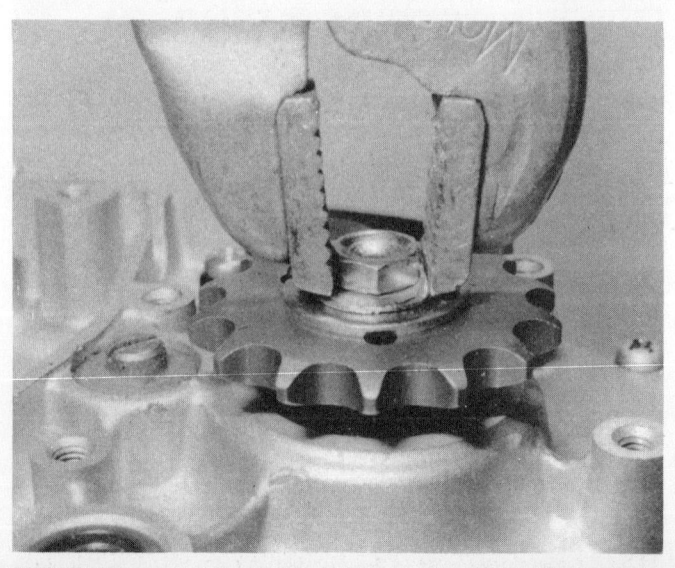

38.3c Tighten sprocket retaining bolt and secure by bending up lock washer tab

38.3d Chain connecting link closed end must always face in direction of chain travel, as shown

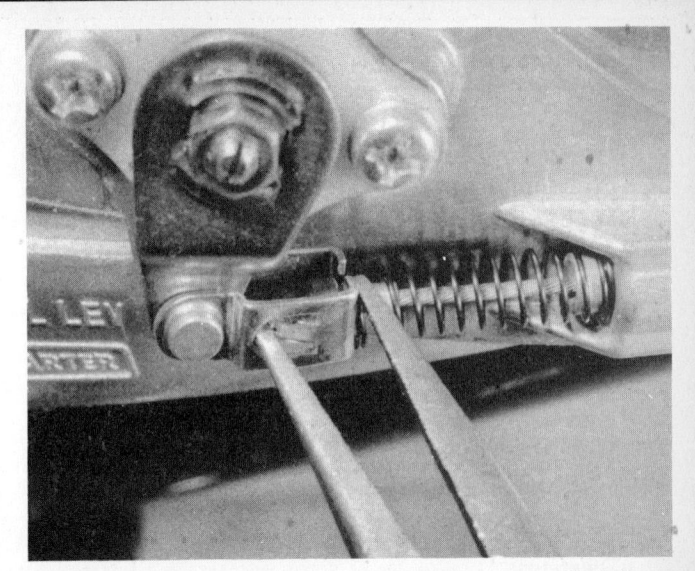

38.7 When refitting clutch cable, do not forget to secure nipple by bending over retaining tang

40 Taking the rebuilt machine on the road

1 Any rebuilt machine will need time to settle down, even if parts have been replaced in their original order. For this reason it is highly advisable to treat the machine gently for the first few miles to ensure oil has circulated throughout the lubrication system and that any new parts fitted have begun to bed down.
2 Even greater care is necessary if the engine has been rebored or if a new crankshaft has been fitted. In the case of a rebore, the engine will have to be run in again, as if the machine were new. This means greater use of the gearbox and a restraining hand on the throttle until at least 500 miles have been covered. There is no point in keeping to any set speed limit; the main requirement is to keep a light loading on the engine and to gradually work up performance until the 500 mile mark is reached. These recommendations can be lessened to an extent when only a new crankshaft is fitted. Experience is the best guide since it is easy to tell when an engine is running freely.

3 Remember that a good seal between the piston and the cylinder barrel is essential for the correct functioning of the engine. A rebored two-stroke engine requires careful running-in. There is a far greater risk of engine seizure during the first hundred miles if the engine is permitted to work hard.
4 If at any time a lubrication failure is suspected, stop the engine immediately and investigate the cause. If the engine is run without oil, even for a short period, irreparable engine damage is inevitable.
5 Do not on any account add oil to the petrol under the mistaken belief that a little extra oil will improve the engine lubrication. Apart from creating excess smoke, the addition of oil will make the mixture much weaker, with the consequent risk of overheating and engine seizure. The oil pump alone should provide full engine lubrication.
6 When the initial run has been completed allow the engine unit to cool and then check all the fittings and fasteners for security. Re-adjust any controls which may have settled down during initial use and check the transmission oil level, topping up if necessary.

Chapter 2 Fuel system and lubrication

Contents

General description 1	Carburettor: adjustment 8
Fuel tank: removal, examination and refitting 2	Air filter: general 9
Fuel tap: removal, dismantling and reassemby 3	Rotary disc valve: removal, examination and refitting 10
Carburettor: removal and refitting 4	Exhaust system: removal, examination and refitting 11
Carburettor: dismantling, examination and reassembly 5	Oil pump: removal, examination and refitting 12
Carburettor: checking the settings 6	Oil pump: bleeding 13
Carburettor adjustment and exhaust emissions: general note 7	

Specifications

Fuel tank capacity

	Litre	Imp gal	US gal
KC100:			
Overall	8.6	1.89	N/App
Reserve	0.2	0.04	N/App
KE100 A (UK), KE100 A5, A6, A7 (US):			
Overall	8.0	1.76	2.11
Reserve	0.8	0.18	0.21
KE100 A8, A9, A10 (US):			
Overall	8.0	1.76	2.11
Reserve	0.5	0.11	0.13
KE100 B:			
Overall	9.0	1.98	2.38
Reserve	1.5	0.33	0.40
KH100:			
Overall	10.0	2.20	N/App
Reserve	1.6	0.35	N/App

Fuel grade .... Unleaded or leaded, minimum octane rating 91 (RON/RM)

Carburettor

	KC100 C1, C2, C3	KC100 C4	KE100 A (UK) KE100 A5, A6, A7 (US)	KE100 A8 (US)
Manufacturer	Mikuni	Mikuni	Mikuni	Mikuni
Type	VM19SC	VM19SC	VM19SC	VM19SC
Main jet	77.5	75	75	77.5
Needle jet	0-2	0-8	0-2	0-2
Jet needle	4EJ7	4EJ34	4EJ7	4EJ12
Clip position – grooves from top	3rd	3rd	3rd	3rd
Throttle valve cutaway	2.0	2.0	2.0	2.0
Pilot air screw – turns out from closed	1½	1½	1½	1½
Pilot jet	17.5	17.5	17.5	15
Float valve seat	N/Av	2.0	2.0	2.0
Fuel level – all models	4.0 ± 1.0 mm	(0.16 ± 0.04 in)		

	KE100 A9, A10, B1, B2, B3 (US)	KE100 B5 to B12 (US)	KE100 B (UK)	KH100 A, G
Manufacturer	Teikei	Mikuni	Mikuni	Mikuni
Type	E19PK-1A	VM19SC	VM19SC	VM19SC
Main jet	82◄	80	90	82.5
Needle jet	2.580	0-1	P-0	0-2
Jet needle	4D21	4EJ39	4M5	4EJ10
Clip position – grooves from top	N/App	N/App	3rd	3rd
Throttle valve cutaway	3.0	2.0	2.0	2.0
Pilot air screw – turns out from closed	1½*	1½*	1½	1½
Pilot jet	38◄	15	17.5	17.5
Float valve seat	N/Av	2.0	2.0	2.0

Fuel level:
  KE100 A9, A10, B1, B2, B3 (US) .... 0.5 ± 1.0 mm (0.02 ± 0.04 in)
  KE100 B5 to B12 (US), KE100 B (UK), KH100 A, G .... 4.0 ± 1.0 mm (0.16 ± 0.04 in)
Idle speed – all models .... Lowest stable speed possible (approx 1300 ± 100 rpm)

* Initial setting only – screws sealed by plugs
◄ Alternative size main jet (80) and pilot jet (36) for B2 and B3 models operating above 4000 ft.

Chapter 2 Fuel system and lubrication

Disc valve

Valve thickness:
- Standard
- Service limit

Valve cover inner surface depth:
- Standard
- Service limit

	KE100 A5	All other KE and KH100 models
Standard	3.00 – 3.20 mm (0.1181 – 0.1260 in)	0.67 – 0.73 mm (0.0264 – 0.0287 in)
Service limit	2.85 mm (0.1122 in)	0.52 mm (0.0205 in)
Standard	3.35 – 3.45 mm (0.1319 – 0.1358 in)	1.10 – 1.20 mm (0.0433 – 0.0472 in)
Service limit	4.00 mm (0.1575 in)	1.50 mm (0.0591 in)

Engine lubrication system

Tank capacity:
- KH100 models .... 1.4 lit (2.46 Imp pint)
- All other models .... 1.2 lit (2.11 Imp pint, 1.27 US qt)

Recommended oil .... Good quality air cooled 2-stroke engine oil

Oil pump output – control lever fully open, @ 2000 rpm:
- KE100 B models .... 3.6 – 4.1 cc in 3 minutes
- All other models .... 2.6 – 3.0 cc in 3 minutes

Transmission oil

Capacity .... 600 cc (1.06 Imp pint, 0.63 US qt)
Recommended oil .... Good quality oil SAE10W/30 or 10W/40 SE or SF

Torque wrench settings

Component	kgf m	lbf ft
Transmission oil drain plug:		
KE100 A	0.7 – 1.0	5 – 7
KE100 B, KH100	2.0	14.5
Oil pipe union banjo bolts	0.4 – 0.5	3 – 3.5

1 General description

Fuel contained in a tank mounted on the frame top tubes is passed via a three-position tap to the conventional concentric float slide-type carburettor. Cold starting is assisted by a separate starting circuit which supplies the correct fuel-rich mixture when the 'choke' control is operated.

Air entering the carburettor first passes through a filter which not only removes any airborne dust which might otherwise enter and damage the engine, but also helps to silence the induction roar, a common problem with two-stroke engines.

The induction system chosen by Kawasaki for these machines is the disc, or rotary valve system in which a flat disc of thin metal (or of a synthetic compound, on early models) rotates with the crankshaft in a narrow chamber next to the crankcase. A cutout in the disc allows the fuel/air mixture to enter the crankcase at the desired moment. This system permits more precise control of the induction timing with the result that more efficient combustion produces an increase in power output and fuel economy.

Engine lubrication is catered for by the Kawasaki Superlube system. Oil from a separate tank is fed by an oil pump to a small injection nozzle in the intake tract. The pump is linked to the throttle twistgrip, and this controls the volume of oil fed to the engine.

All gearbox and primary drive components are lubricated by splash from a supply of oil contained in the reservoir formed by the crankcase castings.

2 Fuel tank: removal and refitting

All models
1 Turn the fuel tap lever to 'Off'. Observe the necessary fire precautions and pull the fuel feed pipe off the tap spigot, draining any fuel from the pipe into a clean container.

KC100
2 Unlatch and raise the seat, then remove the single front mounting bolt and nut, followed by the rear mounting nut. Lift the tank at the rear and manoeuvre it off the machine.

KE100
3 Unlock and raise the seat, disengage the restraining hook and allow the seat to open as far as possible. Unhook the tank rear retaining strap (KE100 A) or remove the single rear mounting bolt (KE100 B), lift the tank at the rear and pull it backwards off its front mountings.

KH100
4 Remove the two bolts which secure the rear end of the seat to the frame, lift the seat at the rear and carefully pull it backwards to remove it. Similarly, lift the tank at the rear to disengage its mounting prong from the rubber socket in the frame, and pull it backwards to remove it.

All models
5 When the tank is removed, place it in safe storage, away from naked flames or sparks. Check for leaks around the tap and cover the tank to protect its paint finish. Renew any damaged mounting rubbers.
6 Reverse the removal procedure to fit the tank. If necessary, wipe the front mounting rubbers with petrol to ease installation. Secure the tank and check for metal-to-metal contact which might split it.
7 Reconnect the pipe, turn on the tap and check for leaks. Fuel leaks will waste petrol and cause a fire hazard.
8 If it is necessary to remove the fuel tank for repairs the following points should be noted. Fuel tank repair, whether necessitated by accident damage or by petrol leaks, is a task for the professional. Welding or brazing is not recommended unless the tank is purged of all petrol vapour; a difficult condition to achieve. Resin-based tank sealing compounds are now available through suppliers who advertise regularly in the motorcycle Press. Accident damage repairs will inevitably involve re-painting the tank; matching of modern paint finishes is a very difficult task not to be lightly undertaken by the average owner. It is therefore recommended that the tank be removed by the owner, and taken to a motorcycle dealer or similar expert for professional attention.
9 Repeated contamination of the fuel tap filter and carburettor by water or rust and paint flakes indicates that the tank should be removed for flushing with clean fuel and internal inspection. Rust problems can be cured by using a resin-based tank sealant.

2.1 Before removing fuel tank switch off fuel supply and disconnect feed pipe

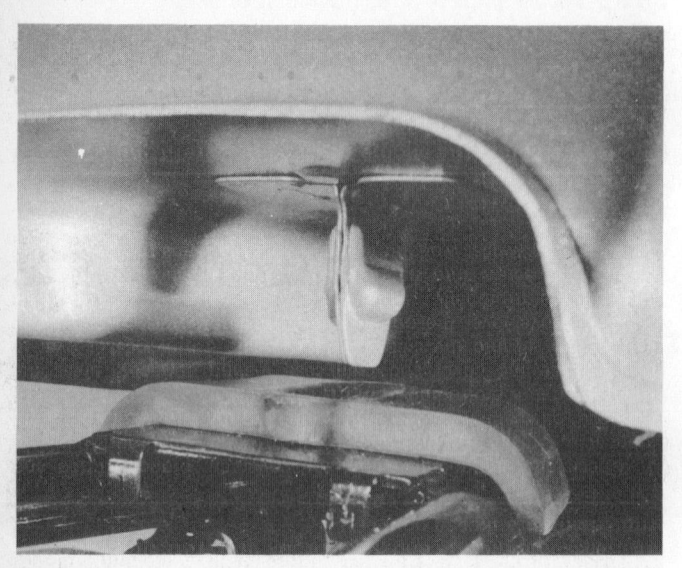

2.4 KH100 models – lift tank at rear to release locating tongue from rubber grommet

2.6 Check front mounting rubbers are in good condition – lubricate to ease tank refitting

3 Fuel tap: removal, dismantling and reassembly

All models

1 If the tap is removed, the tank must first be drained or removed from the machine and placed so that fuel will not leak out. Read the 'Safety First' section of this manual before starting work and take great care to prevent the risk of fire.

KC100, KE100 A

2 Remove the filter bowl as described in Routine Maintenance and unscrew the large gland nut which secures the tap to the tank threaded boss; note the sealing washer inside the gland nut.

3 To remove the lever unscrew the single grub screw from the side of the tap body and pull out the lever complete with its sealing O-ring and spring.

4 Reassembly is the reverse of the dismantling procedure; all seals are available separately and should be renewed as a matter of course. When refitting the tap to the tank, do not omit the sealing washer and be careful to start the threads of the nut simultaneously on those of the tank and tap; the nut will then draw the two together to create a leak-proof joint.

KH100 A, KH100 G2, G3, G4, KE100 B

5 The taps fitted to these machines are retained by two bolts to the underside of the tank. Remove the filter bowl as described in Routine Maintenance, unscrew the two bolts and lift the tap out of the tank noting the O-ring which seals the joint.

6 In most cases a grub screw will be set in the side of the tap body; when unscrewed this will permit the removal of the tap lever with its sealing O-ring and spring. If no grub screw is found, the tap must be regarded as a sealed unit and should be renewed if it is blocked or leaking.

KH100 G5, G6, G7, G8

7 Having drained the tank, remove the two mounting bolts and withdraw the tap, noting the O-ring which seals the joint.

8 Remove the two small screws to release the lever retaining plate, then withdraw the lever with its plate; note the presence of the sealing O-ring next to the tap seal itself.

All models

9 If any leaks have occurred, renew the seal concerned; in the case of the lever seal a temporary repair can be effected sometimes by reversing it (KH100 G5, G6, G7, G8) only.

10 Clean out the tap passages using compressed air and clean the gauze filters with a fine toothbrush, taking great care not to damage the gauze.

11 On reassembly, take care not to overtighten any component: the castings are of a zinc-based alloy which fractures easily. Never overtighten a component in an effort to cure a fuel leak.

3.7a Fuel tap is retained by two bolts

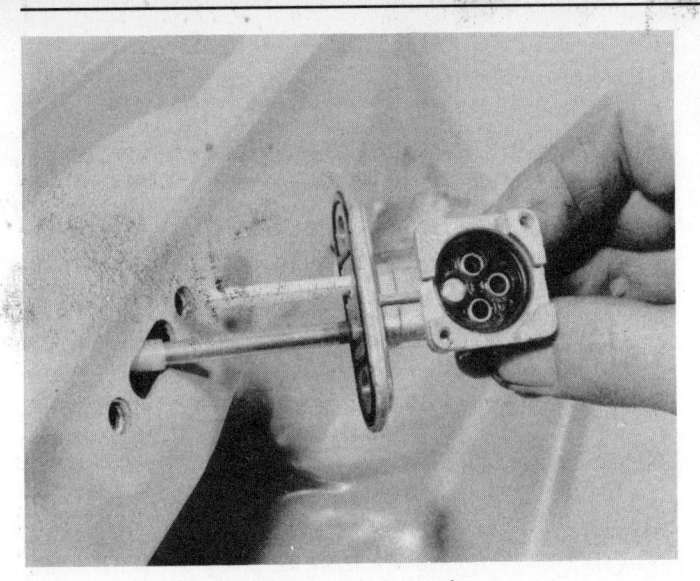

Chapter 2 Fuel system and lubrication 93

3.7b Note O-ring sealing tank/tap mating surface

3.8 Remove two screws and retainer plate to release tap lever – note tap seal and O-ring

4 Carburettor: removal and refitting

1 As a general rule the carburettor should be left alone unless it is in obvious need for overhaul. Before a decision is made to remove and dismantle, ensure that all other possible sources of trouble have been eliminated. This includes the more obvious candidates such as a fouled spark plug, a dirty air filter element or choked exhaust system. If a fault has been traced back to the carburettor, proceed as follows.
2 Firstly, apply a lubricant such as WD40 or similar to the control cables and fuel feed pipe where they pass through the carburettor cover and sealing grommet respectively. This will minimise the risk of their sticking together and tearing.
3 Withdraw the extension piece from the idle speed adjusting screw and, on KC and KE100 models, remove the choke knob by withdrawing the retaining split pin and pulling the knob off the plunger rod. Prise the spring clamp out of its groove and slide the carburettor rubber cover upwards out of the way.
4 Remove the retaining screws and withdraw the carburettor side cover and its gasket. Remove the oil pump cover and disconnect its control cable.
5 Prise the clamp out of its place, check that the fuel tap is in the 'Off' position and pull the feed off the carburettor union.
6 Prise the rubber grommet out of the front of the crankcase cover and use a screwdriver to slacken the carburettor clamp bolt. Work the carburettor off its stub noting the presence of the two sealing rings on the late US KE100 models. Note also the rubber overflow grommet at the bottom of the float bowl; do not allow this to be lost, blocked with dirt or damaged.
7 Refitting is essentially the reverse of the above procedure, but note that great care must be taken to ensure that the carburettor chamber is completely sealed. Renew the cover gasket and any sealing components if they are worn, cracked, split or damaged in the slightest. Remember that not only is there a risk of unfiltered air being allowed into the engine, but that the weakened mixture that will result from a leak in the carburettor chamber will cause overheating and possible severe engine damage.
8 Do not overtighten the carburettor clamp bolt or the body may be distorted, and ensure that the idle adjuster screw extension is correctly refitted, also that the overflow grommet is correctly engaged on its mountings.
9 Before starting the engine after the carburettor has been removed and refitted, check carefully that all control cables are seated correctly in their adjusters and that the throttle, oil pump and choke (where applicable) are working properly. Refer to Routine Maintenance.

4.3 Remove spring clamp and idle adjuster extension piece, then slide rubber cover upwards

4.4 Remove carburettor and oil pump covers to disconnect all components

Chapter 2 Fuel system and lubrication

4.6a Withdraw sealing grommet from crankcase cover ...

4.6b ... and slacken carburettor clamp bolt using a screwdriver

4.8a Do not omit spacer and/or seal inside carburettor mouth when refitting

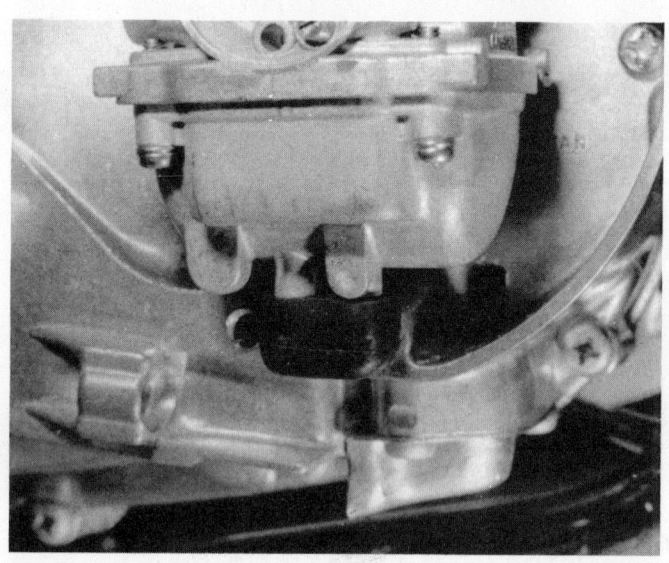

4.8b Ensure overflow grommet is pressed securely into place to prevent leakage

5 Carburettor: dismantling, examination and reassembly

1 With the carburettor removed from the machine as described in the previous Section, unscrew the choke plunger assembly and put it to one side. On KH100 models the cable can be disconnected by compressing the spring and sliding the end nipple out of the plunger.
2 Unscrew the carburettor top and carefully withdraw the throttle valve (slide) assembly; be very careful not to bend the valve rod. To release the rod, withdraw the split pin from its upper end and allow the rod to drop down through the valve; the adjuster screw and spring need not be adjusted. On KH100 models, prise out the wire clip which secures the throttle cable into the adjuster.
3 Compress the throttle return spring against the carburettor top, leaving the valve clear for the retainer to be tipped out. The cable inner wire can then be pressed down and its end nipple slipped into the longer slot so that the two can be separated.
4 On all UK models and early US models the jet needle is now free to be tipped out of the valve. On all US models from the KE100 A8 onwards remove the retaining screw and lock washer from the base of the valve then lift out the retainer plate followed by the spacer, the needle and circlip, the spring seat and the spring.

5 On reassembly, be very careful to check that all components are correctly installed and that none are damaged; be especially careful with the jet needle and the valve rod. On installing the valve assembly in the carburettor, note that the groove in the side of the valve must engage with the projecting pin set in the carburettor body.
6 To dismantle the remainder of the carburettor, first peel the overflow grommet off the base of the float chamber, then prepare a clean working area covered with clean paper to prevent small components from being lost.
7 Remove the float chamber retaining screws. If necessary, tap around the chamber to body joint with a soft-faced hammer to free the chamber. Remove the float pivot pin and detach the twin floats. Displace the float needle and unscrew the float needle seat with sealing washer.
8 Use a narrow-bladed screwdriver to unscrew the pilot jet, then a close-fitting spanner to unscrew the main jet. Do not forget to remove the washer beneath the main jet and be very careful not to damage the soft brass jets. Using a slim wooden rod such as a pencil, tap the needle jet out from below, ie upwards through the throttle valve bore.
9 On all US models from the KE100 A9 onwards the pilot air screw is sealed by a blanking plug. If it is necessary to disturb the screw, use a small punch or similar to pierce the plug until it can be displaced. The

Chapter 2 Fuel system and lubrication

screw can now be removed in the same way as for all other models, ie screw it **inwards** until it seats lightly and count the exact number of turns necessary to do this, before unscrewing it completely and removing it with its spring. On reassembly, simply screw it in until it seats lightly, then unscrew it by the previously noted number of turns to return it to its original position. On late US models no further adjustment is necessary and a new sealing plug should be tapped into place and secured with a small smear of bonding agent. On all other models the screw position will usually serve as a good basis for subsequent adjustment; the setting should be very close to that specified.

10 Before examination, thoroughly clean each part in clean petrol, using a soft nylon brush to remove stubborn contamination and a compressed air jet to blow dry. Avoid using rag because lint will obstruct jet orifices. Do not use wire to clear blocked jets; this will enlarge the jet and increase petrol consumption. If an air jet fails, use a soft nylon bristle. Observe the necessary fire precautions and wear eye protection against blow-back from the air jet.

11 Check carefully for any distorted or cracked castings, renew all O-rings, gaskets, fatigued or broken springs and flattened spring washers.

12 Wear of the float needle takes the form of a groove around its seating area; renew if worn. Check for similar wear of the needle seat and, if possible, renew when worn. Where fitted, check the needle end pin is free to move and is spring-loaded.

13 Check the floats for damage and leakage. Renew if damaged; it is not advisable to attempt a repair.

14 Examine the choke assembly, renewing any worn or damaged parts. Renew hardened drain or fuel feed pipes.

15 Wear of the throttle valve will be indicated by polished areas on its external diameter, causing air leaks which weaken the mixture and produce erratic slow running. Examine the carburettor body for similar wear and renew each component as necessary.

16 Examine the needle for scratches or wear along is length and for straightness. If necessary, it must be renewed, in conjunction with the needle jet. Do not attempt to straighten a bent needle; they fracture very easily.

17 Renew the seal within the mixing chamber top if damaged. The throttle return spring must be free of fatigue or corrosion.

18 Before assembly, clean all parts and replace them on clean paper in a logical order. Do not use excessive force during reassembly, it is easy to shear a jet or damage a casting.

19 Reassembly is the reverse of dismantling. If in doubt, refer to the accompanying figures or photographs.

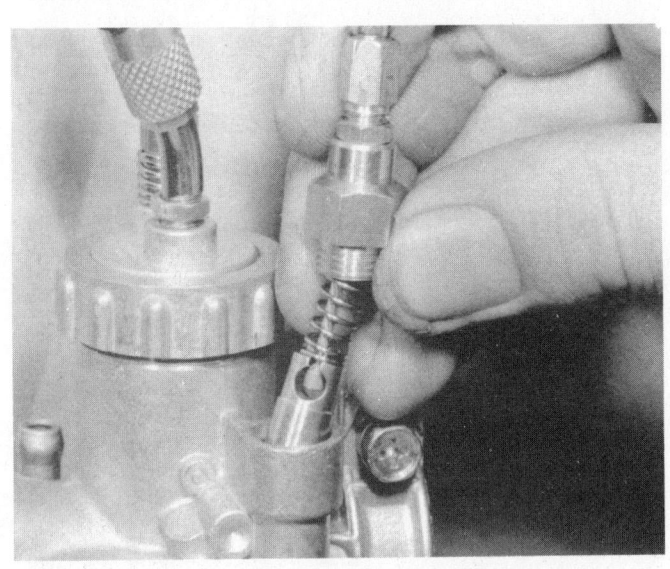

5.1 KH100 models – unscrew choke plunger assembly and disconnect from cable if required

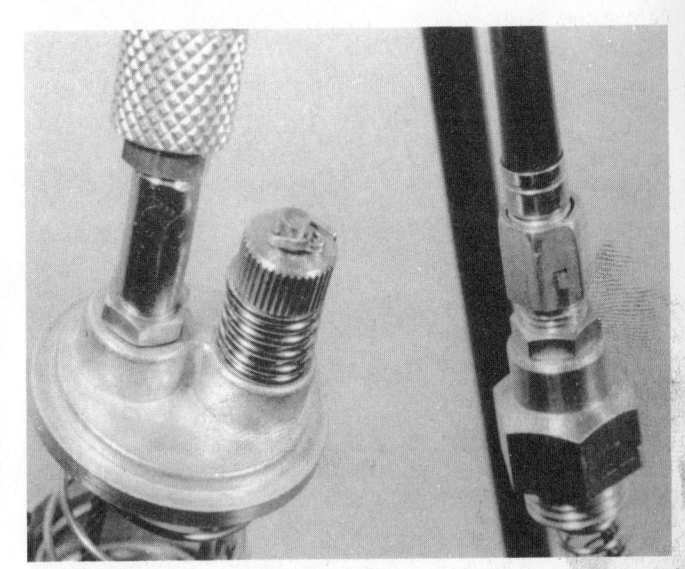

5.2a Straighten and remove split pin from above idle speed adjuster ...

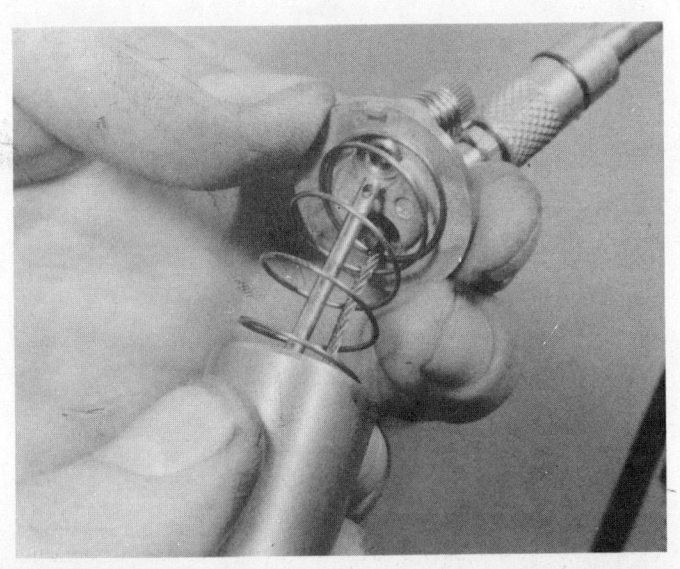

5.2b ... so that valve rod can be lowered away from valve assembly

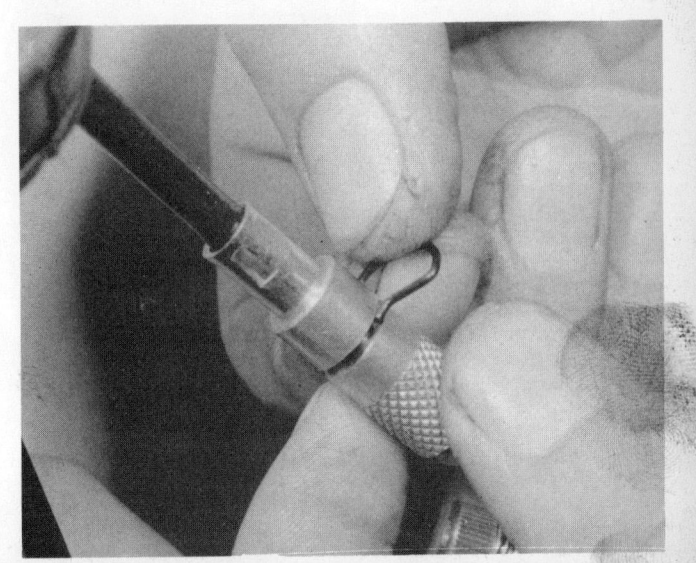

5.2c Where fitted, remove spring clip to release throttle cable outer from carburettor top

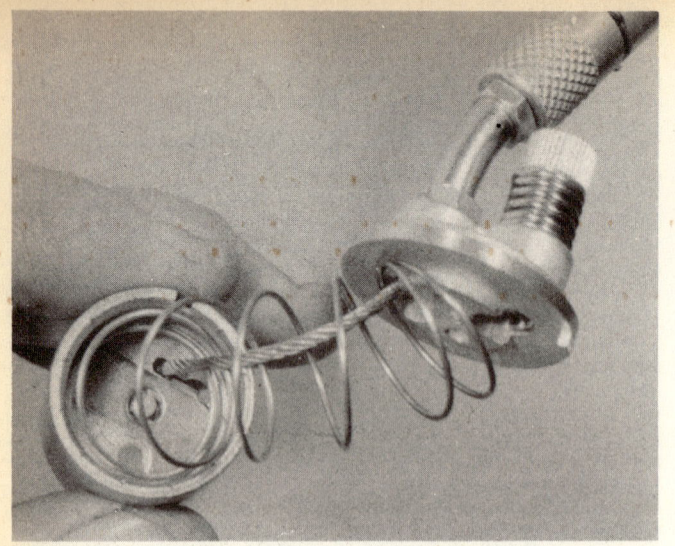

5.3a Compress return spring against carburettor top and tip out retainer ...

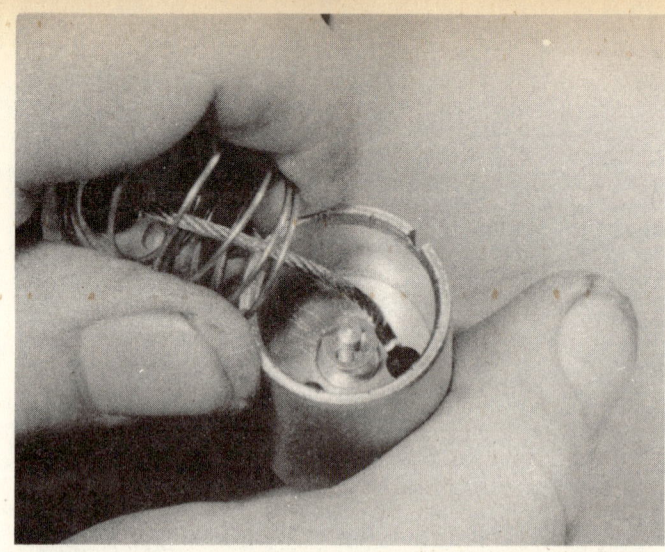

5.3b ... so that cable end nipple can be slipped into larger slot and removed

5.4 Jet needle assembly is now free to be removed – do not disturb clip position unless necessary

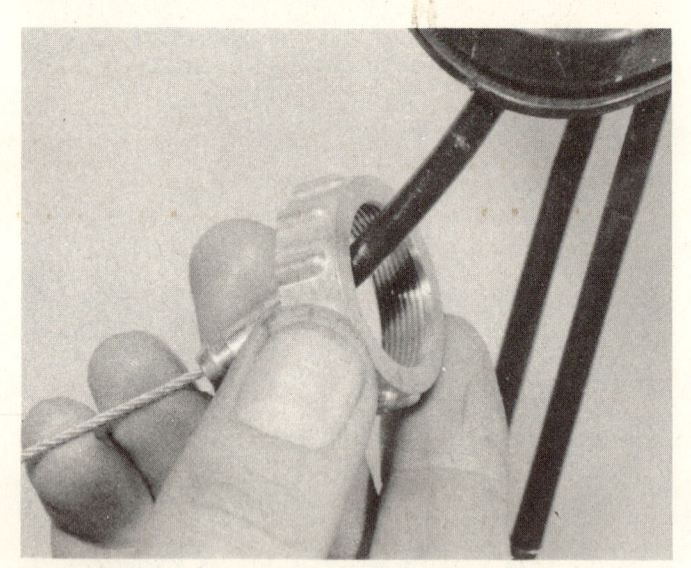

5.5a On reassembly do not forget to refit knurled ring ...

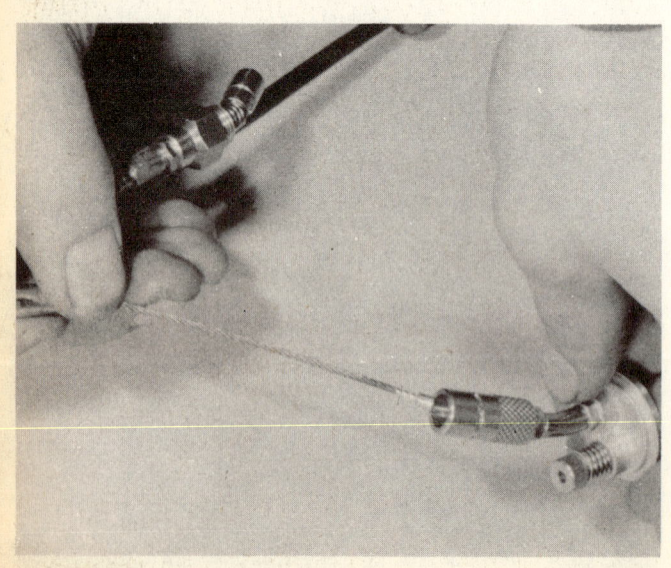

5.5b ... before fitting carburettor top to throttle cable

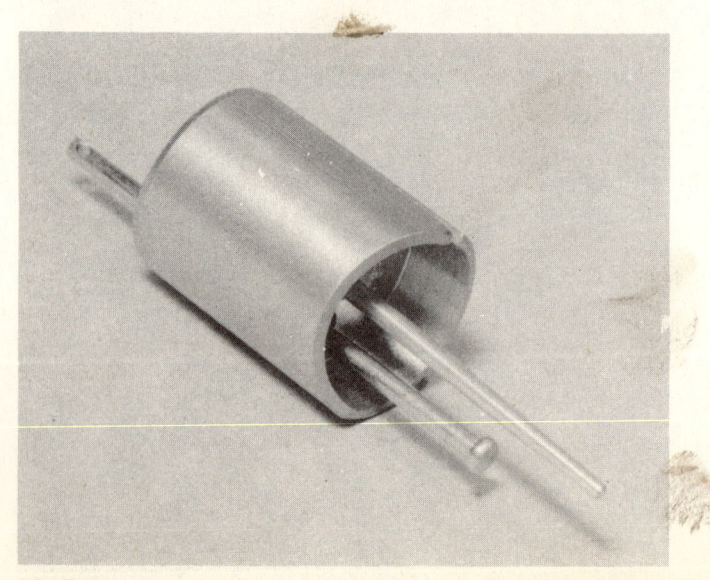

5.5c Take great care not to bend jet needle or throttle valve rod on refitting

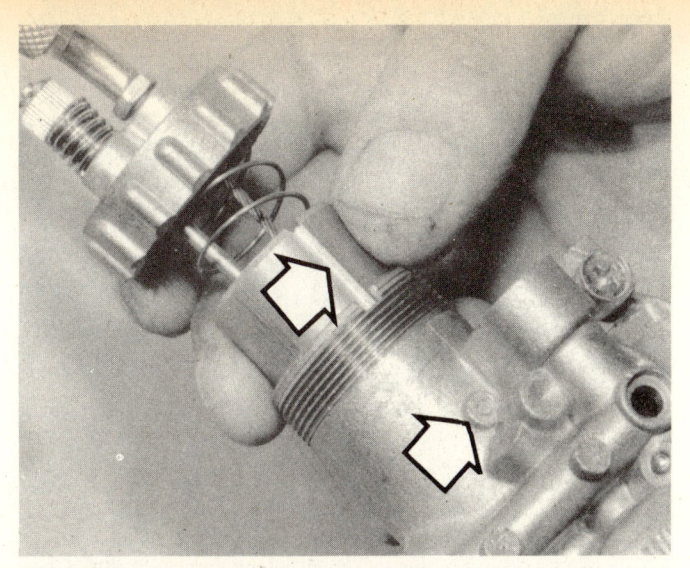

5.5d Ensure locating pin in carburettor body engages with slot (arrowed) as throttle valve is refitted

5.7a Remove retaining screws to release float chamber ...

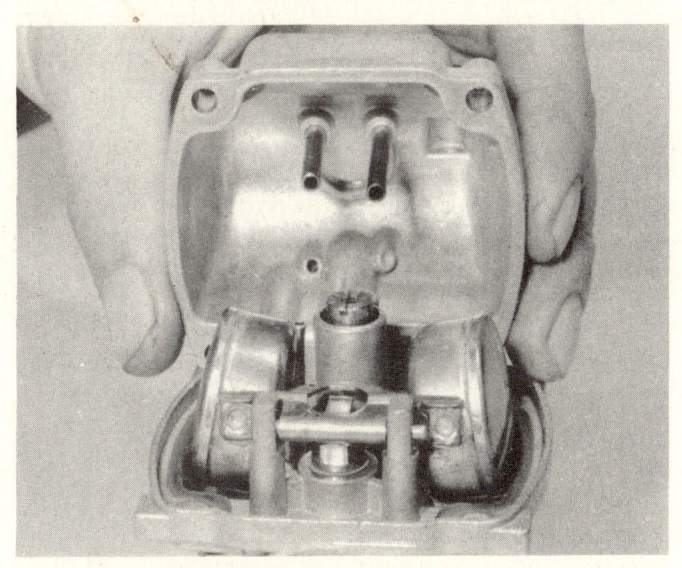

5.7b ... and expose float assembly and jets

5.7c Remove pivot pin to release float assembly ...

5.7d ... withdraw float needle ...

5.7e ... and unscrew float needle seat – note sealing washer

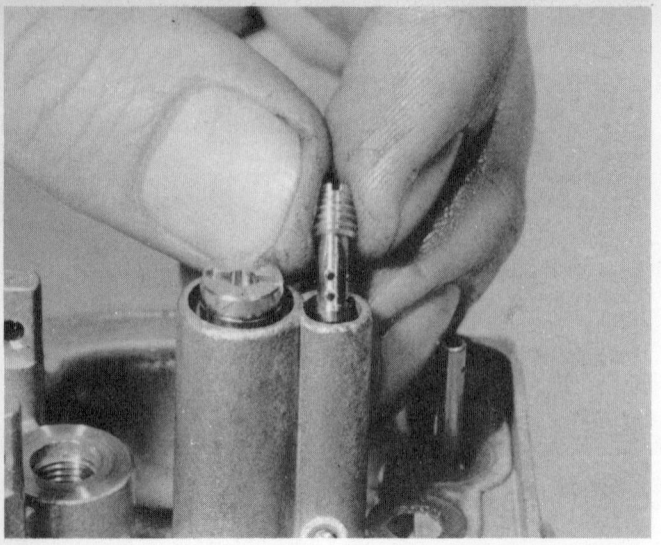

5.8a Unscrew pilot jet ...

5.8b ... followed by main jet

5.8c Note flat washer underneath main jet

5.8d Needle jet must be pressed out from below using a wooden rod or similar

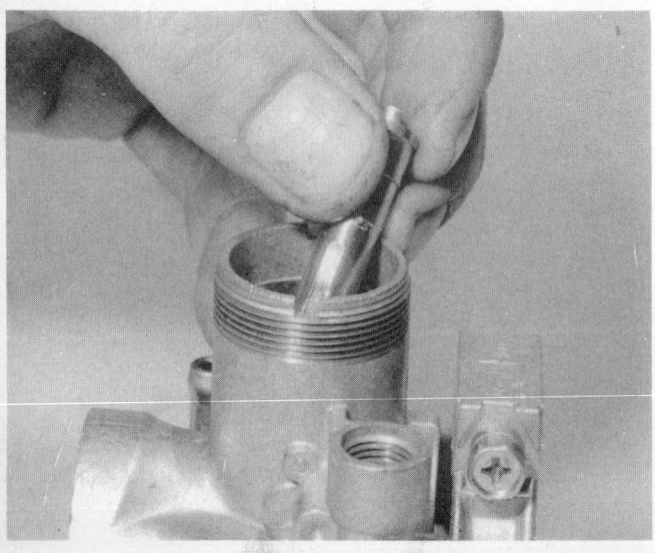

5.8e When refitting needle jet note that it is located by a groove ...

5.8f ... which must engage with locating peg in jet pillar

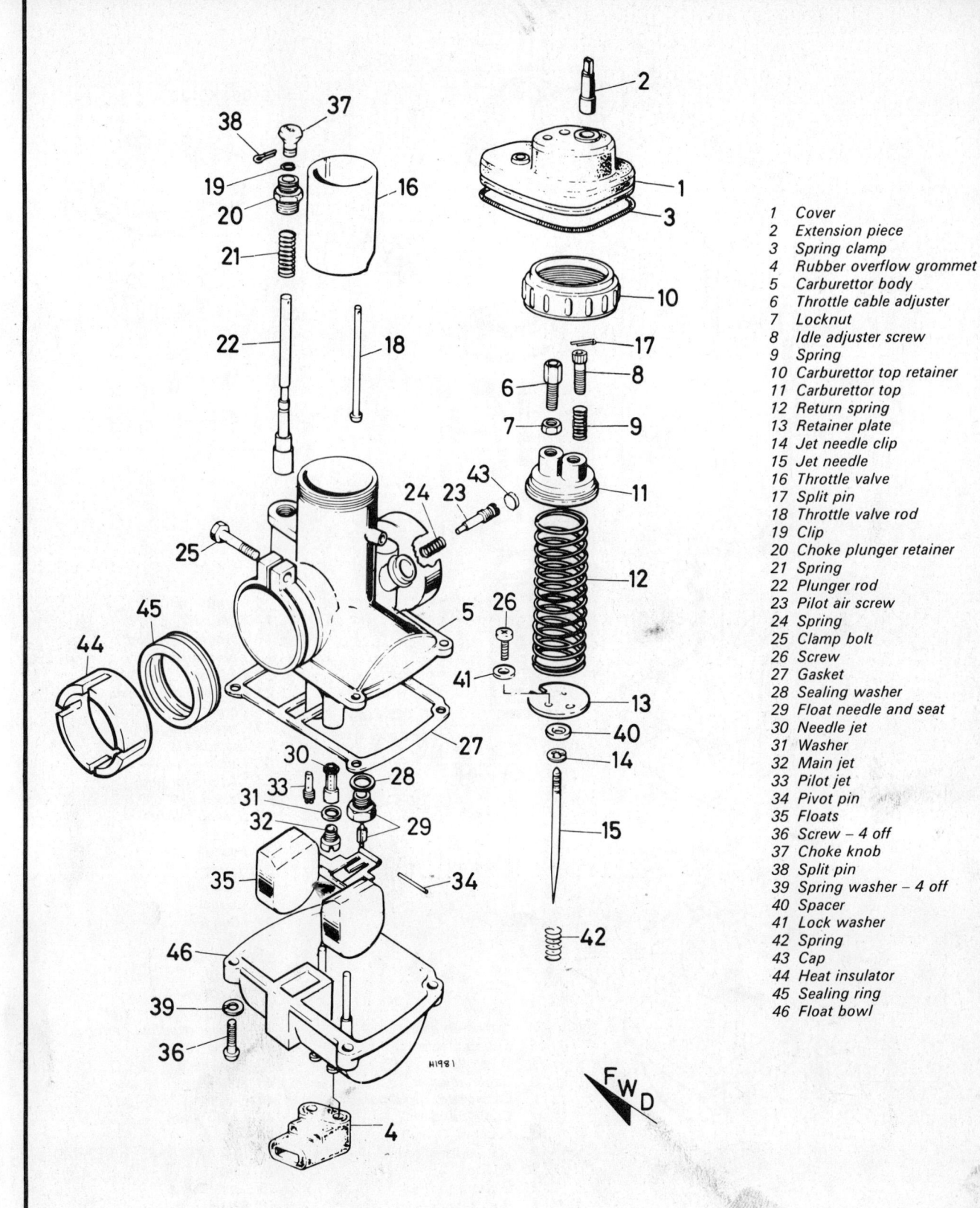

1. Cover
2. Extension piece
3. Spring clamp
4. Rubber overflow grommet
5. Carburettor body
6. Throttle cable adjuster
7. Locknut
8. Idle adjuster screw
9. Spring
10. Carburettor top retainer
11. Carburettor top
12. Return spring
13. Retainer plate
14. Jet needle clip
15. Jet needle
16. Throttle valve
17. Split pin
18. Throttle valve rod
19. Clip
20. Choke plunger retainer
21. Spring
22. Plunger rod
23. Pilot air screw
24. Spring
25. Clamp bolt
26. Screw
27. Gasket
28. Sealing washer
29. Float needle and seat
30. Needle jet
31. Washer
32. Main jet
33. Pilot jet
34. Pivot pin
35. Floats
36. Screw – 4 off
37. Choke knob
38. Split pin
39. Spring washer – 4 off
40. Spacer
41. Lock washer
42. Spring
43. Cap
44. Heat insulator
45. Sealing ring
46. Float bowl

Fig. 2.1 Carburettor – KE100 A9, A10, B1, B2, B3 (US) models

16 Throttle valve
17 Split pin
18 Throttle valve rod
19 Cable adjuster
20 Locknut
21 Choke plunger retainer
22 Spring
23 Choke plunger
24 Clamp bolt
25 Nut
26 Carburettor body
27 Gasket
28 Sealing washer
29 Float needle seat
30 Pilot air screw
31 Spring
32 Needle jet
33 Washer
34 Main jet
35 Pilot jet
36 Pivot pin
37 Floats
38 Float bowl
39 Spring washer – 4 off
40 Rubber overflow grommet
41 Screw – 4 off
42 Float needle
43 Split pin
44 Cover
45 Choke knob
46 Clip
47 Cap – KE100 B5 to B12 (US)
48 Screw*
49 Lock washer*
50 Retainer plate*
51 Spacer*
52 Spring seat*
53 Spring*

1 Cover
2 Extension piece
3 Spring clamp
4 Wire clip
5 Cable adjuster
6 Carburettor top retainer
7 Cable guide
8 Locknut
9 Idle adjuster screw
10 Spring
11 Carburettor top
12 Return spring
13 Retainer plate
14 Jet needle clip
15 Jet needle

Fig. 2.2 Carburettor – all other models

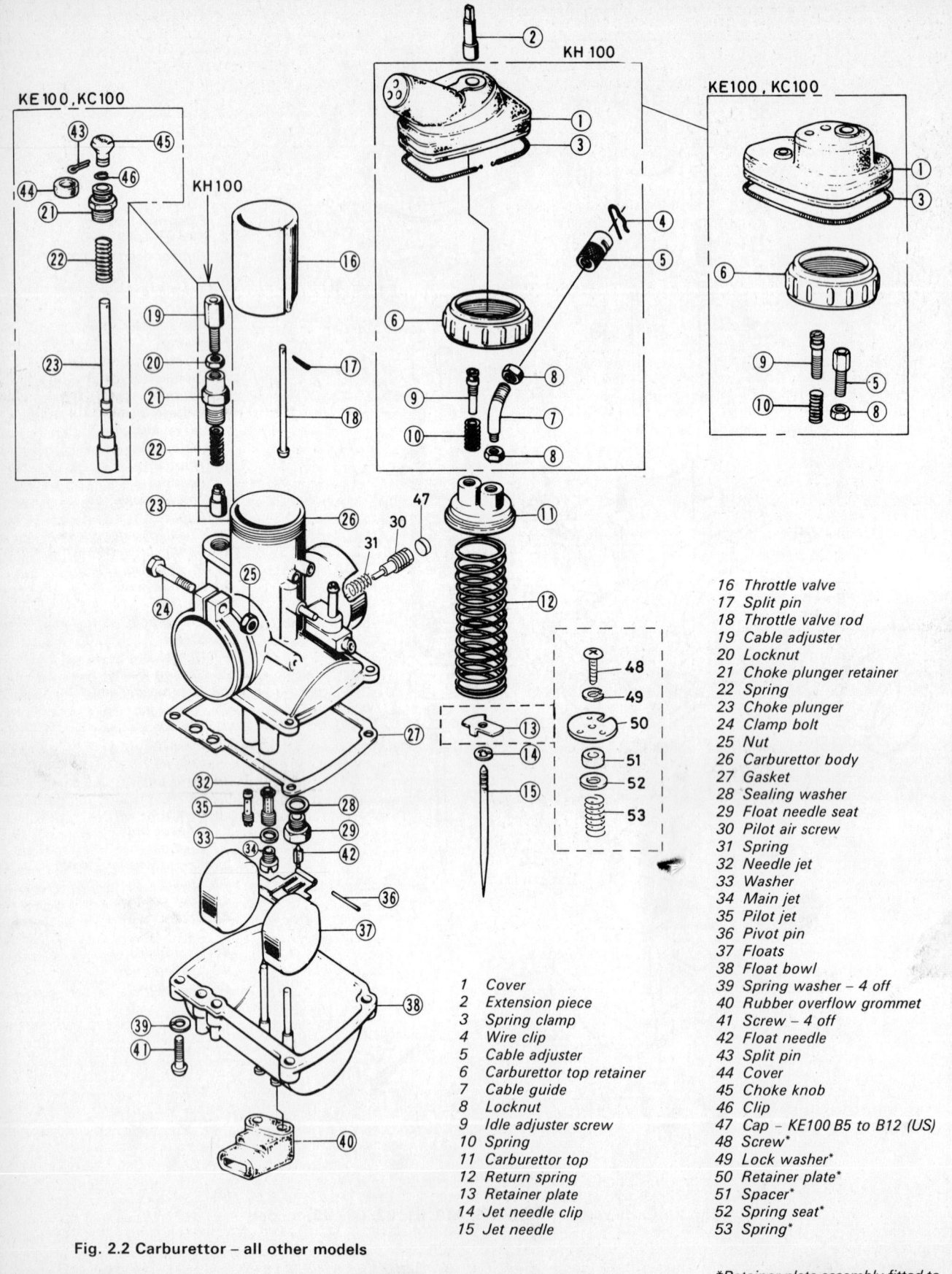

*Retainer plate assembly fitted to US KE100 A8 to B12 models

6 Carburettor: checking the settings

1 The various jet sizes and throttle valve cutaway are predetermined by the manufacturer and should not require modification. Check with the Specifications list at the beginning of this Chapter if there is any doubt about the types fitted. If a change appears necessary it can often be attributed to a developing engine fault unconnected with the carburettor. Although carburettors do wear in service, this process occurs slowly over an extended length of time and hence wear of the carburettor is unlikely to cause sudden or extreme malfunction. If a fault does occur check first other main systems, in which a fault may give similar symptoms before proceeding with carburettor examination or modification.

2 Where non-standard items such as an exhaust system or an air filter has been fitted to a machine, some alterations to carburation may be required. Arriving at the correct settings often requires trial and error, a method which demands skill borne of previous experience. In many cases the manufacturer of the non-standard equipment will be able to advise on correct carburation changes.

3 As a rough guide up to ⅛ throttle is controlled by the pilot jet, ⅛ to ¼ by the throttle valve cutaway, ¼ to ¾ throttle by the needle position and from ¾ to full by the size of the main jet. These are only approximate divisions, which are by no means clear cut. There is a certain amount of overlap between various stages.

4 If alterations to the carburation must be made, always err on the side of the slightly rich mixture. A weak mixture will cause the engine to overheat which, particularly on two-stroke engines, may cause engine seizure. Reference to Routine Maintenance will show how, after some experience has been gained, the condition of the spark plug electrodes can be interpreted as a reliable guide to mixture strength.

7 Carburettor adjustment and exhaust emissions: general note

1 In some countries legal provision is made for describing and controlling the types and levels of toxic emission from motor vehicles.

2 In the USA exhaust emission legislation is administered by the Environmental Protection Agency (EPA) which has introduced stringent regulations relating to motor vehicles. The Federal law entitled The Clean Air Act, specifically prohibits the removal (other than temporary), or modification, of any component incorporated by the vehicle manufacturer to comply with the requirements of the law. The law extends the prohibition to any tampering which includes the addition of components, use of unsuitable replacement parts or maladjustment of components which allow the exhaust emissions to exceed the prescribed levels. Violations of the provisions of this law may result in penalties of up to $10 000 for each violation. It is strongly recommended that appropriate requirements are determined and understood prior to making any change to or adjustments of components in the fuel, ignition, crankcase breather or exhaust systems.

3 To help ensure compliance with the emission standards some manufacturers have fitted to the relevant systems fixed or pre-set adjustment screws as anti-tamper devices. In most cases, this is restricted to plastic or metal limiter caps fitted to the carburettor pilot adjustment screws, which allow normal adjustment only within narrow limits. Occasionally the pilot screw may be recessed and sealed behind a small metal blanking plug, or locked in position with a thread-locking compound, which prevents normal adjustment.

4 It should be understood that none of the various methods of discouraging tampering actually prevent adjustment, nor, in itself, is re-adjustment an infringement of the current regulations. Maladjustment, however, which results in the emission levels exceeding those laid down, is a violation. It follows that no adjustments should be made unless the owner feels confident that he can make those adjustments in such a way that the resulting emissions comply with the limits. For all practical purposes a gas analyser will be required to monitor the exhaust gases during adjustment, together with EPA data of the permissible Hydrocarbon and CO levels. Obviously, the home mechanic is unlikely to have access to this type of equipment or the expertise required for its use, and, therefore, it will be necessary to place the machine in the hands of a competent motorcycle dealer who has the equipment and skill to check the exhaust gas content.

5 For those owners who feel competent to carry out correctly the various adjustments, specific information relating to the anti-tamper components fitted to the machines covered in this manual is given in the relevant Sections of this Chapter.

8 Carburettor: adjustment

1 Before any dismantling or adjustment is undertaken, eliminate all other possible causes of running problems, checking in particular the spark plug, ignition timing, air cleaner and the exhaust. Checking and cleaning these items as appropriate will often resolve a mysterious flat spot or misfire.

2 The first step in carburettor adjustment is to ensure that the jet sizes, needle position and fuel level are correct, which will require the removal and dismantling of the carburettor as described in Sections 4 and 5 of this Chapter.

3 If the carburettor has been removed for the purpose of checking jet sizes, the fuel level should be measured at the same time. It is unlikely that once this is set up correctly there will be a significant amount of variation, unless the float needle or seat have worn. These should be checked and renewed, if necessary, as described in Section 5.

4 Unfortunately measuring the fuel level requires the use of a special service tool, Kawasaki Part Number 57001-203 which consists of a modified float chamber with a long clear plastic pipe attached to it; the end of the pipe is finely graduated. This is used on all machines fitted with Mikuni carburettors and is indispensible as these carburettors do not have float chamber drain plugs to which other tools could be attached. For those late US KE100 models with Teikei carburettors, owners must obtain a float bowl from a local Kawasaki dealer under Part Number 16020-1003. This must be modified by cutting off the overflow pipe inside it and attaching a length of clear plastic tubing to the overflow pipe outer end. The Kawasaki service tool Part Number 57001-1017 (a short length of finely graduated clear plastic tubing) is then attached to this tubing and the whole assembly is fitted to the carburettor when in place on the machine.

5 Since this tool is not likely to be required very often, it is assumed that most owners will take their machines to a competent Kawasaki dealer for the fuel level to be checked. For those who wish to check it themselves, the procedure is outlined below. Reference to the accompanying illustrations will show the basic principles of the check.

6 Having acquired the necessary equipment, remove the carburettor from the machine, substitute the service tool for the original float chamber and refit the carburettor. The graduated tube should have a thicker ring around its mid-part which is the 'O' mark for zeroing the tool. Hold the tube against the side of the carburettor with the 'O' mark above the carburettor body bottom edge, being careful that the tube is not kinked, and switch on the fuel supply.

7 When the fuel level has stabilised in the tube, very carefully move the tube downwards until the 'O' mark is level with the carburettor body bottom edge. Hold the tube steady; do not lower it below the specified point and raise it again to take a reading, or the level will be inaccurate. Use the tube graduations to measure the fuel level, which should be the specified distance (or at least within tolerances) **below** the bottom edge of the carburettor body.

8 If adjustment is necessary, remove the carburettor and withdraw the float chamber. Invert the carburettor and carefully measure the height of the two floats, taking measurements from the carburettor body bottom edge to the bottom surface of each float, at a point in line with the main jet idle pillar. If there is any difference in height bend very gently and carefully the bridge piece which joins the two floats until they are at exactly the same height; it is advisable to remove the float assembly from the carburettor before attempting this.

9 To adjust the fuel level itself, when the floats are known to be at the same height remove the float assembly and bend very carefully the small brass tang which actually bears on the float valve needle end; bending the tang towards the float valve will lower the fuel level, bending it away from the valve will raise the fuel level. Be very careful to bend the tang only slightly; the tiniest alteration will have a large effect on the fuel level. Refit the float chamber, install the carburettor again and re-check the level. It may be necessary to repeat the procedure several times until the correct level is established; if care and patience is exercised the task will prove easier.

10 When the fuel level has been checked and re-set if necessary, check all the remaining carburettor settings and jet sizes given in the

Specifications Section of this Chapter; unless some other aspect of the machine has been modified, such as the exhaust, always return to standard settings to act as a sound basis for subsequent adjustment. Fit the carburettor to the machine. On KH100 models, check that the choke cable has the specified free play. See Routine Maintenance.

11 Except for later US models, screw in the pilot air screw until it seats lightly, unscrew it by exactly the number of turns specified, then refit the carburettor cover using a new gasket to ensure an airtight seal. Owners of US models should note that the air screw is sealed by a blanking plug and will have been set at the factory. Subsequent adjustment is not possible as Kawasaki do not provide the data necessary even for those owners with the skill, qualifications and equipment necessary to carry out adjustments. Owners should therefore return the screw to its exact original position, as established at the factory, if the screw is ever removed for cleaning, and should not attempt adjustment. Note that the initial setting given in the Specifications Section of this Chapter is only to be used when renewing the screw.

12 The normal tuning procedure is to warm up the engine and, starting from the basic position specified to turn the screw inwards by ¼ turn at a time noting the effect on the engine at each turn, then outwards by ¼ turn at a time until the position is found where the engine runs fastest and smoothest. However on these machines the carburettor cover must be removed to reach the air screw, thus completely upsetting the mixture balance and making accurate adjustment impossible. This means that adjustment will become a frustrating and repetitive task involving the removal of the cover, altering the screw position, refitting the cover and starting the engine again to note the effect on it. This will require great skill and patience if accurate adjustment is to be achieved.

13 In practice, it is suggested that the specified setting will be close enough for most owners and machines in normal working conditions; even when the full procedure is followed on a 'normal' machine the screw setting rarely varies by more than ¼ turn from that specified by the makers. If any large adjustment is required it will be because the machine has been altered from standard specification in some way or because a fault, possibly not in the carburettor, has developed. In the latter case of course, the fault must be found and rectified before the carburettor setting is altered from standard.

14 In view of this, once the carburettor has been checked and reset as described in paragraphs 1-11 of this Section, it is suggested that most owners confine their carburettor tuning to adjusting the idle speed. To do this, take the machine on a journey of at least 10 minutes duration or until the engine is fully warmed up to normal operating temperature. Using the extension piece which projects from the idle adjusting screw through the black rubber carburettor cover, set the tickover to the lowest speed that the engine will maintain while running smoothly and steadily. For machines with rev-counters, this is usually between 1200-1400 rpm. Check that the engine does not falter and stop after the twistgrip has been opened and closed a few times.

15 When the carburettor settings are correct, check the throttle and oil pump cable settings and routing as described in Routine Maintenance.

8.11 Idle speed mixture is set using pilot air screw

8.14 Idle speed is adjusted by rotating idle speed adjuster extension piece

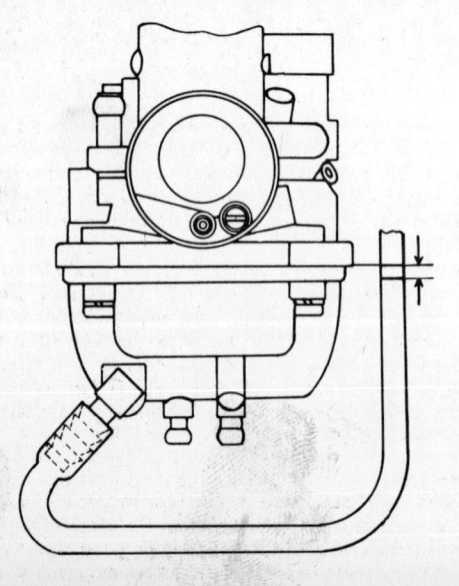

Fig. 2.3 Measuring the fuel level

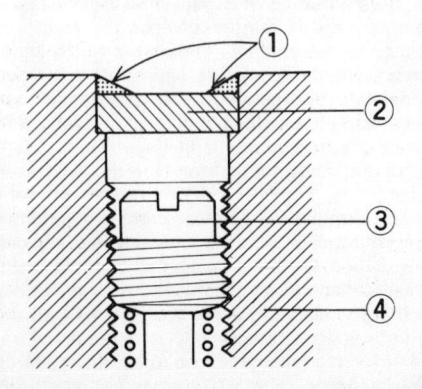

Fig. 2.4 Pilot air screw plug – all US models 1980 on

1 Bonding agent
2 Plug
3 Pilot air screw
4 Carburettor body

9 Air filter: general

1 The care and maintenance of the filter element(s) is described in Routine Maintenance. Never run the engine with the air filter disconnected or the element removed. Apart from the risk of increased engine wear due to unfiltered air being allowed to enter, the carburettor is jetted to compensate for the presence of the filter and a dangerously weak mixture will result if the filter is omitted.
2 US owners should note that the air filter is subject to the anti-tampering legislation currently in force, which means the machine must never be run with the filter element removed or rendered inoperative, or with the assembly altered in any way. Furthermore, only genuine Kawasaki replacement parts may be used if the renewal of any component is necessary.

10 Rotary disc valve: removal, examination and refitting

1 Since the valve forms such an integral part of the engine/gearbox unit its removal and refitting is outlined only here. Refer to this Chapter, Chapter 1 and Routine Maintenance for full details of the individual operations. The procedure is as follows:
2 Drain the transmission oil
3 Remove the kickstart lever
4 Remove the carburettor and withdraw both the fuel feed pipe and the oil pump control cable from their respective sealing grommets. On early KE100 models remove the metal carburettor overflow drain cover from underneath the carburettor chamber.
5 Disconnect the clutch cable.
6 Remove the oil pump.
7 On early KE100 models, it may be necessary to remove the right-hand footrest assembly (three bolts). Similarly on KH100 models it may be necessary to remove the exhaust system.
8 Remove the crankcase right-hand cover; take care not to lose the oil transfer pipe or the clutch release piece.
9 Remove the clutch assembly and primary drive gear.
10 Remove the rotary valve cover and withdraw the valve, noting the comments made in Chapter 1, Section 10 about making your own timing marks before disturbing the disc.
11 Reassembly is the reverse of the above, following the instructions given for the relevant operation.
12 Examine the large O-ring which seals the disc cover and the crankshaft oil seal fitted in the centre of the cover. If either of these are worn or damaged, air may be admitted, upsetting the mixture.
13 The disc is actually free-floating on the sleeve (KE100 A5) or splined collar (all other models) so that it can be moved from one face of its chamber to the other under the effect of the variations in crankcase pressure and thus act as a more effective seal across the inlet tract. While this means that there should be very little actual wear, the contact faces of the valve cover, disc and crankcase must be checked carefully for signs of wear or damage such as scoring. Cases have occured of discs being very badly scored or even cut through by particles of foreign matter. This is most likely to occur if the machine is used with a poorly-sealed carburettor chamber or with a damaged, badly-maintained or missing air filter element.
14 If the equipment is available, the disc can be checked for wear by measuring its thickness at points all around its circumference; if it is worn to the specified service limit, or beyond at any point, it must be renewed. Check it carefully, using a magnifying glass if necessary, for any signs of wear or damage such as hairline cracks, and check that it is a reasonably close fit on the sleeve or splined collar. Renew any component that is found to be worn or damaged in any way.
15 The crankcase surface must be completely smooth and flat. If any serious scoring is found, or distortion is suspected, the crankcase must be renewed. Unfortunately the only alternative is to have the complete right-hand face skimmed, which would bring too many associated problems with the consequent alteration in working positions and tolerances.
16 The valve cover must be renewed if it is found to be cracked or heavily scored. To measure the depth of the disc recess machined in its inner (left-hand) face, place a straightedge across the cover gasket surface and pass feeler gauges of the necessary thickness underneath it. The cover must be renewed if it is found to be worn at any point to the specified service limit or beyond.

11 Exhaust system: removal, examination and refitting

KC100
1 Remove the retaining nut from the swinging-arm pivot shaft right-hand end noting the presence of any washer behind it, then remove the single bolt which secures the silencer front end to the footrest bracket. Remove the two bolts securing the exhaust pipe clamp to the cylinder barrel and withdraw the system, depressing the rear brake pedal to gain as much movement as possible. Prise the gasket out of the exhaust port and discard it.
2 Note that the system can be separated into two parts by slackening the clamp screws and pulling the pipe out of the silencer. It is possible to remove either the pipe or the silencer as separate items from the machine if required. The baffle tube is retained by a single screw and is removed as described in Routine Maintenance.
3 Check the condition of the seal fitted to the pipe/silencer joint and renew it if it is split, damaged, or distorted so badly as to leak.
4 On refitting, always renew the gasket of the exhaust port. If the pipe and silencer have been separated fit them together loosely (not forgetting the two clamps) and offer up the system as a single unit. Fit loosely the mounting bolts and nut, not forgetting their plain and/or lock washers, then check that the system is settled securely and without stress on its mountings. To avoid distortion, tighten first the clamp screws, ensuring that the seal shoulder is correctly engaged against the silencer end and that the clamp is correctly fitted over the top of the seal. Next tighten evenly and securely the two front clamp bolts followed by the silencer front mounting bolt, then push the machine off its stand and have an assistant sit astride it so that the suspension is in its normal working position as the swinging arm pivot shaft nut is securely fastened.

KE100
5 Remove the two nuts or bolts securing the exhaust front clamp to the cylinder barrel. The system is mounted by one bolt to the frame just above the swinging arm pivot and by one bolt in front of the left-hand suspension unit top mounting. Remove these two bolts and withdraw the system from the machine.
6 If the front clamp is to be removed from the system, grip the split insert with large self-locking pliers or vice-grips and tap the clamp gently with a hammer to separate the two; if they are well rusted it will be necessary to soak them in penetrating fluid. When the two are separated open up the split with a screwdriver and work the insert over the pipe front end, followed by the clamp. Use a wire brush to clean thoroughly both components before refitting them.
7 Details of the baffle tube vary according to model, but all are secured at the rear end of the main silencer section by two bolts. Remove these bolts and pull the tube out of the silencer. Some US models have a separate spark arrester unit which is retained in the baffle tube by a single bolt. Note the presence of the O-ring sealing the baffle tube joint; this must be renewed if worn or damaged.
8 Refitting is the reverse of the removal procedure. With the system loosely assembled on its mountings tighten securely the retaining nuts or bolts working from the front to the rear.

KH100
9 The exhaust system fitted to the models is removed and refitted exactly as described above for the KC100 models; refer to paragraphs 1 and 4 above for details. Note that the swinging arm pivot shaft nut should be tightened to the specified torque settings.
10 The baffle tube is removed and refitted as described in Routine Maintenance.
11 The two parts of the system are joined by a tubular seal which is clamped inside the silencer front end by a threaded sleeve. Use a C-spanner to unscrew the sleeve when separating the pipe from the silencer. If worn or damaged the seal must be dug out of the silencer and renewed. On refitting the seal, note that its chamfered end must face to the rear of the machine. When the engine has first been run after refitting the exhaust system, allow the exhaust to cool down and re-tighten the sleeve.

All models
12 Whenever the exhaust system is disturbed, renew the exhaust port seal as a matter of course to prevent leaks. If the system is found to be so badly corroded that it is holed, or if it is damaged badly enough to leak, it must be renewed to prevent excessive noise and to preserve the performance and reliability of the engine.

Chapter 2 Fuel system and lubrication

13 Do not attempt to modify the exhaust system in any way. The exhaust system is designed to give the maximum power possible yet meet legal requirements and produce the minimum noise level. It is very unlikely that an unskilled person could improve the performance of any of the machines by working on the exhaust. If an aftermarket accessory system is being considered, check very carefully that it will maintain or increase performance when compared with the standard system. Very few 'performance' exhaust systems live up to the claims made by manufacturers, and even fewer offer any performance increase at all over the standard component.

14 For KE100 models, the final point to be borne in mind when considering the exhaust system is the finish. The matt-black painted finish employed is cheaper to renovate but less durable than a conventional chrome-plated system. It is inevitable that the original finish will deteriorate to the point where the system must be removed from the machine and repainted, therefore some thought must be given to the type of paint to be used. Reference to the advertisements in the national motorcycle press, or to a local Kawasaki dealer will help in selecting the most effective finish. The best are those which require the paint to be baked on, although some aerosol sprays are almost as effective. Whichever finish is decided upon, ensure that the surface is properly prepared according to the paint manufacturer's instructions and that the paint itself is correctly applied.

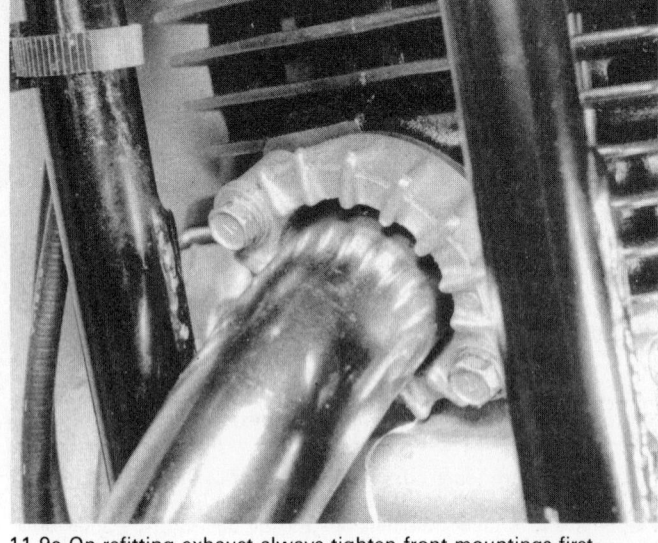

11.9a On refitting exhaust always tighten front mountings first ...

11.9b ... then work towards the rear, tightening mountings in succession

11.11a Use a C-spanner to unscrew threaded sleeve so that pipe can be separated from silencer ...

11.11b ... always renew tubular seal to prevent exhaust leaks

11.12 Always fit a new exhaust port gasket to prevent leaks

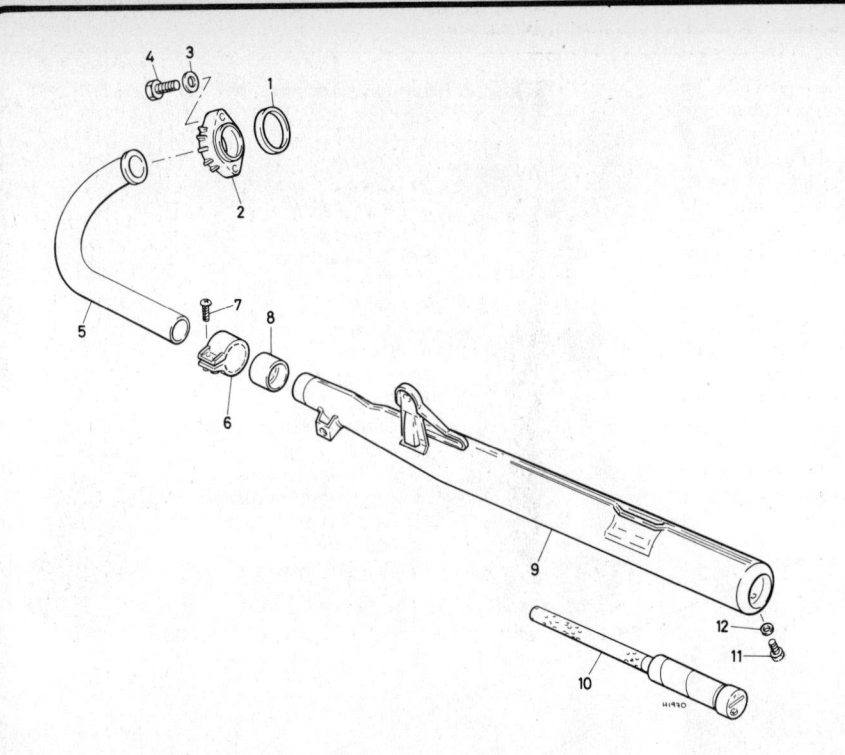

Fig. 2.5 Exhaust system – KC100 model

1 Gasket
2 Clamp
3 Spring washer – 2 off
4 Bolt – 2 off
5 Exhaust pipe
6 Clamp
7 Screw
8 Seal
9 Silencer
10 Baffle
11 Screw
12 Spring washer

Fig. 2.6 Exhaust system – KE100 models (typical)

1 Exhaust pipe
2 Clamp
3 Split insert
4 Gasket
5 Bolt – 2 off △
6 Spring washer – 2 off △
7 Nut – 2 off □
8 Stud – 2 off □
9 O-ring*
10 Baffle
11 Bush*
12 Spark arrester*
13 Bolt*
14 Spring washer*
15 Bolt – 2 off
16 Washer – 2 off
17 Bolt
18 Spring washer
19 Washer
20 Heat shield
21 Screw – 2 off
22 Lock washer – 2 off
23 Damping rubber
24 Bolt
25 Spring washer
26 Washer

△ All UK models and A5 to A7 US models
□ A8 to B12 US models
* KE100 A US models

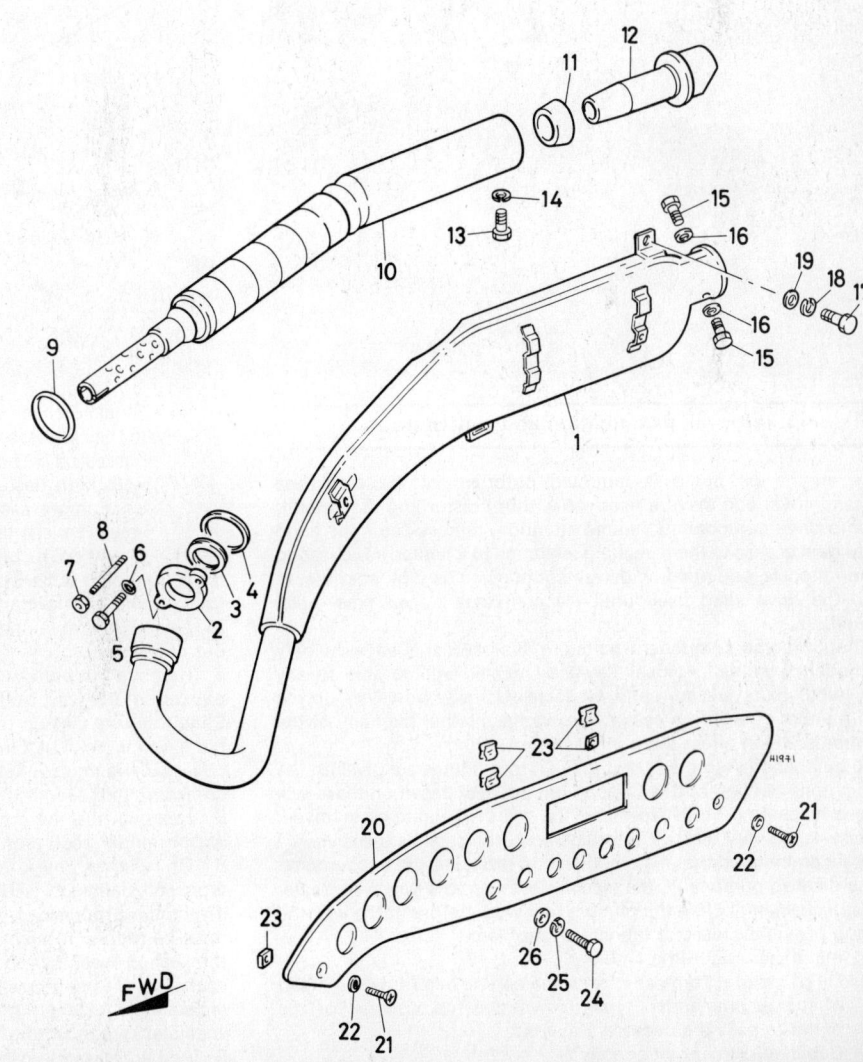

Fig. 2.7 Exhaust system – KH100 model

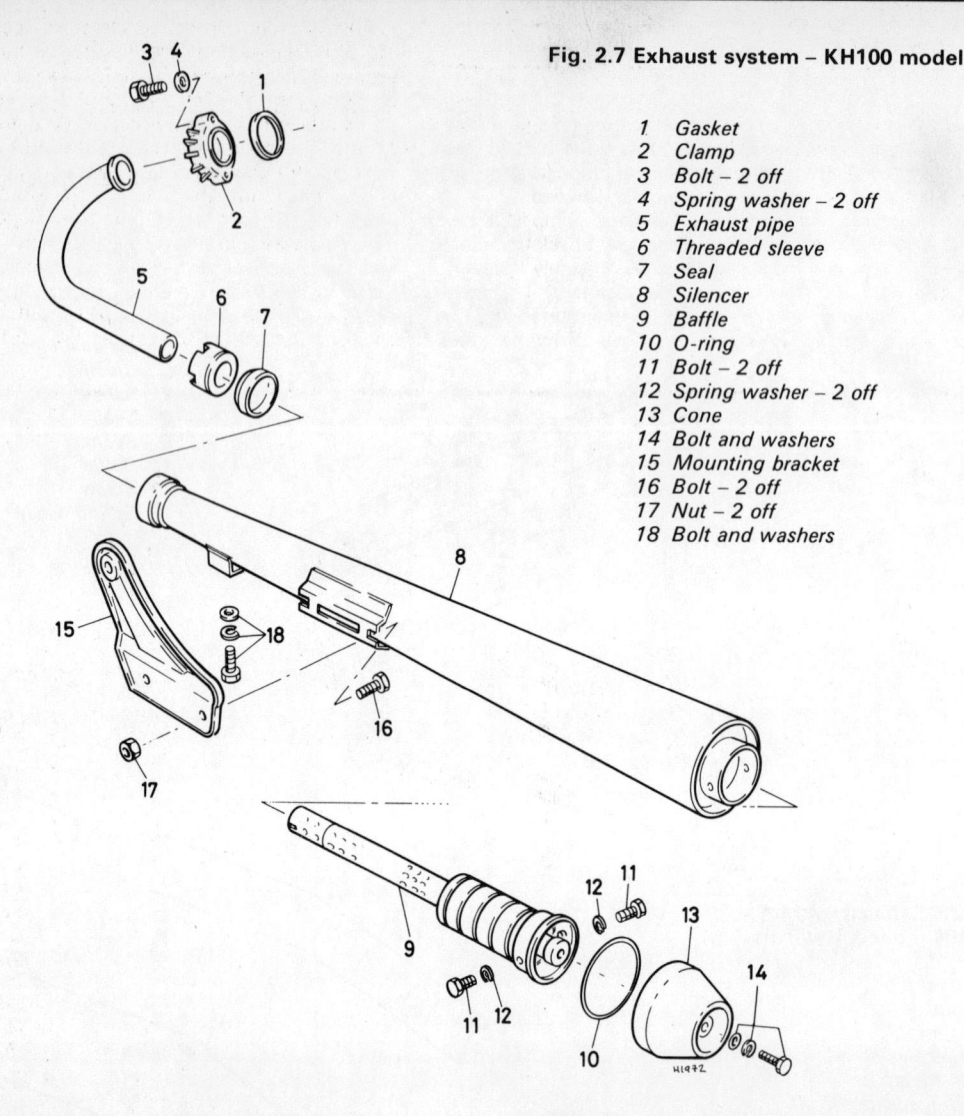

1 Gasket
2 Clamp
3 Bolt – 2 off
4 Spring washer – 2 off
5 Exhaust pipe
6 Threaded sleeve
7 Seal
8 Silencer
9 Baffle
10 O-ring
11 Bolt – 2 off
12 Spring washer – 2 off
13 Cone
14 Bolt and washers
15 Mounting bracket
16 Bolt – 2 off
17 Nut – 2 off
18 Bolt and washers

12 Oil pump: removal, examination and refitting

1 The engine oil pump is mounted outboard of the crankcase right-hand cover and may be removed without disturbing the latter. If the pump drive components require attention, remove the right-hand cover to gain access to them, making reference to Chapter 1 and noting that, for models equipped with rev-counters, it is not possible to remove the drive shaft itself until the crankcase halves have been separated.

2 Although some component parts are available, this varies widely from model to model. A local Kawasaki dealer will be able to say exactly what parts are currently available. Usually, however, only a limited number of seals are available separately, rather than any of the components which affect the pump's output.

3 The pump is therefore a sealed unit. Output figures are given in the Specifications Section of this Chapter but any test based on these will require a measuring vessel to show the minute quantities involved; such vessels are very difficult to obtain and are usually expensive.

4 It is recommended that if any lubrication problems are encountered which indicate a pump fault, the machine is taken to a Kawasaki dealer for careful checking before the pump is renewed. Before this, check the following possible causes of lubrication problems:
a) Insufficient oil in the tank
b) Oil of incorrect type in use – some engine oils (if used instead of the recommended type) may prove too viscous for the pump to pass the necessary quantity
c) Cable adjustment incorrectly set
d) Blocked, kinked or otherwise damaged oil tank/pump and pump/engine feed pipes
e) Blocked check valve – a spring-loaded one-way valve is fitted in the crankcase cover union adjacent to the carburettor. If this passes oil in both directions or in none at all, try flushing it with high flash-point solvent to clear the blockage. **Do not** use compressed air which may damage the valve spring. If the valve remains faulty, it must be renewed.
f) Faulty drive components

5 If the pump is faulty it must be renewed. Remove the pump cover from behind the kickstart lever. Bend up the control lever retaining tang to release the cable end nipple and disconnect the cable, securing it out of the way.

6 Remove the pump/engine feed pipe banjo bolt and place the feed pipe out of the way; note the sealing washer on each side of the union. Slide the wire clip up the tank/pump feed pipe until it is clear of the pump union, then pull the pipe off the union and plug it with one of the cover screws to prevent the loss of oil. Pull the pipe grommet out of the crankcase and secure the pipe out of the way.

7 Remove the two pump mounting screws, noting the sealing washer under each and withdraw the pump.

8 On refitting, check first the sealing washers and renew any that are worn or damaged. Always renew the pump gasket. Reverse the dismantling procedure to fit the pump, but note that its drive tongue must be rotated to align with the notch in the driveshaft. Do not omit the sealing washers, one on each pump mounting screw and one on each side of the pump engine banjo union. Ensure that both feed pipes are securely fastened, but do not overtighten the banjo bolts; note the specified torque setting.

9 Adjust the pump cable as described in Routine Maintenance, then bleed the pump as described in the next Section of this Chapter. Check carefully for any signs of oil leakage after the pump has been disturbed.

Chapter 2 Fuel system and lubrication

13 Oil pump: bleeding

1 It is necessary to bleed the oil pump every time the feed pipe from the tank is removed and refitted or when refilling if the oil tank has been allowed to run dry. This is because air will be trapped in the oil line, no matter what care is taken when the pipe is removed.

2 Remove the pump cover and carefully pull the tank/pump feed pipe grommet out of its recess in the pump chamber wall. Check that there is sufficient oil in the tank and that the feed pipe is securely fastened.

3 Slacken through two full turns the (usually slotted) hexagon-headed screw which is threaded into the upper rear surface of the pump body; this is the bleed screw. Place a container below the pump to collect the oil expelled as the pump is bled. Allow the oil to trickle out of the bleed hole, checking for air bubbles. The bubbles should eventually disappear as the air is displaced by fresh oil. When clear of air, tighten the bleed screw. DO NOT replace the pump cover until the pump setting has been checked, as described in Routine Maintenance.

4 It will be necessary to ensure that the oil pump/engine feed pipe is primed if this was disturbed. Unless this is checked, the engine will be starved of oil until the pipe fills. To avoid this, start the engine and allow it to idle for a few minutes whilst holding the pump control lever in its fully open position by pulling the pump cable. The excess oil will make the exhaust smoke heavily for a while, indicating that the pump is delivering oil to the engine. Continue the process until no more air bubbles can be seen in the pipe. Refit all disturbed components, checking that the pump control lever is restored to normal operation.

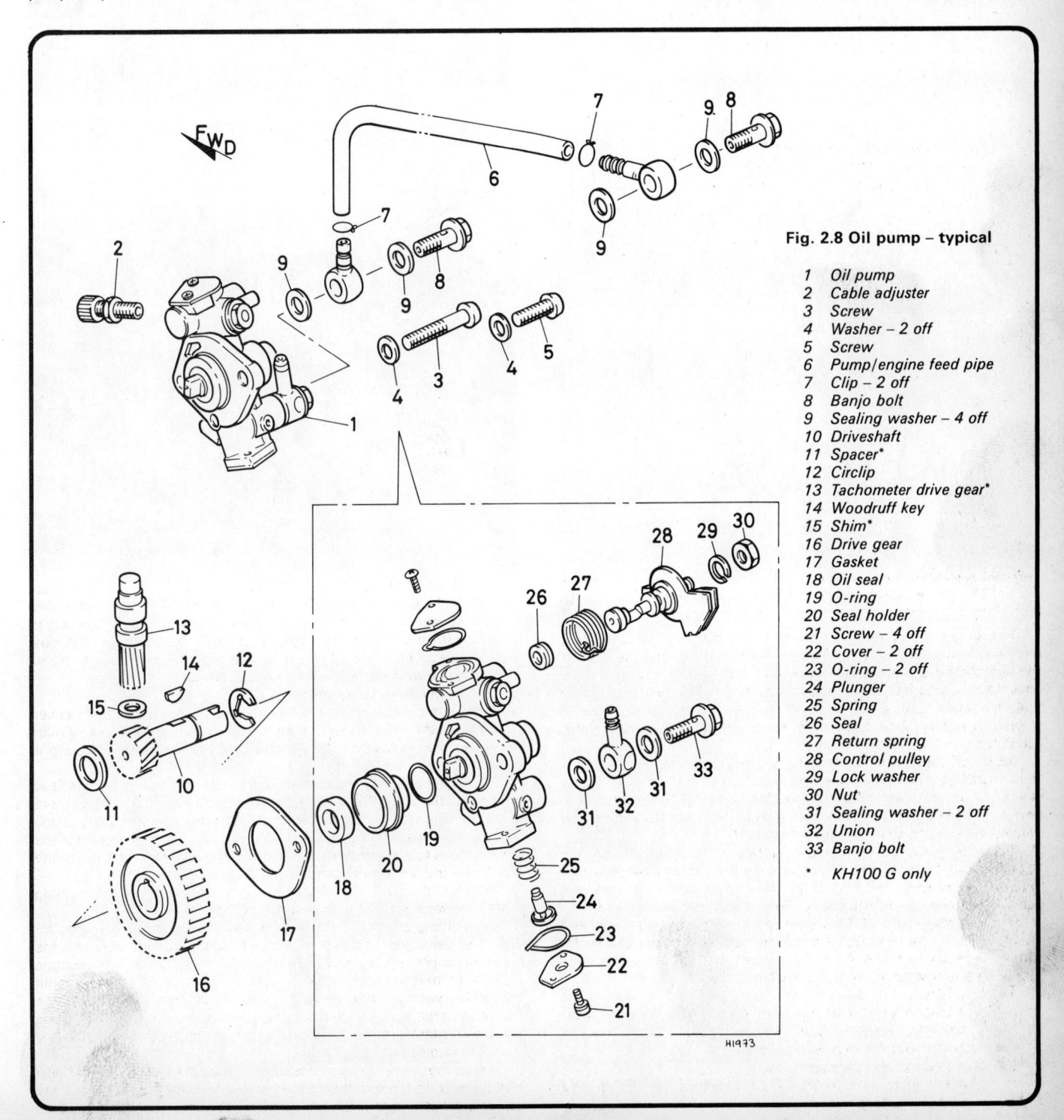

Fig. 2.8 Oil pump – typical

1 Oil pump
2 Cable adjuster
3 Screw
4 Washer – 2 off
5 Screw
6 Pump/engine feed pipe
7 Clip – 2 off
8 Banjo bolt
9 Sealing washer – 4 off
10 Driveshaft
11 Spacer*
12 Circlip
13 Tachometer drive gear*
14 Woodruff key
15 Shim*
16 Drive gear
17 Gasket
18 Oil seal
19 O-ring
20 Seal holder
21 Screw – 4 off
22 Cover – 2 off
23 O-ring – 2 off
24 Plunger
25 Spring
26 Seal
27 Return spring
28 Control pulley
29 Lock washer
30 Nut
31 Sealing washer – 2 off
32 Union
33 Banjo bolt

* KH100 G only

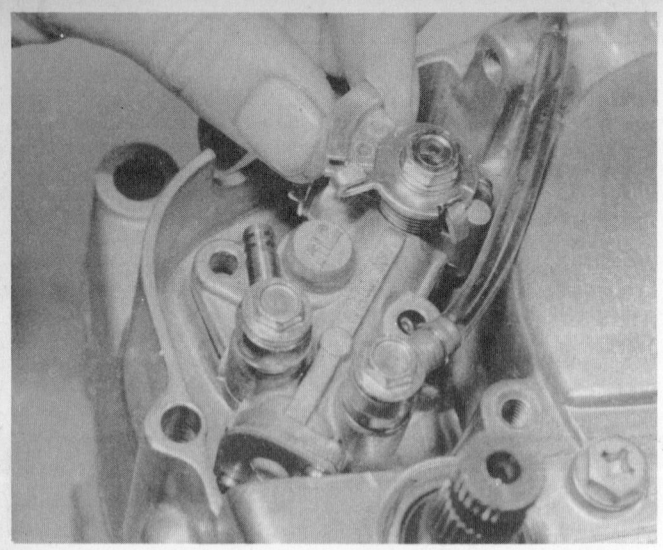

12.5 Oil pump should be regarded as a sealed unit and must be renewed if faulty

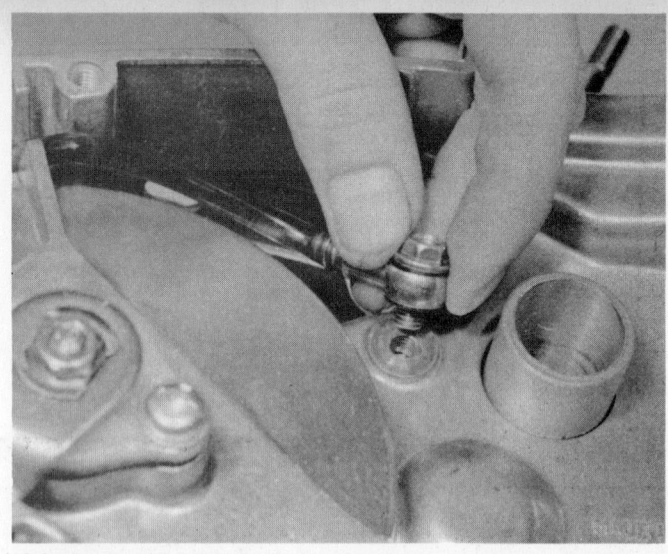

12.6a Disconnect oil pump/engine feed pipe – note sealing washer on each side of union

12.6b Disconnect black rubber oil tank/pump feed pipe and quickly plug with cover screw

12.8 Ensure cable retaining tang is correctly clamped on refitting

13.3 Slacken bleed screw to clear air from oil tank/pump feed pipe

Chapter 3 Ignition system

Contents

General description 1	Ignition source coil: testing 4
Condenser (capacitor): testing and renewal 2	HT lead and suppressor cap: examination 5
Ignition HT coil: location and testing 3	

Specifications

Ignition timing – BTDC | **Crankshaft position** | **Piston position**
KE100 A8 to A10, B1 to B12 (US) .... | 23° | 2.58 mm (0.1016 in)
All other models .... | 20° | 1.96 mm (0.0772 in)

Contact breaker
Gap .... 0.35 mm (0.014 in)
Tolerance – for ignition timing .... 0.30 – 0.40 mm (0.012 – 0.016 in)

Ignition source coil resistance
KE100 A5, A6, A7 (US and UK), KE100 A8, A9, A10 (UK) .... 1.1 – 1.7 ohm
KE100 A8, A9, A10 (US), all KE100 B models .... 2.0 – 3.1 ohm
KH100 A2, A3 .... 2.4 – 3.6 ohm
KH100 A4, all KH100 G models .... 2.0 – 3.0 ohm

Ignition HT coil*
Minimum spark gap – at cranking speed .... 5.0 mm (0.20 in)
Winding resistances: | **Primary** | **Secondary**
KH100 G5, G6, G7, G8 .... | 1.2 – 1.8 ohm | 6.8 – 10.2 K ohm
KE100 B3, B4, B5, B6, B7, B10, B11 (UK) .... | 1.2 – 2.2 ohm | 6.5 – 10.0 K ohm
KE100 B3, B5, B6, B7, B8, B9, B10, B11, B12 (US) .... | 1.9 – 2.5 ohm | 8.0 – 12.0 K ohm
All other models .... | 1.7 – 2.5 ohm | 8.0 – 12.0 K ohm

Information not available for KC100 models

Condenser
Capacity .... 0.25 ± 0.03 microfarad

Spark plug | **KE100 A5, A6** | **All other models**
Make | NGK | NGK
Type | B8HS | B8ES*
Gap .... | 0.6 – 0.7 mm | 0.7 – 0.8 mm
 | (0.024 – 0.028 in) | (0.028 – 0.032 in)

BR8ES on UK KE100 B11 model

Torque wrench settings
Component | kgf m | lbf ft
Spark plug .... | 2.5 – 3.0 | 18 – 22

1 General description

A conventional contact breaker ignition system is fitted. As the generator rotor moves, alternating current (ac) is generated in the ignition source coil of the stator. With the contact breaker closed, the current runs to earth. When the breaker opens, the current transfers to the ignition coil primary windings. A high voltage is thus produced in the coil secondary windings (by mutual induction) and fed to the spark plug via the HT lead.

As energy flows to earth across the plug electrodes, a spark is produced and the combustible gases in the cylinder ignite. A condenser prevents arcing across the contact breaker points which helps reduce erosion due to burning.

This Chapter is concerned only with the testing and repair of the system; refer to Routine Maintenance for details of the regular attention needed for the spark plug, contact breaker points and ignition timing. Refer to the relevant Sections of Chapter 6 for details of how to check the wiring and switches.

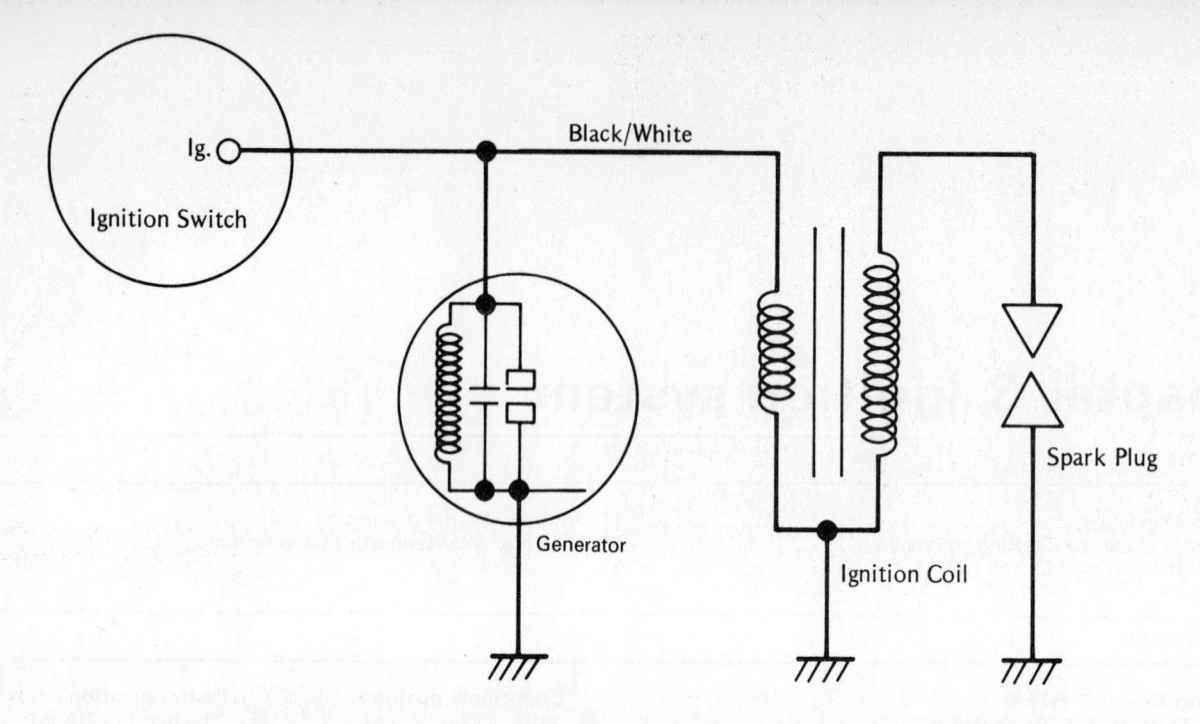

Fig. 3.1 Ignition system circuit diagram

2 Condenser (capacitor): testing and renewal

1 The condenser prevents arcing across the contact breaker points as they separate; it is connected in parallel with the points. If misfiring occurs or starting proves difficult, especially with the engine hot, it is possible that the condenser is faulty. To check, watch the points via the rotor apertures when the engine is running. If the points spark excessively and appear burnt, the condenser requires renewal.
2 The condenser can be tested in place, after the generator rotor has been removed as described in Chapter 1, but its leads must first be disconnected and a capacitor tester used, although a good quality ohmmeter can also be used for the first part of the test.
3 Using the tester or ohmmeter, measure the resistance between the condenser centre terminal and its outer casing; a value of at least 5 M ohms should be obtained (ie very close to infinite resistance).
4 Switch the tester to the condenser capacity test scale and connect its leads as described in the preceding paragraph. The needle should deflect and then return as the condenser is charged; note the reading obtained when the needle stops. A value of 0.25 microfarad should be obtained. After this test the condenser must be discharged by touching a length of thick insulated wire to its centre terminal and outer casing.
5 If the results of either test are not satisfactory, the condenser is faulty and must be renewed. If the test equipment is not available, the checks in paragraph 1 above must be assumed to be sufficiently accurate; if the symptoms described are found, the condenser should be renewed.
6 Remove the stator plate and unscrew the condenser mounting screw, then press out the condenser, once its leads have been unsoldered. On refitting, ensure that the leads are soldered securely to the centre terminal and that they do not touch the outer casing. Check that all screws are securely fastened.

3 Ignition HT coil: location and testing

1 The ignition HT coil is mounted on the frame top tubes and requires no attention other than to ensure its mounting bolts and connections are clean and tight. If found to be faulty, the coil must be renewed as it is a sealed unit which cannot be repaired.

2 The most accurate test of a coil of this type is to check its performance on a spark-gap tester; a sound coil should produce a strong blue spark across a certain minimum gap (see Specifications) for at least 5 minutes. This can be approximated as a quick check when tracing faults by removing the suppressor cap and holding the end of the HT leads approximately 1/4 in from the cylinder head finning while turning the engine over on the kickstart. Be very careful; the spark may find it easier to travel to earth by way of your hand, thus producing a most unpleasant shock.
3 The condition of the coil windings can be assessed by using a multimeter or ohmmeter set to the x 1 ohm scale for the primary windings and to the kilo ohms scale for the secondary windings. Make the meter connections for each test as shown in the accompanying illustration. If the readings obtained differ markedly from those given in the Specifications section, the coil must be taken to an authorized Kawasaki dealer for testing as described in paragraph 2.

2.1 Ensure condenser leads are securely soldered on refitting

Chapter 3 Ignition system

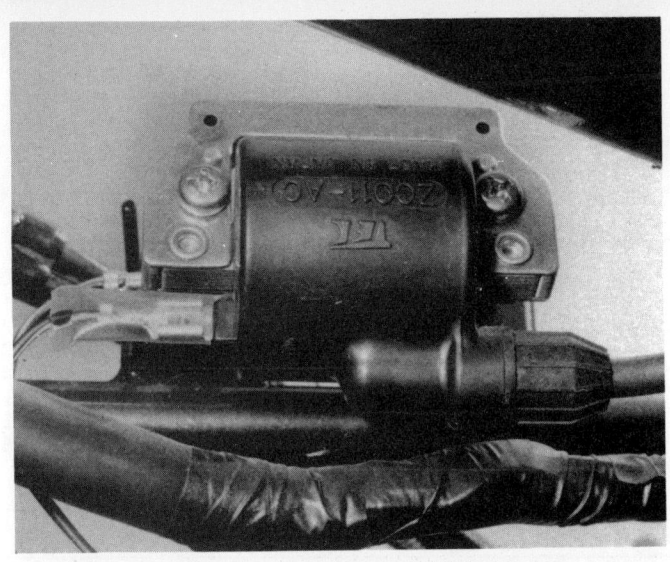

3.1 Ignition HT coil is mounted underneath the fuel tank

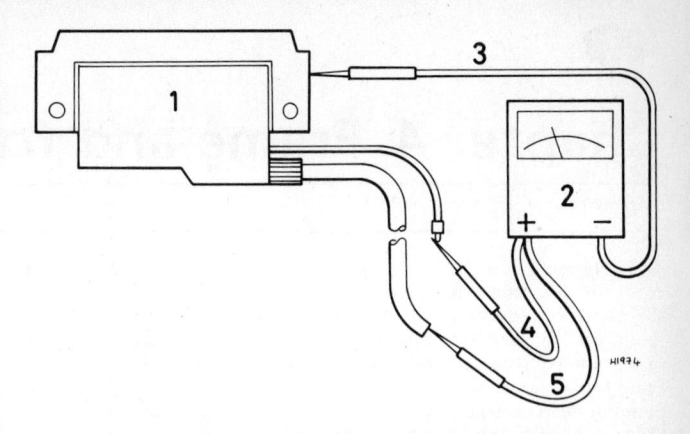

Fig. 3.2 Ignition HT coil resistance tests

1 HT coil
2 Meter
3 Earth
4 Primary windings test
5 Secondary windings test

4 Ignition source coil: testing

1 Remove the left-hand side panel or unlock and raise the seat, as appropriate, to reach the connectors between the generator stator and the main wiring loom.
2 Disconnect the black or black/white wire and connect a meter set to the x 1 ohm resistance range between the terminal of the wire leading to the generator coil and a good earth point such as the crankcase. Rotate the generator rotor until the points open and note the reading obtained.
3 If the test produces a result different from that given, the coil is defective, but note that a reading of infinite resistance would indicate a broken wire (open circuit) and a reading of very little resistance would indicate a trapped wire (short circuit), either of which may be easily repaired by the owner. If the coil must be renewed, remember that an auto-electrician specialist may be able to rewind it, saving some of the cost of a new item.
4 To renew the coil, the generator stator must be removed as described in Chapter 1 so that the coil connections can be unsoldered and the mounting screws removed to release the coil. Note carefully at which point the wires are connected and ensure that the new soldered joint is secure and well insulated on refitting; if in any doubt take the stator to an expert. Check carefully on refitting that the coil does not foul the rotor at any point, and that the air gap is the same between each coil pole and the rotor magnets.
5 If, on the other hand, the coil is proved sound by this test, the fault may be due to loss of flywheel magnetism, which can be cured only by the renewal of the rotor. Check with a competent Kawasaki dealer before taking such an expensive source of action.

5 HT lead and suppressor cap: examination

1 Erratic running faults can sometimes be attributed to leakage from the HT lead and spark plug. With this fault, it will often be possible to see tiny sparks around the lead and suppressor cap at night. One cause is dampness and the accumulation of road salts around the lead. It is often possible to cure the problem by drying and cleaning the components and spraying them with an aerosol ignition sealer.
2 If the system has become swamped with water, use a water dispersant spray. Renew the cap seals if defective. If the lead or cap is suspected of breaking down internally, renew the component concerned.
3 Where the lead is permanently attached to the ignition coil, entrust its renewal to an auto-electrician who will have the expertise to solder on a new lead without damaging the coil windings.

Chapter 4 Frame and forks

Contents

General description 1	Swinging arm: removal 10
Front fork legs: removal 2	Swinging arm: examination and renovation 11
Front fork legs: dismantling 3	Swinging arm: refitting 12
Front fork legs: examination and renovation 4	Rear suspension units: removal, examination and refitting 13
Front fork legs: reassembly 5	Footrests, stands and controls: examination and renovation 14
Front fork legs: refitting 6	Speedometer and tachometer: removal, examination and
Steering head assembly: removal and refitting 7	refitting 15
Steering head bearings: examination and renovation 8	Instrument drive components: location and examination 16
Frame: examination and renovation 9	

Specifications

Front forks
Wheel travel:
- KC100 .... 90 mm (3.54 in)
- KE100 .... 145 mm (5.71 in)
- KH100 .... 110 mm (4.33 in)

Fork oil: Capacity – per leg   Oil level*
- KC100 .... 130 cc (4.58 Imp fl oz)   315 ± 2 mm (12.40 ± 0.08 in)
- KE100 A .... 162 ± 4 cc (5.70 ± 0.14   408 ± 2 mm
  Imp fl oz, 5.48 ± 0.13 US fl oz)   (16.06 ± 0.08 in)
- KE100 B .... 182 ± 4 cc (6.41 ± 0.14   408.5 ± 2 mm
  Imp fl oz, 6.15 ± 0.13 US fl oz)   (16.08 ± 0.08 in)
- KH100 A .... 140 ± 4 cc   355 ± 2 mm
  (4.93 ± 0.14 Imp fl oz)   (13.98 ± 0.08 in)
- KH100 G .... 72 ± 4 cc   396 ± 2 mm
  (2.53 ± 0.14 Imp fl oz)   (15.59 ± 0.08 in)

Recommended fork oil:
- KC100, KH100 A .... SAE 30 fork oil or SAE 10W/20 engine oil
- KE100 A .... SAE 10 fork oil or SAE 10W/20 engine oil
- KE100 B, KH100 G .... SAE 10W/20 engine oil

Fork oil level is measured from top of stanchion to top of oil, forks fully extended and (internal) springs removed

Fork spring free length: Standard   Service limit
- KC100 .... N/Av   N/Av
- KE100 A (UK), KE100 A5, A6, A7, A8 (US) .... 405.0 mm (15.94 in)   395.0 mm (15.55 in)
- KE100 A9, A10 (US) .... 410.0 mm (16.14 in)   402.0 mm (15.83 in)
- KE100 B .... 420.5 mm (16.56 in)   412.0 mm (16.22 in)
- KH100 A .... 187.0 mm (7.36 in)   177.0 mm (6.97 in)
- KH100 G .... 378.5 mm (14.90 in)   368.5 mm (14.51 in)

Rear suspension
Wheel travel:
- KC100 .... 70 mm (2.76 in)
- KE100 A .... 90 mm (3.54 in)
- KE100 B .... 110 mm (4.33 in)
- KH100 A2, A3 .... 67 mm (2.64 in)
- KH100 A4, KH100 G .... 73 mm (2.87 in)

Swinging arm pivot shaft – maximum runout .... 0.14 mm (0.0055 in)

Torque wrench settings
Component	kgf m	lbf ft
Handlebar clamp bolts:		
KE100	1.6 – 2.2	11.5 – 16
KH100	1.9	14
Steering stem head bolt:		
KE100 A (UK), KE100 A5, A6, A7 (US)	3.0 – 3.5	22 – 25
KE100 A8, A9, A10 (US), KE100 B	3.0 – 5.0	22 – 36
KH100	1.9	14
Steering stem clamp bolt – KE100	1.6 – 2.2	11.5 – 16
Top yoke pinch bolts – KE100	1.6 – 2.2	11.5 – 16
Fork leg top bolt/plug:		
KE100	1.5 – 2.0	11 – 14.5
KH100	1.9	14

Chapter 4 Frame and forks

Component	kgf m	lbf ft
Damper rod Allen screw – KH100 G	1.6	1.5
Bottom yoke pinch bolts:		
KE100	2.6 – 3.5	19 – 25
KH100	1.7	12
Swinging arm pivot bolt nut:		
KE100	4.0 – 6.0	29 – 43
KH100	4.0	29
Rear suspension unit mounting nuts or bolts:		
KE100	2.6 – 3.5	19 – 25
KH100	3.1	22.5

1 General description

The front forks are of the telescopic type, with coil springs and hydraulic damping. Those fitted to KC100 and KH100 A models have external springs while those fitted to all other models have internal springs.

The frame is constructed of welded tubular steel and is of the full cradle type.

Rear suspension is by two hydraulically-damped, coil spring suspension units acting on a pivoted fork-type swinging arm.

2 Front fork legs: removal

1 Referring to Chapter 5, Section 2, remove the front wheel.
2 On KH100 G models only, remove the caliper from the fork lower leg, release the clamps securing the hydraulic hose and tie the caliper to the frame out of harm's way. Take care not to twist the hose. Wedge a piece of wood between the caliper's pads to prevent the risk of damage should the brake lever be accidentally applied.
3 On KC and KH 100 models, remove the four mounting bolts and carefully withdraw the front mudguard.
4 Remove the fork leg top bolts, slacken fully the bottom yoke pinch bolts and pull each leg downwards, out of the yokes.
5 On KE100 models slacken the top bolt while the fork legs are still securely held, if the legs are to be dismantled. Prise off the top plastic cap (later models only) and unscrew the top bolt/plug.
6 Slacken fully the top and bottom yoke pinch bolts and pull each leg downwards out of the yokes.
7 On all models, if the fork legs are stuck in the yokes, apply penetrating fluid and attempt to rotate the legs by hand to free them. It may be necessary to completely remove the pinch bolts and to spring the clamps apart slightly with a large, flat-bladed screwdriver. Great care must be taken not to distort or to break the clamp, as this will necessitate renewal of the complete yoke. If the leg is still reluctant to move, push a metal bar of suitable diameter through the spindle lug in the fork lower leg and tap firmly downwards on the protruding end of the bar to drive the fork leg from the yokes.
8 Once the legs have been removed, put them to one side to await stripping and examination. If they are not to be dismantled, ensure that they remain upright so that no fork oil is lost.

2.3 KC and KH100 models – remove mounting bolts and withdraw mudguard ...

2.4a ... unscrew fork leg top bolts ...

2.4b ... and slacken bottom yoke pinch bolts to release fork legs

3 Front fork legs: dismantling

1 Dismantle and rebuild the fork legs separately so that there is no chance of interchanging components, thus promoting undue wear.

KC100, KH100 A

2 Displace the O-ring from the top of the stanchion, then slide off the spring top seat and spring.
3 Invert the leg over a suitable container and tip out the oil, pumping the leg several times to expel as much oil as possible.
4 Using a vice with soft jaw covers to protect the chrome, clamp the lower leg by its wheel spindle lug. Wrap the chrome seal holder with a piece of rubber (cut from an old inner tube, for example) and use this to protect the chrome while the holder is unscrewed, using a strap wrench or chain wrench. Slide the dust seal and seal holder off the stanchions.
5 Pull sharply on the stanchion to tug it out of the lower leg, complete with bushes. Tap the oil seal out of the seal holder, noting which way up it was fitted.

KE100 A

6 The fork springs are internal on these models, and must therefore be pulled out of the stanchion after the top bolt has been removed. The forks are very similar to those fitted to KC and KH100 A models and are dismantled as described in paragraphs 3 – 5 above.

KE100 B

7 These forks are very similar to those described above but are fitted with internal springs and a modified oil seal retaining arrangement.
8 With the spring removed and the fork oil drained as described above, prise the rubber dust-seal off the fork lower leg and displace the seal retaining circlip underneath it. Withdraw the dust seal, circlip and back-up washer, clamp the fork lower leg by the wheel spindle lug in a soft-jawed vice (as described above) and pull the stanchion sharply out of the lower leg. Several hard tugs may be necessary before the slide hammer action of the bushes extracts the oil seal from the lower leg.

KH100 G

9 Withdraw the spacer, spring seat and spring from the fork leg, then invert the leg over a suitable container and tip out the oil; pump the leg several times to expel as much oil as possible.
10 Remove the damper rod Allen screw from the bottom of the fork lower leg and pull the stanchion out of the lower leg. It may be necessary to insert a length of wooden dowel inside the stanchion to bear down on the head of the damper rod, thus locking it while the screw is removed. To remove the oil seal, displace the dust seal and circlip and lever the seal out of the lower leg, taking care not to scratch or distort the seal housing. Pouring boiling water over the lower leg will help to release a particularly tight seal, but take great care to prevent personal injury. Invert the stanchion to tip out the damper rod assembly.

3.9 KH100 G – removal of top bolt reveals spacer, spring seat and spring

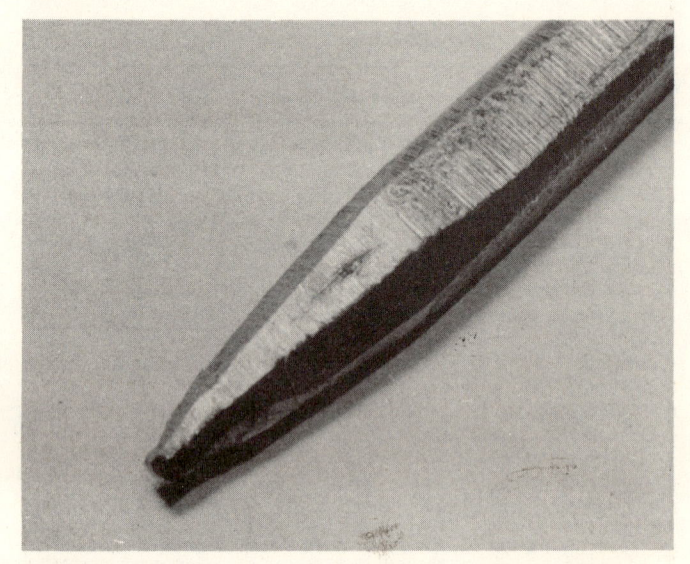

3.10a Damper rod must be held by length of wooden dowel or metal bar ...

3.10b ... so that Allen screw can be unscrewed ...

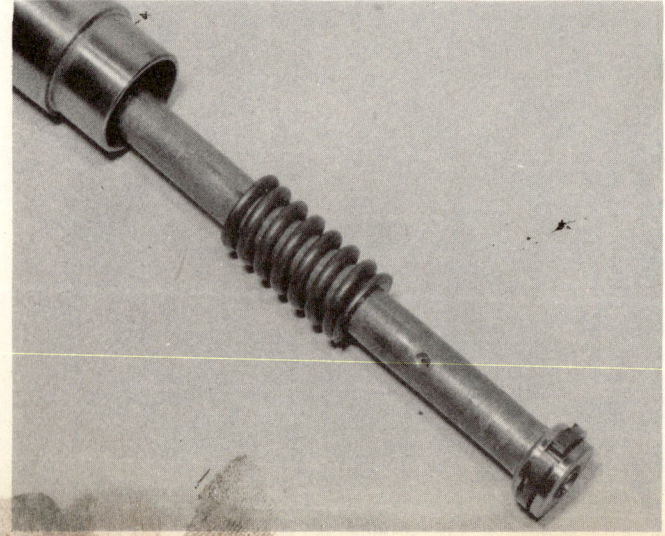

3.10c ... to release damper rod assembly

3.10d Displace retaining circlip ...

3.10e ... take care not to damage lower leg when removing oil seal

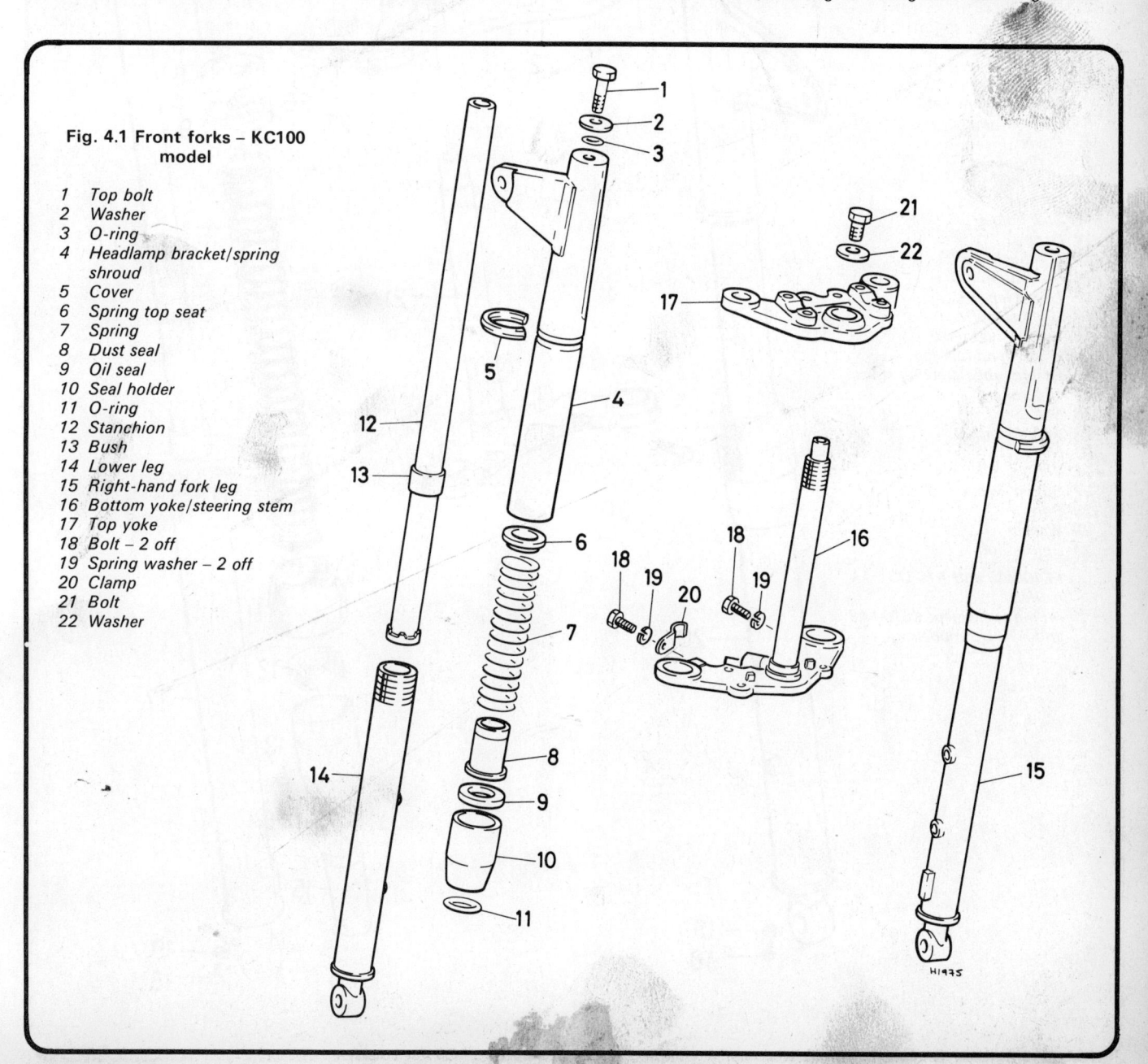

Fig. 4.1 Front forks – KC100 model

1. Top bolt
2. Washer
3. O-ring
4. Headlamp bracket/spring shroud
5. Cover
6. Spring top seat
7. Spring
8. Dust seal
9. Oil seal
10. Seal holder
11. O-ring
12. Stanchion
13. Bush
14. Lower leg
15. Right-hand fork leg
16. Bottom yoke/steering stem
17. Top yoke
18. Bolt – 2 off
19. Spring washer – 2 off
20. Clamp
21. Bolt
22. Washer

Fig. 4.2 Front forks – KE100 A model

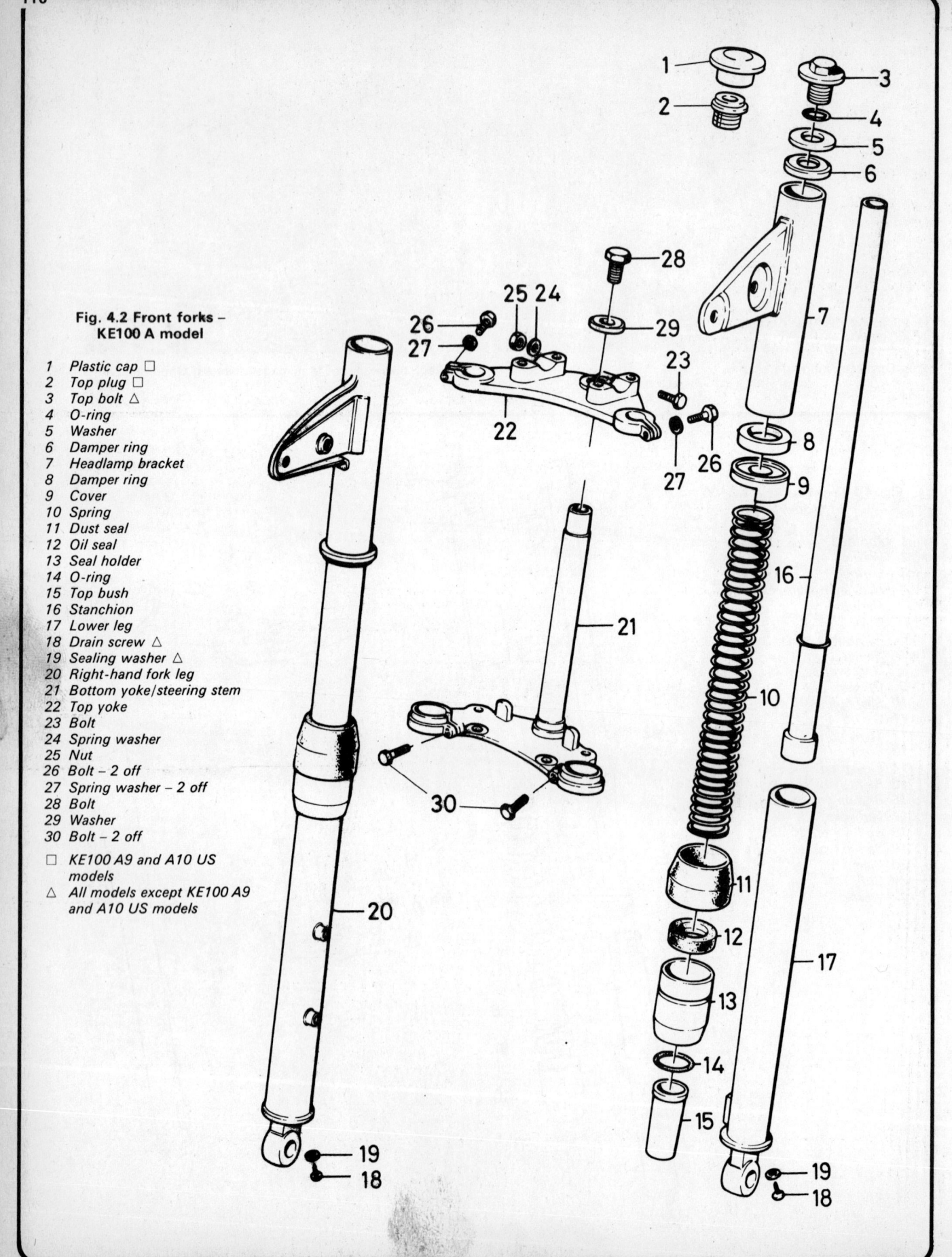

1 Plastic cap □
2 Top plug □
3 Top bolt △
4 O-ring
5 Washer
6 Damper ring
7 Headlamp bracket
8 Damper ring
9 Cover
10 Spring
11 Dust seal
12 Oil seal
13 Seal holder
14 O-ring
15 Top bush
16 Stanchion
17 Lower leg
18 Drain screw △
19 Sealing washer △
20 Right-hand fork leg
21 Bottom yoke/steering stem
22 Top yoke
23 Bolt
24 Spring washer
25 Nut
26 Bolt – 2 off
27 Spring washer – 2 off
28 Bolt
29 Washer
30 Bolt – 2 off

□ KE100 A9 and A10 US models
△ All models except KE100 A9 and A10 US models

Fig. 4.3 Front forks – KE100 B model

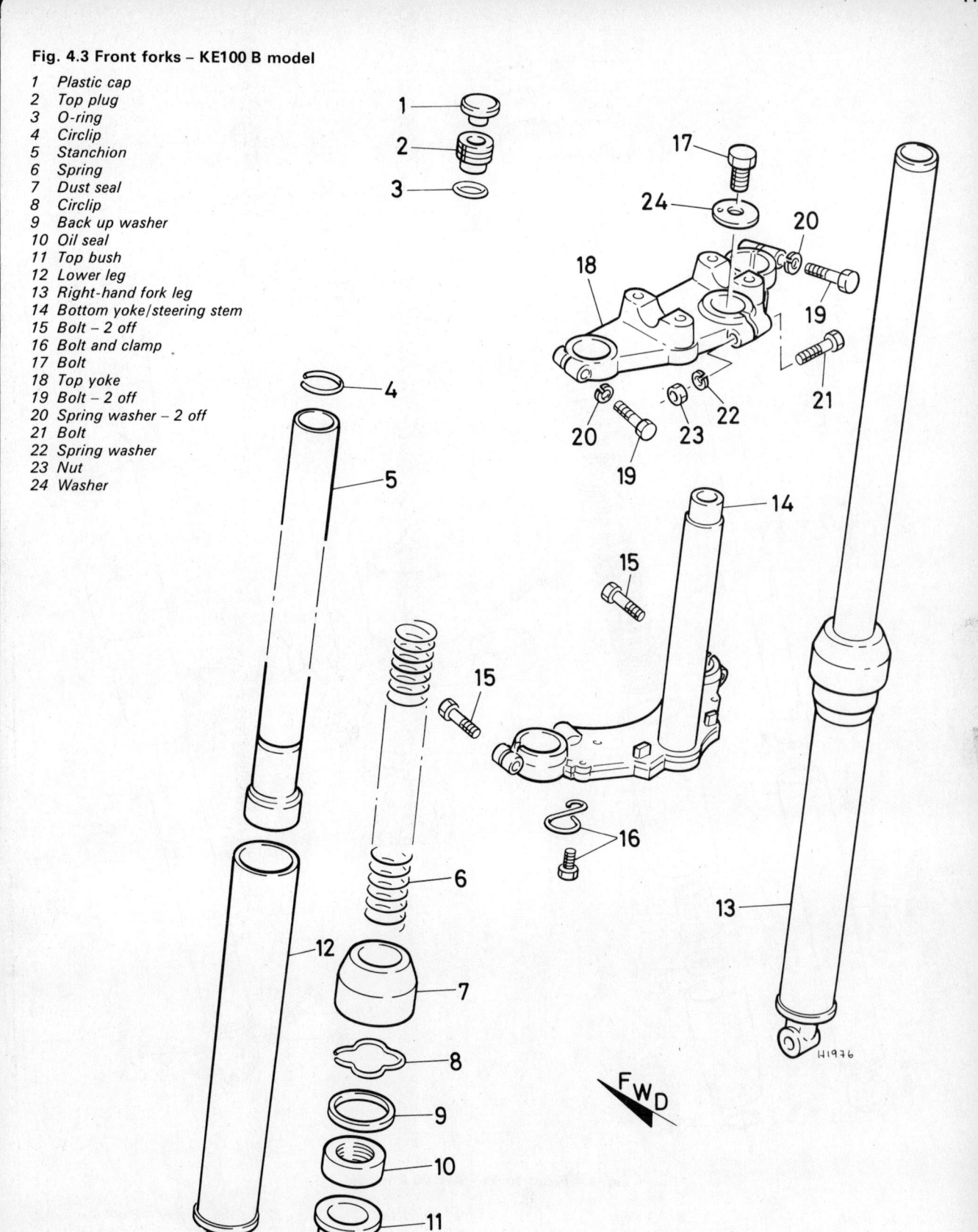

1. Plastic cap
2. Top plug
3. O-ring
4. Circlip
5. Stanchion
6. Spring
7. Dust seal
8. Circlip
9. Back up washer
10. Oil seal
11. Top bush
12. Lower leg
13. Right-hand fork leg
14. Bottom yoke/steering stem
15. Bolt – 2 off
16. Bolt and clamp
17. Bolt
18. Top yoke
19. Bolt – 2 off
20. Spring washer – 2 off
21. Bolt
22. Spring washer
23. Nut
24. Washer

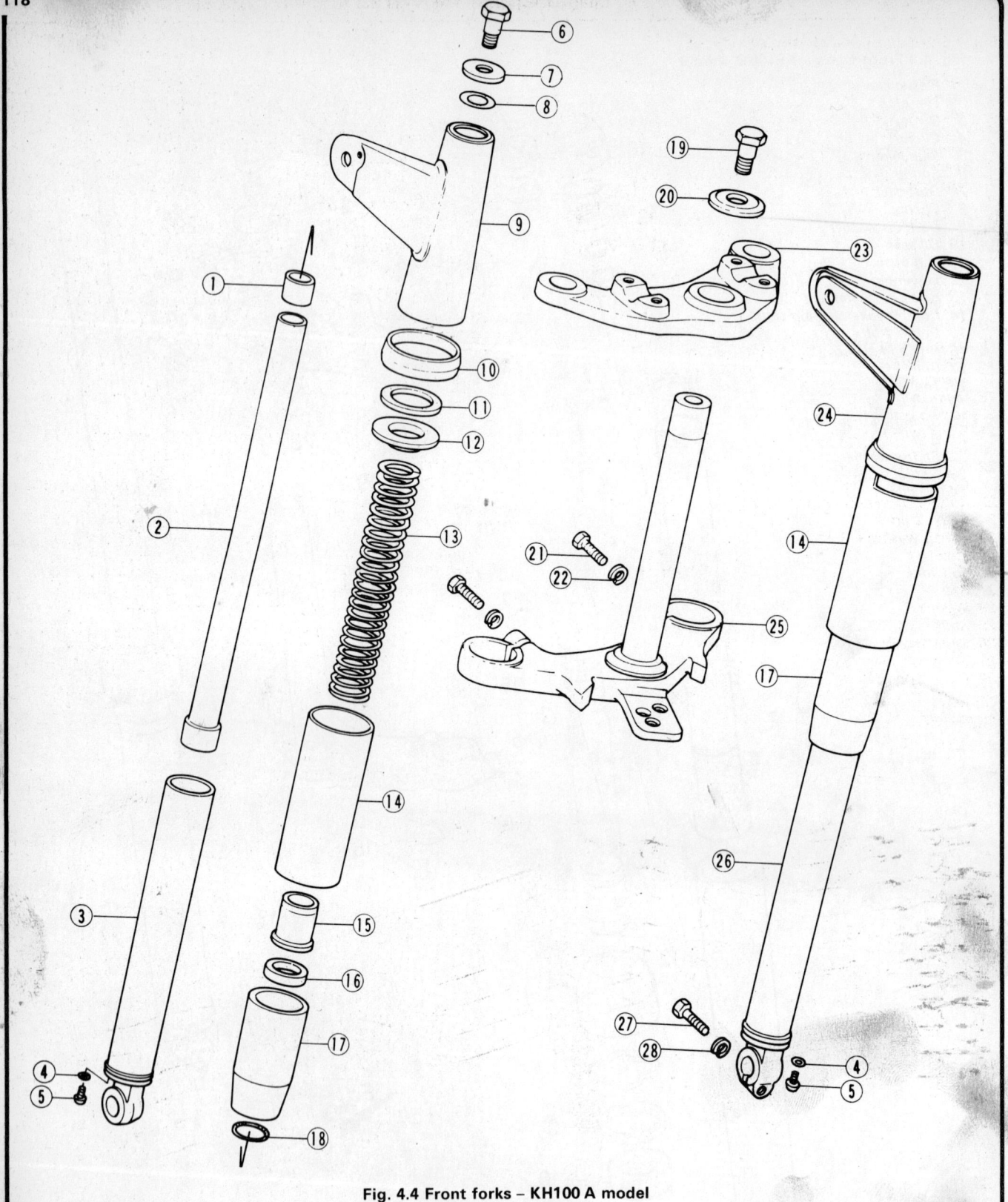

Fig. 4.4 Front forks – KH100 A model

1 Bush	8 O-ring	15 Dust seal	22 Spring washer – 2 off
2 Stanchion	9 Headlamp bracket	16 Oil seal	23 Top yoke
3 Lower leg	10 Cover	17 Seal holder	24 Headlamp bracket
4 Sealing washer	11 Damper ring	18 O-ring	25 Bottom yoke/steering head
5 Drain screw	12 Spring top seat	19 Bolt	26 Right-hand fork leg
6 Top bolt	13 Spring	20 Washer	27 Bolt
7 Washer	14 Spring shroud	21 Bolt – 2 off	28 Spring washer

Chapter 4 Frame and forks 119

Fig. 4.5 Front forks – KH100 G model

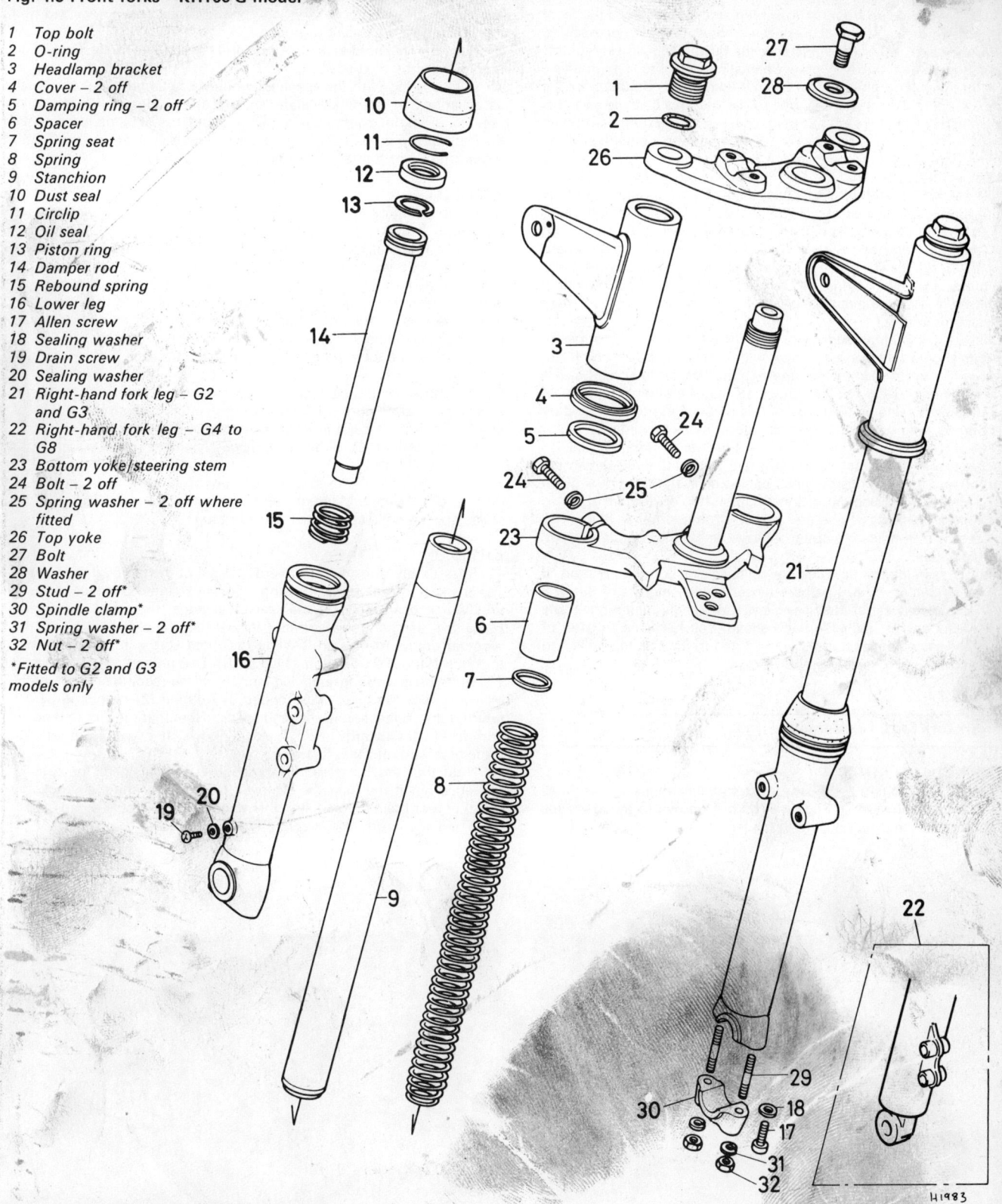

1. Top bolt
2. O-ring
3. Headlamp bracket
4. Cover – 2 off
5. Damping ring – 2 off
6. Spacer
7. Spring seat
8. Spring
9. Stanchion
10. Dust seal
11. Circlip
12. Oil seal
13. Piston ring
14. Damper rod
15. Rebound spring
16. Lower leg
17. Allen screw
18. Sealing washer
19. Drain screw
20. Sealing washer
21. Right-hand fork leg – G2 and G3
22. Right-hand fork leg – G4 to G8
23. Bottom yoke/steering stem
24. Bolt – 2 off
25. Spring washer – 2 off where fitted
26. Top yoke
27. Bolt
28. Washer
29. Stud – 2 off*
30. Spindle clamp*
31. Spring washer – 2 off*
32. Nut – 2 off*

*Fitted to G2 and G3 models only

4 Front fork legs: examination and renovation

1 Carefully clean and dry all the components of the fork leg. Lay them out on a clean work surface and inspect each one, looking for excessive wear, cracks or other damage. All traces of oil, dirt and swarf should be removed, and any damaged or worn components renewed.

2 Examine the sliding surface of the stanchion or bushes, as applicable, and the internal surface of the lower leg, looking for signs of scuffing which will indicate that excessive wear has taken place. Slide the stanchion into the lower leg so that it seats fully. Any wear present will be easily found by attempting to move the stanchion backwards and forwards, and from side to side, in the bore of the lower leg. It is inevitable that a certain degree of slackness will be found,

especially when the test is repeated at different points as the stanchion is gradually withdrawn from the lower leg, and it is largely a matter of experience to assess with accuracy the amount of wear necessary to justify renewal of either the stanchion, the bush, or the lower leg. It is recommended that the two components be taken to a motorcycle dealer for an expert opinion to be given if there is any doubt about the degree of wear found. Note that while wear will only become a serious problem after a high mileage has been covered, it is essential that such wear be rectified by the renewal of the components concerned if the handling and stability of the machine are impaired.

3 Check the outer surface of the stanchion for scratches or roughness; it is only too easy to damage the oil seal during reassembly if these high spots are not eased down. The stanchions are unlikely to bend unless the machine is damaged in an accident. Any significant bend will be detected by eye, but if there is any doubt about straightness, roll the stanchion tubes on a flat surface such as a sheet of plate glass. If the stanchions are bent, they must be renewed. Unless specialised repair equipment is available it is rarely practicable to effect a satisfactory repair.

4 Check the stanchion sliding surface for pits caused by corrosion. Such pits should be smoothed down with fine emery paper and filled, if necessary, with Araldite. Once the Araldite has set fully hard, use a fine file or emery paper to rub it down so that the original contour of the stanchion is restored. Damage of this nature can be eliminated by the fitting of gaiters available from any good motorcycle dealer.

5 After an extended period of service, the fork springs may take a permanent set. If the spring lengths are suspect, they should be measured and the readings obtained compared with the lengths given in the Specifications Section of this Chapter. It is always advisable to fit new fork springs where the length of the original items has decreased by a significant amount. Always renew the springs as a set, never separately.

6 Closely examine the dust seal for splits or signs of deterioration. If found to be defective, it must be renewed as any ingress of dirt will rapidly accelerate wear of the oil seal and fork stanchion. It is advisable to renew any gasket washers fitted beneath bolt heads as a matter of course. The same applies to the O-rings fitted to the fork top bolts and especially to the fork seals themselves.

5 Front fork legs: reassembly

KC100, KE100 A, KH100 A

1 Smear the stanchion and bushes with fork oil and insert it as far as possible into the lower leg. Slide the top bush down to the lower leg and gently tap it into its housing in the leg.

2 Using a tubular drift such as a socket spanner which bears only on the seal hard outer edge, tap a new seal into the seal holder and fit a new O-ring to the locating groove.
3 Slide the seal holder into position and tighten it securely. On KE100 models refit the dust seal.
4 Refill the leg with the specified amount of oil and check the level, as described in Routine Maintenance. If this is done with the forks removed from the machine, take care to avoid loss of oil on reassembly.
5 Refit the spring with the closer-spaced coils at the top. On KE 100 models refit and tighten the top bolt.

KE100 B

6 Fit the stanchion and bushes to the lower leg as described in paragraph 1 above.
7 Smear it liberally with fork oil to protect the sealing lips and slide the oil seal down the stanchion. Press it as far as possible into the lower leg by hand, then refit the back up washer. Either use a hammer and a small drift to tap the seal gently and evenly into the lower leg, using the washer for protection, or obtain a length of tubing of the same outside diameter as the seal and which will fit over the stanchion, and use this as a tubular drift to tap the seal into place. Be very careful not to damage the seal.
8 Whichever method is used, cease tapping as soon as the circlip groove is exposed above the back-up ring; attempting to force the seal any further will damage it. Refit the circlip and the dust seal.
9 Be very careful to tap the seal only gently and to keep it square in its housing at all times.
10 Refill the leg with the specified amount of oil and check the level as described in Routine Maintenance before refitting the spring (with the closer spaced coils at the top) and top plug.

KH100 G

11 Reassembly is a straightforward reversal of the removal procedure. Tap the seal into its housing using a socket spanner or similar tubular drift which bears only on the seal outer edge, then refit the circlip. Place the damper rod assembly inside the stanchion and use the wooden dowel to restrain it while the Allen screw is tightened to a torque setting of 21.6 kgf m (11.5 lbf ft). See paragraph 13 below.
12 Fill the leg with the specified amount of the recommended type of oil and check the level as described in Routine Maintenance before refitting the spring, spring seat and spacer. Note that the spring tapered end faces downwards. Pack grease above the seal as additional protection and refit the dust seal.
13 Note that the full assembly procedure is given here for ease of reference but if the forks are removed from the machine it would be easier to refit them before filling them with oil; this would avoid all risk of oil spillage and of the loss of any components.

5.11a Refit fork oil seal as shown ...

5.11b ... and refit circlip to secure seal

Chapter 4 Frame and forks 121

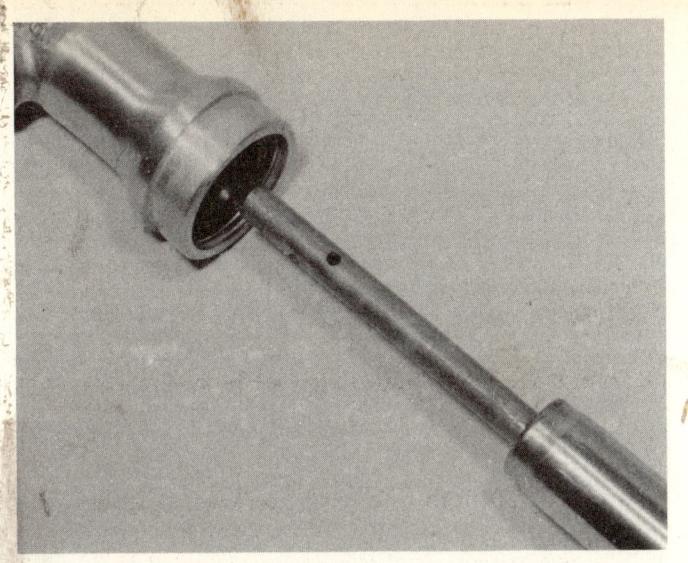

5.11c Insert stanchion and damper rod into lower leg ...

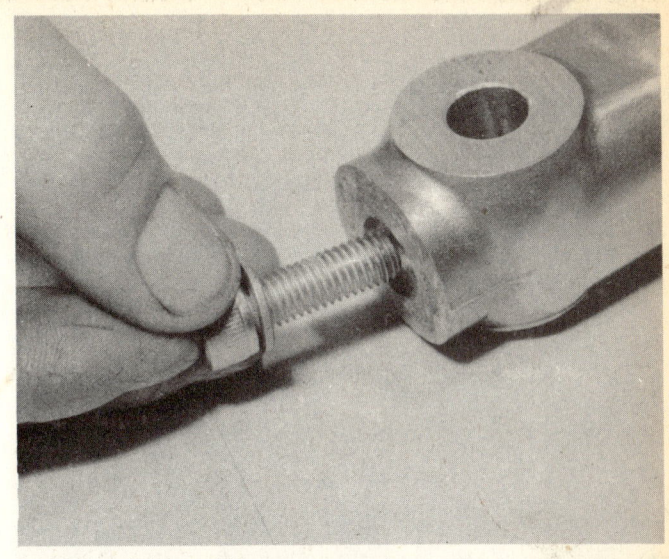

5.11d ... and tighten Allen screw securely

5.12a Fill fork leg with specified amount and grade of oil, then check oil level ...

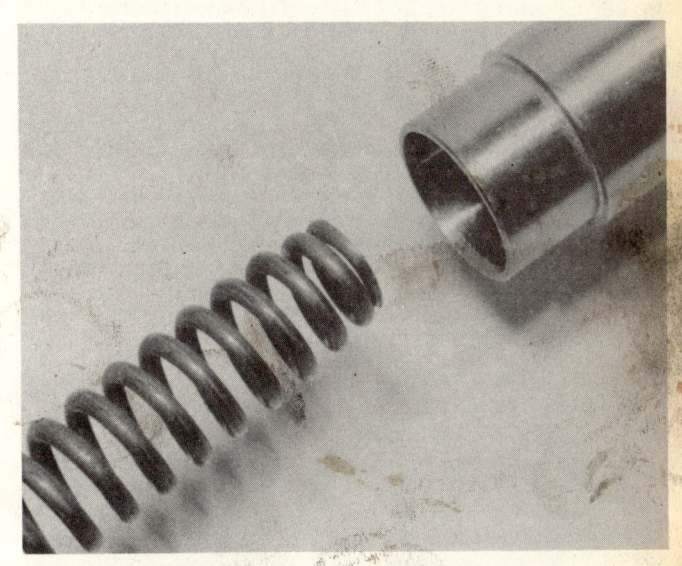

5.12b ... before refitting fork spring – note tapered end points downwards

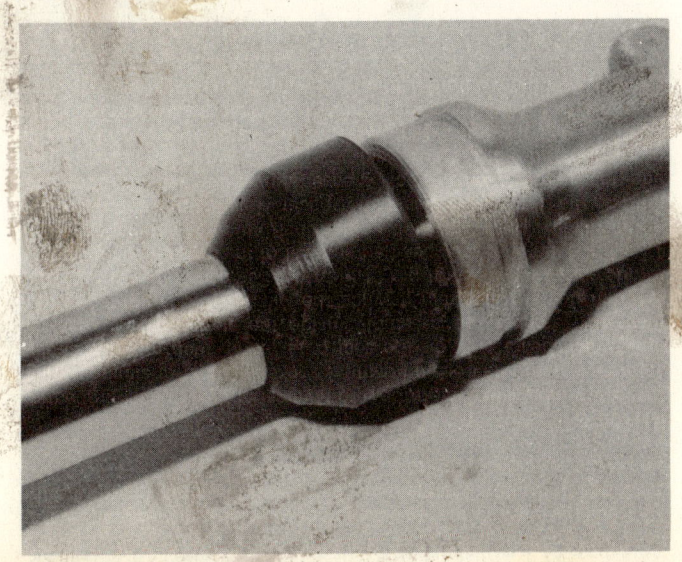

5.12c Refit dust seal to top of lower leg

6 Front fork legs: refitting

1 Refitting is a straightforward reversal of the removal sequence, noting the following points.
2 On KC and KH100 models smear the stanchion upper end with oil or grease, check that all components have been refitted to the leg, and pass it up through the bottom yoke and headlamp bracket to the underside of the top yoke. Ensuring that the sealing O-ring has been refitted, tighten down the fork leg top bolt to secure the leg.
3 If this procedure proves difficult it will be necessary, on KC100 and KH100 A models, to find a long bolt which will pass down through the yokes and headlamp bracket far enough for it to be screwed into the stanchion to provide a grip by which the fork leg can be pulled firmly into place. The Kawasaki service tool for this job is part number 57001-177. Where appropriate, fill the fork legs with oil. See Section 5.
4 On KE100 models slide the stanchion up through the yokes until it is flush with the top yoke top surface and tighten lightly the top and bottom yoke pinch bolts to retain it.
5 Refit the front mudguard (where removed) and the front wheel, then assemble the brake components again. Tighten the spindle nut

Chapter 4 Frame and forks

and other mounting bolts only lightly at first and take the machine off its centre stand or box. Apply the front brake and push down on the handlebars several times so that the operation of the fork legs settles each component in its correct place. Using a torque wrench if possible, tighten all nuts and bolts from the wheel spindle upwards to the recommended torque settings, where specified. This will ensure that the fork components can operate freely and easily, with no undue strain imposed from an overtightened bolt or an awkwardly positioned part.

6 Be very careful to check the fork operation, front brake operation and that all nuts and bolts are securely fastened before taking the machine out on the road.

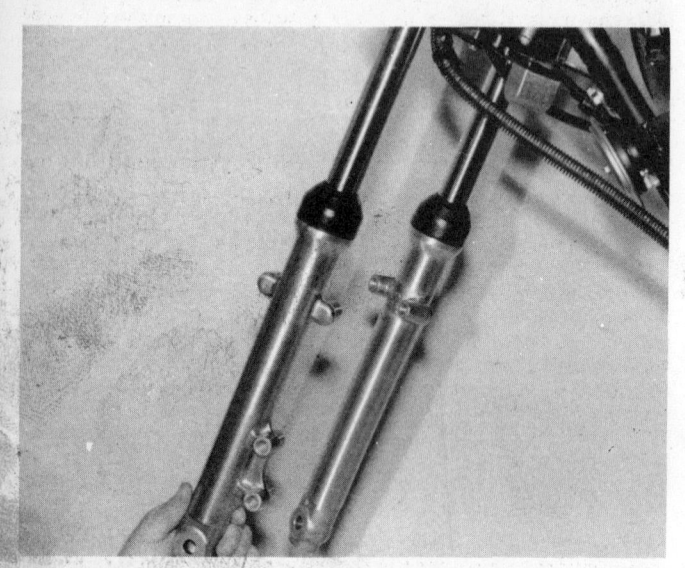

6.2 Smear stanchion with oil or grease to aid refitting

6.5 Refit all disturbed components – tighten fasteners to recommended torque wrench settings, where given

7 Steering head assembly: removal and refitting

1 Working as described in Section 2 of this Chapter, remove the front fork legs, then remove the fuel tank as described in Section 2 of Chapter 2.
2 Remove the four handlebar clamp bolts and the handlebar clamps. Move the handlebars backwards as far as possible without straining the control cables, brake hose, or wiring and secure them clear of the steering head area. Remove or disconnect all control cables, wiring and the instrument drive cables which might hinder the removal of the fork yokes. Components such as the horn and instrument panel can remain in place on the yokes.
3 Slacken its clamp bolt, where fitted, then unscrew the large chromium-plated bolt from its location through the centre of the top yoke. Using a soft-faced hammer, give the top yoke a gentle tap to free it from the steering head and lift it from position, ensuring that the headlamp brackets do not drop away. Carry out a final check around the bottom yoke to ensure that all components have been removed which might prevent its release.
4 Support the weight of the bottom yoke and, using a C-spanner of the correct size, remove the steering head bearing adjusting ring. If a C-spanner is not available, a soft metal drift may be used in conjunction with a hammer to slacken the ring.
5 Remove the dust excluder and the cone of the upper bearing. The bottom yoke, complete with steering stem, can now be lowered from position. Ensure that any balls that fall from the bearings are caught and retained as the bearing races separate. It is quite likely that only the balls from the lower bearing will drop free, since those of the upper bearing will remain seated in the bearing cup. Full details of examining and renovating the steering head bearings are given in Section 8 of this Chapter.
6 Fitting of the steering head assembly is a direct reversal of the removal procedure, whilst noting the following points. It is advisable to position all twenty-three balls of the lower bearing around the bearing cone before inserting the steering stem fully into the steering head. Retain these balls in position with grease of the recommended type and fill both bearing cups with the same type of grease.
7 With the bottom yoke pressed fully home into the steering head, place the twenty-three balls into the upper bearing cup and fit the bearing cone followed by the dust excluder. Refit the adjusting ring and tighten it, finger-tight. The ring should now be tightened firmly to seat the bearings, using a C-spanner only; do not apply excessive force. Turn the bottom yoke from lock-to-lock five or six times to settle the balls, then slacken the adjusting ring until all pressure is removed.
8 To provide the initial setting for steering head bearing adjustment, tighten the adjusting ring carefully until resistance is felt, then loosen it by $1/8$ to $1/4$ of a turn. Remember to check that the adjustment is correct, as described in Routine Maintenance, when the steering head assembly has been reassembled and the forks and front wheel refitted.
9 Finally, whilst refitting and reconnecting all disturbed components, take care to ensure that all control cables, drive cables, electrical leads, etc are correctly routed and that reference is made to the list of torque wrench settings given in the Specifications Section of this Chapter and of Chapter 5. Check that the headlamp beam height has not been disturbed and ensure that all controls and instruments function correctly before taking the machine on the public highway.

8 Steering head bearings: examination and renovation

1 Before commencing reassembly of the steering head component parts, take care to examine each of the steering head bearings. The ball bearing tracks of their respective cup and cone bearings should be polished and free from any indentations or cracks. If wear or damage is evident, the cups and cones must be renewed as complete sets.
2 Carefully clean and examine the balls contained in each bearing assembly. These should also be polished and show no signs of surface cracks or blemishes. If any one ball is found to be defective, the complete set should be renewed. Remember that a complete set of these balls is relatively cheap and it is not worth the risk of refitting items that are in doubtful condition.
3 Note that a total of forty-six balls, all of $3/16$ in diameter, are fitted; twenty-three in each race. This arrangement will leave a gap between any two balls but an extra ball **must not** be fitted, otherwise the balls will press against each other thereby accelerating wear and causing the steering action to be stiff.
4 The bearing cups are a drive fit in the steering head and may be removed by passing a long drift through the inner bore of the steering head and drifting out the defective item from the opposite end. The drift must be moved progressively around the race to ensure that it leaves the steering head evenly and squarely.
5 The lower of the two cones fits over the steering stem and may be

Chapter 4 Frame and forks

removed by carefully drifting it up the length of the stem with a flat-ended chisel or a similar tool. Again, take care to ensure that the cone is kept square to the stem.

6 Fitting of the new cups and cone is a straightforward procedure whilst taking note of the following points. Ensure that the cup locations within the steering head are clean and free of rust; the same applies to the steering stem. Lightly grease the stem and head locations to aid fitting and drift each cup or cone into position whilst keeping it square to its location. Fitting of the cups into the steering head will be made easier if the opposite end of the head to which the cup is being fitted has a wooden block placed against it to absorb some of the shock as the drift strikes the cup.

9 Frame: examination and renovation

1 The frame is unlikely to require attention unless accident damage has occurred. In some cases, renewal of the frame is the only satisfactory remedy if it is badly out of alignment; repairs are a task for the specialist only.

2 At regular intervals, examine the frame closely for signs of cracking or splitting at the welded joints. Rust can also cause weakness at these joints. Minor damage can be repaired by welding or brazing depending on the extent and nature of the damage.

3 Remember that a frame which is out of alignment will cause handling problems and may even promote a 'speed wobble'. If misalignment is suspected, it will be necessary to strip the machine completely so that the frame can be checked, and if necessary, renewed.

10 Swinging arm: removal

1 Referring to Chapter 5, remove the rear wheel. On KC100 and KH100 models, while its removal is not strictly necessary, work is a great deal easier if the exhaust system is first removed. Refer to Chapter 2.11.

2 Remove, if required, the brake torque arm and the chainguard from the swinging arm, then slacken all four suspension unit mounting nuts. Remove the two suspension unit bottom mounting nuts and their washers, then pull the units sideways off their swinging arm mounting lugs. Note carefully the position and number of the plain washers at each mounting.

3 Remove the swinging arm pivot bolt securing nut, then withdraw the pivot bolt. If the bolt proves stubborn, apply a good quantity of penetrating fluid, allow time for it to work, then displace the bolt using a hammer and a metal drift. Withdraw the swinging arm.

11 Swinging arm: examination and renovation

1 Thoroughly clean all components, removing all traces of dirt, corrosion and old grease.

2 Inspect closely all components, looking for obvious signs of wear such as heavy scoring, or for damage such as cracks or distortion due to accidental impact. Any obviously damaged or worn component must be renewed. If the painted finish has deteriorated it is worth taking the opportunity to repaint the affected area, ensuring that the surface is correctly prepared beforehand.

3 Check the pivot bolt for wear. If the shank of the bolt is seen to be stepped or badly scored, it must be renewed. Remove all traces of corrosion and hardened grease from the bolt before checking it for straightness by rolling it on a flat surface, such as a sheet of plate glass; if the bolt is not perfectly straight it must be renewed. Check also that its threads and those of the retaining nut are in good condition.

4 Detach the nylon buffer from the left-hand pivot lug and inspect it for signs of excessive wear. If the buffer no longer protects the pivot lug from the final drive chain, it must be renewed.

5 The bonded rubber bushes are an interference fit in the swinging arm mounting bosses, and are unlikely to wear or deteriorate until a high mileage has been covered, but wear may occur between the bush inner sleeves and the pivot, especially if the retaining nut has come loose. In normal service there should be no relative movement between the inner bushes and shaft; the fork movement is allowed by flexing of the bonded rubber.

6 Inspect the bush rubber for damage or separation from both inner and outer sleeves. Check also that the inner sleeves have not been moving on the shaft. If damage is evident, the bushes must be renewed. Driving the bushes from position is unlikely to prove successful, particularly if they have been in position for a long time and corrosion has taken place. Removal is accomplished most easily using a fabricated puller as shown in the accompanying illustration. The puller sleeve should have an internal diameter slightly greater than the outside diameter of the bush outer sleeve. If possible, use a high tensile nut and bolt because these will be better able to take the strain during use. New bushes may be drawn into place using the same puller. If difficulty is encountered in removing the old bushes it is recommended that the swinging arm be returned to an authorised Kawasaki dealer.

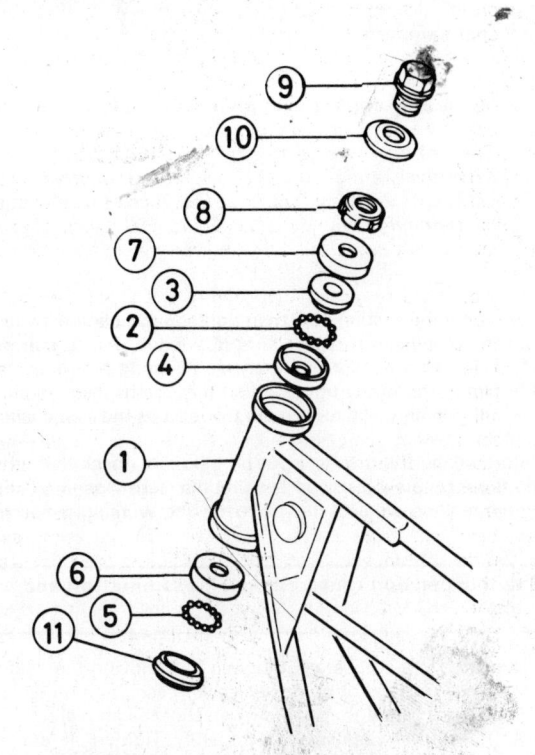

Fig. 4.6 Steering head bearing assembly

1 Frame
2 Upper bearing balls
3 Upper bearing cone
4 Upper bearing cup
5 Lower bearing balls
6 Lower bearing cup
7 Dust excluder
8 Adjuster nut
9 Bolt
10 Washer
11 Lower bearing cone

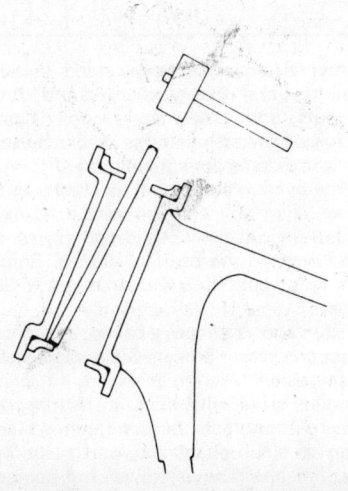

Fig. 4.7 Method of removing steering head bearing cups

124 Chapter 4 Frame and forks

11.6 Swinging arm pivot bushes should be checked regularly for wear

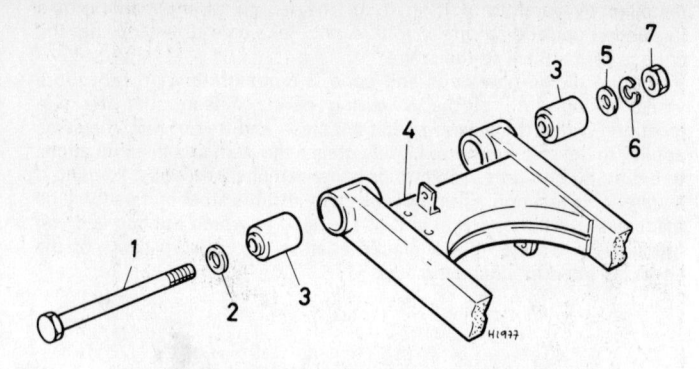

Fig. 4.8 Swinging arm pivot assembly – typical

1 Pivot bolt
2 Washer
3 Bush – 2 off
4 Swinging arm
5 Washer
6 Spring washer
7 Nut

twice to settle the suspension, then tighten securely the swinging arm pivot bolt retaining nut; use a torque wrench if a torque setting is specified. **Note:** it is essential that the swinging arm is in its normal position when the nut is tightened so that the bushes are clamped in the position which will distort them the least as the suspension moves through its travel.

5 Before taking the machine out on the road, check that all the nuts and bolts are securely fastened and that the rear suspension, rear brake, and chain tension are adjusted correctly and working properly.

13 Rear suspension units: removal, examination and refitting

1 The suspension units are withdrawn by removing the retaining cap nuts and plain washers or bolts then pulling each unit off its mounting lugs. Refitting is a straightforward reversal of the above, but care must be taken to refit the plain washers correctly. Tighten all retaining nuts or bolts to the specified torque setting.

2 The suspension units are sealed; if any wear or damage is found they must be renewed as a matched pair of complete units. Look for signs of oil leakage around the damper rod, worn mounting bushes, and any other signs of damage. Note that the mounting bushes can be renewed separately if necessary. Press out the old bushes and fit the new ones using a smear of soapy water or similar as a lubricant.

14 Footrests, stands and controls: examination and renovation

1 At regular intervals all footrests, the stand, the brake pedal and the gearchange lever pivots should be checked and lubricated. Check that all mounting nuts and bolts are securely fastened, using the recommended torque wrench settings where these are given. Check that any securing split pins are correctly fitted.

2 Check that the bearing surfaces at all pivot points are well greased and unworn, renewing any component that is excessively worn. If lubrication is required, dismantle the assembly to ensure that grease can be packed fully into the bearing surface. Return springs, where fitted, must be in good condition with no traces of fatigue and must be securely mounted.

3 If accident damage is to be repaired, check that the damaged component is not cracked or broken. Such damage may be repaired by welding, if the pieces are taken to an expert, but since this will destroy the finish, renewal is usually the most satisfactory course of action. If a component is merely bent it can be straightened after the affected area has been heated to a full cherry red, using a blowlamp or welding torch. Again the finish will be destroyed, but painted surfaces can be repainted easily, while chromed or plated surfaces can only be replated, if the cost can be justified.

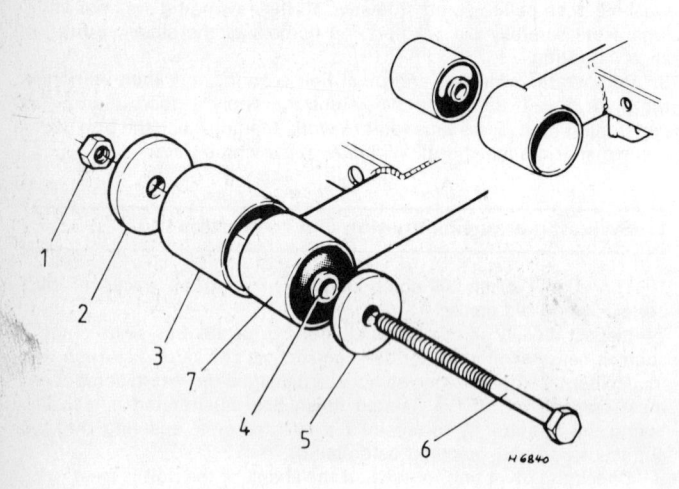

Fig. 4.9 Swinging arm bush removal tool

1 Nut
2 Washer
3 Sleeve
4 Bush
5 Washer
6 Bolt
7 Swinging arm

12 Swinging arm: refitting

1 Smear grease over the length of the pivot bolt and pack a small amount into the inner sleeves of the pivot bushes and into the frame passages through which the bolt passes. Grease also the suspension unit mounting lugs.

2 Pass the left-hand fork end through the drive chain and manoeuvre the swinging arm into place. Fit the pivot bolt from left to right and refit the suspension units to their mounting lugs, ensuring that the plain washers are refitted in their original locations. Tighten all four mounting nuts securely, using the recommended torque setting where specified.

3 Refit the exhaust system (if removed) and lightly tighten the swinging arm pivot bolt retaining nut. Refit the brake torque arm and the chainguard (if removed). Refit the rear wheel.

4 Push the machine off its stand and bounce the rear end once or

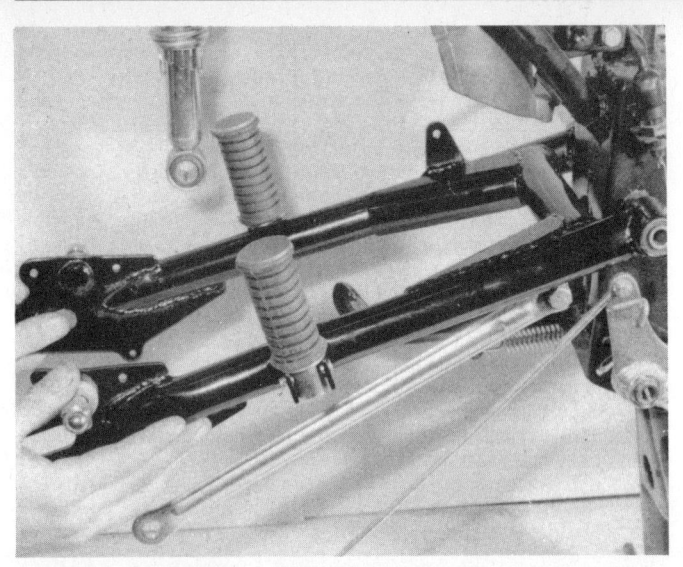

Chapter 4 Frame and forks

12.2a Offer up swinging arm assembly and grease pivot bolt before refitting

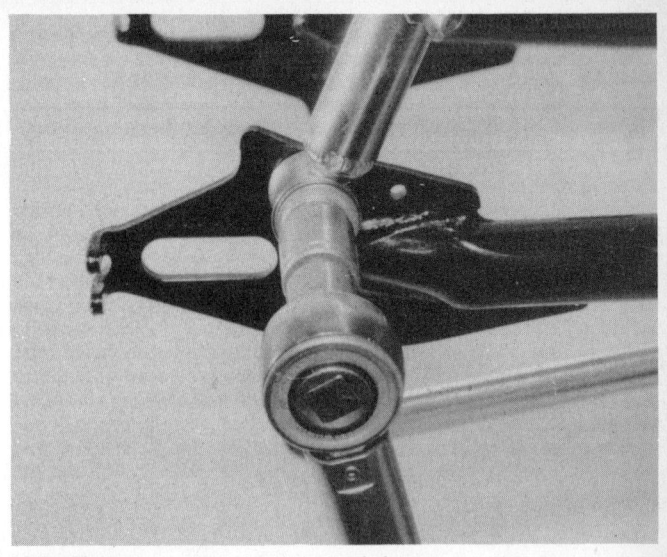

12.2b Tighten securely suspension unit fasteners – use torque wrench settings where available

12.4 Swinging arm pivot bolt nut should be tightened only with suspension in normal position

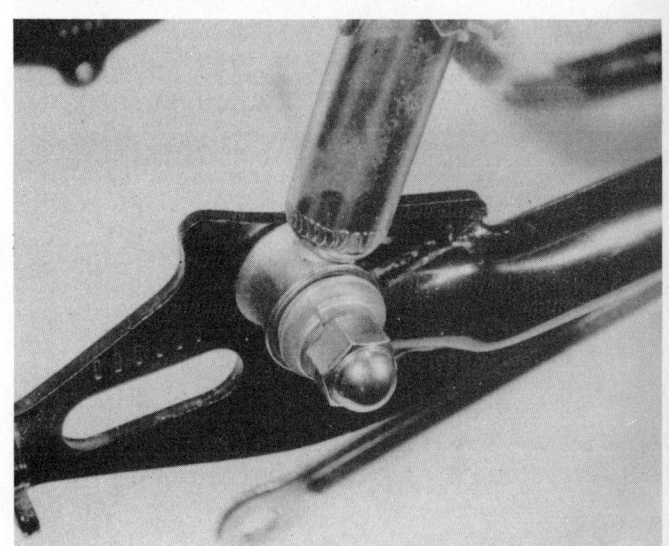

13.1 Note fitted position of washers before removing suspension unit fasteners

14.1 Check and grease all control and stand pivots at regular intervals

15 Speedometer and tachometer: removal, examination and refitting

1 To remove the instruments, disconnect the drive cables and remove the headlamp rim and reflector unit so that the wires can be disconnected.

2 Remove the two nuts retaining each instrument and remove the shaped washers, noting which way up each is fitted, then lift the instrument out of its mounting. Refitting is the reverse of the removal procedure.

3 The instruments must be carefully handled at all times and must never be dropped or held upside down. Dirt, oil, grease and water all have an equally adverse effect on them and so a clean working area must be provided if they are to be removed.

4 The instrument heads are very delicate and should not be dismantled at home. In the event of a fault developing, an instrument should be entrusted to a specialist repairer or a new unit fitted. If a replacement unit is required, it is well worth trying to obtain a good secondhand item from a motorcycle breaker in view of the high cost of a new replacement.

5 Remember that a speedometer in correct working order is a statutory requirement in the UK. Apart from this legal necessity,

reference to the odometer readings is the most satisfactory means of keeping pace with the maintenance schedules.

16 Instrument drive components: location and examination

Speedometer

1 The speedometer drive is formed by a gearbox, mounted on the right-hand side of the front wheel on KH100 G models, and set in the brake backplate on all other models. It must be removed and packed with fresh grease at regular intervals, as described in Routine Maintenance.

2 If the unit is thought to be faulty, first check that the raised tangs of the drive ring in the wheel hub are not flattened or damaged. If this does not cure the fault, check with an authorised Kawasaki dealer as to what replacement parts are available. For some models there are almost none listed while others can be fully reconditioned.

3 For machines with front drum brakes remove the front wheel and withdraw the brake backplate. The drive ring and pinion in the hub can be simply pulled out, provided that care is taken to prevent damage to the grease seal; this can be renewed separately, if necessary.

4 To remove the driven pinion, the roll pin must be removed. Passing a 1 mm drill bit up through the pin, drill through the brake backplate, then use this as a pilot hole to drill a hole (using a 2 mm drill bit) through to the rear of the pin which can then be tapped out, using a punch. The cable bush, thrust washers and pinion can then be extracted. On reassembly, always use a new pin to retain the cable bush, and stake it in place to prevent it from dropping out. The same method can be used to extract the drive gearbox components on KH100 G models, but note that the drive ring is retained by a circlip in some cases and that the driven pinion is retained by a grub screw.

Tachometer (rev-counter) – KH100 G

5 This instrument is driven from the crankshaft via the primary drive, the kickstart idler gear and a white nylon drive gear. It is removed and refitted as described in Chapter 1. If any component is found to be worn or damaged, it must be renewed.

Chapter 5 Wheels, brakes and tyres

Contents

General description 1	Brake disc: examination and renovation 8
Front wheel: removal and refitting 2	Brake caliper: examination and renovation 9
Rear wheel: removal and refitting 3	Bleeding the hydraulic brake system 10
Wheel bearings: removal, examination and refitting 4	Drum brake: examination and renovation 11
Rear wheel cush drive: examination and renovation 5	Tyres: removal, repair and refitting 12
Brake master cylinder: removal, overhaul and refitting 6	Valve cores and caps: general 13
Hydraulic hoses: examination 7	

Specifications

Wheels

	Front	Rear
Rim size:		
KC100	1.40 x 18	1.40 x 18
KE100 A	1.40 x 19	1.60 x 18
KE100 B	1.40 x 19	1.60 x 17
KH100 A	1.40 x 18	1.60 x 18
KH100 G2, G3, G4	N/Av	N/Av
KH100 G5, G6, G7, G8	1.40 x 18	1.85 x 18
Rim maximum warpage:	Wire-spoked wheels	Cast alloy wheels
Radial	2.0 mm (0.079 in)	0.5 mm (0.020 in)
Axial	2.0 mm (0.079 in)	0.8 mm (0.032 in)
Spindle maximum warpage	0.2 mm (0.008 in)	

Drum brakes – except KC100

	Standard	Service limit
Front brake drum ID:		
KE100	110.000 – 110.087 mm (4.3307 – 4.3341 in)	110.750 mm (4.3602 in)
KH100 A	130.000 – 130.160 mm (5.1181 – 5.1244 in)	130.750 mm (5.1476 in)
Rear brake drum ID:		
KE100, KH100 A	110.000 – 110.087 mm (4.3307 – 4.3341 in)	110.750 mm (4.3602 in)
KH100 G	N/Av	N/Av
Brake shoe return spring free length	30.8 – 31.2 mm (1.2126 – 1.2283 in)	34.0 mm (1.3386 in)
Brake camshaft OD	11.957 – 11.984 mm (0.4708 – 0.4718 in)	11.830 mm (0.4658 in)
Backplate ID at camshaft	12.000 – 12.027 mm (0.4724 – 0.4735 in)	12.180 mm (0.4795 in)
Brake shoe friction material thickness:		
KH100 A front, KH100 G5, G6, G7, G8 rear	4.0 mm (0.16 in)	2.0 mm (0.08 in)
KE100 front and rear, KH100 A, KH100 G2, G3, G4, rear	3.0 mm (0.12 in)	1.5 mm (0.06 in)

Front hydraulic disc brake

	KH100 G2, G3	KH100 G4, G5, G6, G7, G8
Master cylinder:		
Cylinder standard ID	14.000 – 14.063 mm (0.5512 – 0.5537 in)	12.700 – 12.743 mm (0.500 – 0.5017 in)
Service limit	14.080 mm (0.5543 in)	12.760 mm (0.5024 in)
Piston standard OD	13.957 – 13.984 mm (0.5495 – 0.5506 in)	12.657 – 12.684 mm (0.4983 – 0.4994 in)
Service limit	13.900 mm (0.5472 in)	12.630 mm (0.4972 in)
Primary cup standard OD	14.200 – 14.600 mm (0.5591 – 0.5748 in)	13.300 – 13.700 mm (0.5236 – 0.5394 in)
Service limit	14.100 mm (0.5551 in)	13.100 mm (0.5157 in)
Secondary cup standard OD	14.650 – 15.150 mm (0.5768 – 0.5965 in)	13.500 – 13.900 mm (0.5315 – 0.5472 in)
Service limit	14.500 mm (0.5709 in)	13.300 mm (0.5236 in)
Spring free length	N/Av	37.800 – 39.800 mm (1.4882 – 1.5669 in)
Service limit	N/Av	36.000 mm (1.4173 in)
Caliper:	KH100 G2, G3	KH100 G4, G5, G6, G7, G8
Cylinder standard ID	38.100 – 38.150 mm (1.4999 – 1.5020 in)	33.960 – 34.010 mm (1.3370 – 1.3390 in)
Service limit	38.400 (1.5118 in)	34.030 mm (1.3398 in)
Piston standard OD	37.970 – 38.020 mm (1.4949 – 1.4969 in)	33.878 – 33.928 mm (1.3338 – 1.3358 in)
Service limit	37.700 (1.4843 in)	33.850 mm (1.3327 in)

Chapter 5 Wheels, brakes and tyres

	KH100 G2, G3, G4	KH100 G5, G6, G7, G8
Disc:		
Diameter	240 mm (9.5 in)	240 mm (9.5 in)
Thickness	4.8 – 5.1 mm (0.1890 – 0.2008 in)	3.8 – 4.1 mm (0.1496 – 0.1614 in)
Service limit	4.5 mm (0.1772 in)	3.5 mm (0.1378 in)
Maximum runout	0.3 mm (0.0118 in)	0.3 mm (0.0118 in)
	KH100 G2, G3	KH100 G4, G5, G6, G7, G8
Brake pad:		
Friction material standard thickness	Not Available	4.5 mm (0.1772 in)
Service limit	To wear limit marks	1.0 mm (0.0394 in)

Tyres

Size:	Front	Rear
KC100 C1, C2	2.50 – 18 4PR	2.50 – 18 6PR
KC100 C3, C4	2.50 – 18 4PR	2.50 – 18 4PR
KE100 A	2.75 – 19 4PR	3.00 – 18 4PR
KE100 B	2.75 – 19 4PR	3.00 – 17 4PR
KH100 A	2.50 – 18 4PR	2.75 – 18 6PR
KH100 G	2.50 – 18 4PR	2.75 – 18 4PR
Manufacturers recommended minimum tread depth	2.0 mm (0.08 in)	2.0 mm (0.08 in)

Tyre pressures – tyres cold

	Front	Rear
KE100 A:		
Up to 97.5 kg (215 lb) load	25 psi (1.75 kg/cm²)	25 psi (1.75 kg/cm²)
97.5 kg (215 lb) up to maximum permissible load	25 psi (1.75 kg/cm²)	32 psi (2.25 kg/cm²)
All other models:		
Up to 97.5 kg (215 lb) load	21 psi (1.50 kg/cm²)	28 psi (2.00 kg/cm²)
97.5 kg (215 lb) up to maximum permissible load	21 psi (1.50 kg/cm²)	32 psi (2.25 kg/cm²)

Loads quoted represent total weight of rider, passenger and any accessories or luggage. Refer to 'Model dimensions and weights' for maximum permissible loads. Pressures apply to OE tyres only – if different tyres are fitted check with tyre manufacturer or supplier whether different pressures are necessary.

Torque wrench settings

Component	kgf m	lbf ft
Front wheel spindle nut:		
KE100	3.4 – 4.6	24.5 – 33
KH100	6.0	43
Front wheel spindle clamp nut or bolt – KH100	1.9	14
Brake disc mounting nuts or bolts:		
KH100 G2, G3, G4	1.0	7
KH100 G5, G6, G7, G8	2.0	14.5
Bleed nipple – KH100 G	0.8	6
Brake union banjo bolts – KH100 G	3.0	22
Stop lamp front pressure switch – KH100 G2, G3	2.8	20
Brake caliper mounting bolts:		
KH100 G2, G3	3.0	22
KH100 G4, G5, G6, G7, G8	2.3	16.5
Master cylinder clamp bolt:		
KH100 G2, G3	0.8	6
KH100 G4, G5, G6, G7, G8	0.9	6.5
Brake lever pivot bolt locknut – KH100 G	0.6	4
Three-way joint mounting bolts – KH100 G2, G3	0.8	6
Rear wheel spindle nut:		
KE100	3.4 – 4.6	24.5 – 33
KH100	6.0	43
Rear sprocket mounting nuts:		
KE100	2.0 – 2.2	14.5 – 16
KH100	3.1	22.5
Rear brake torque arm nuts:		
KE100	2.6 – 3.5	19 – 25
KH100	3.1	22.5

1 General description

KH100 G models are fitted with cast alloy wheels with tubed tyres. An hydraulically-operated disc brake is fitted to the front wheel and a rod-operated single leading shoe drum brake to the rear.

All other models are fitted with wheels in which full width light alloy drums are laced to steel rims by steel spokes. Both brakes are of the single leading shoe drum type, the front being cable operated and the rear being rod-operated. All tyres are of the tubed type.

This Chapter covers only the removal, refitting and overhaul of the wheels, brakes and tyres; for details of regular checks and maintenance such as brake adjustment, refer to Routine Maintenance.

2 Front wheel: removal and refitting

KH100 G

1 Place the machine on its centre stand, then straighten and remove the split pin and slacken the spindle nut. Disconnect the speedometer cable at its lower end. Wedge a wooden box or other strong support under the crankcase so that the wheel is raised clear of the ground.

Fig. 5.1 Front wheel – KH100 G model

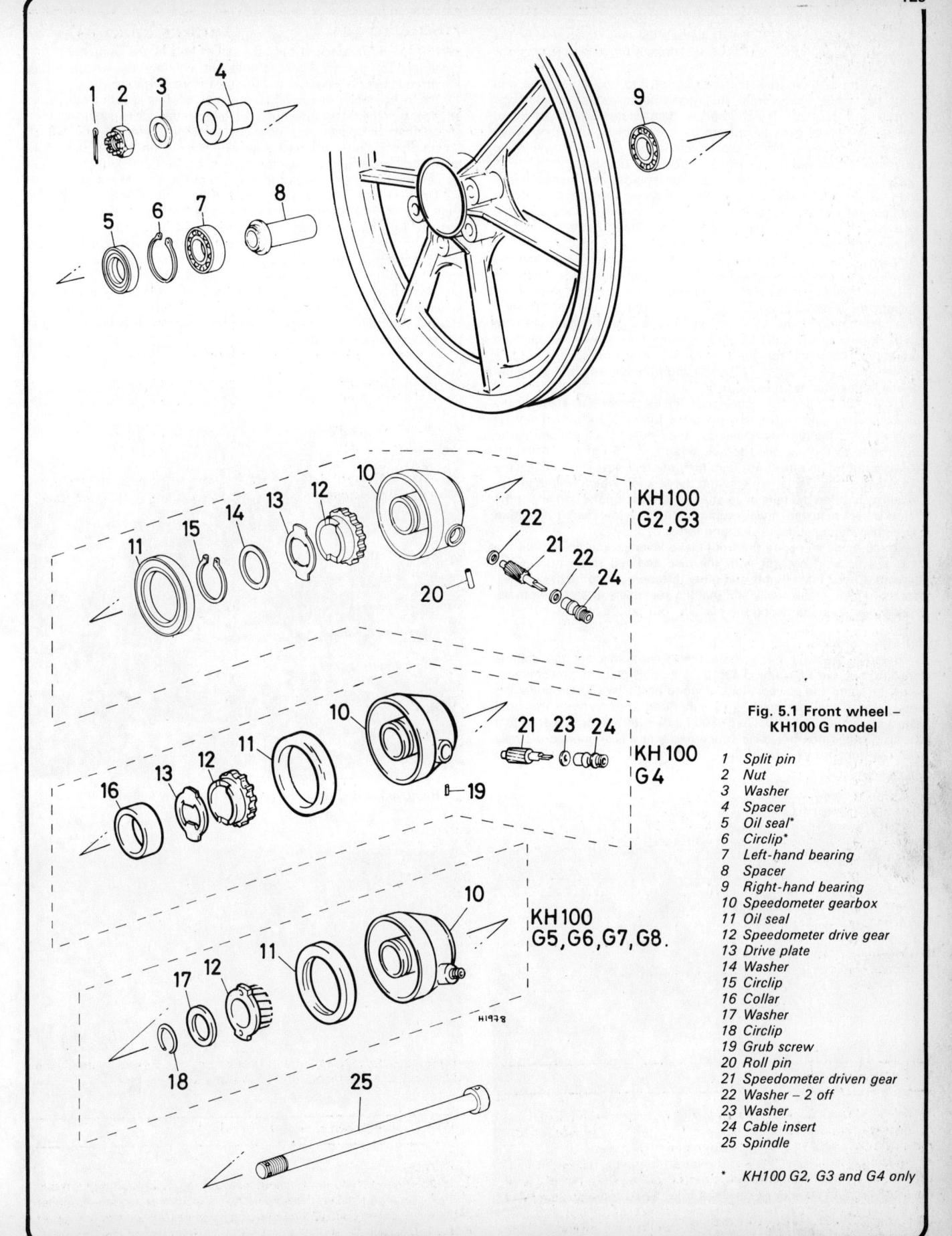

1 Split pin
2 Nut
3 Washer
4 Spacer
5 Oil seal*
6 Circlip*
7 Left-hand bearing
8 Spacer
9 Right-hand bearing
10 Speedometer gearbox
11 Oil seal
12 Speedometer drive gear
13 Drive plate
14 Washer
15 Circlip
16 Collar
17 Washer
18 Circlip
19 Grub screw
20 Roll pin
21 Speedometer driven gear
22 Washer – 2 off
23 Washer
24 Cable insert
25 Spindle

* KH100 G2, G3 and G4 only

2 On KH100 G2 and G3 models slacken the two clamp nuts at the bottom of the fork right-hand lower leg; there is no need to remove the clamp completely.

3 On all models unscrew the spindle nut, tap out the spindle and lower the wheel away from the forks. Remove the speedometer gearbox and the hub left-hand spacer, then wedge a piece of wood between the brake pads to prevent them from being disturbed while the wheel is removed. **Note:** do not apply the front brake lever while the wheel is removed or the caliper piston may be ejected.

4 On reassembly, thoroughly clean the spindle and smear grease along its length to prevent corrosion. Fit the hub spacer and the speedometer gearbox to the hub, ensuring that the tongues of the speedometer drive ring engage the hub slots. Remove the wood from between the brake pads.

5 Offer up the wheel, ensuring that the brake disc fits between the pads and that the lug on the speedometer gearbox engages with the lug on the fork lower leg. Refit the spindle and tap it into place.

6 Refit the washer and spindle nut and connect the speedometer drive cable, spinning the wheel to help the cable inner engage with the drive. Tighten securely the knurled retaining ring. Lower the machine so that the front wheel touches the ground, tighten the spindle nut to a torque setting of 6.0 kgf m (43 lbf ft) and fit a new split pin to secure the nut, spreading its ends securely.

7 On KH100 G2 and G3 models push the machine off its stand, apply the front brake and vigorously pump the forks up and down several times to settle the right-hand leg on the spindle before tightening the clamp nuts to the specified torque setting of 1.9 kgf m (14 lbf ft). **Always** tighten the front nut first, then the rear nut. This will leave a gap between the clamp rear end and the fork leg which is intentional. Never overtighten the nuts in an attempt to close up the gap and if the clamp is ever removed, always refit it with the arrow mark (indicating the higher mating surface) pointing forward.

8 On all models operate the front brake lever several times until the pads are back in contact with the disc and full brake pressure is restored. Check that all nuts and other fasteners are correctly secured, that the wheel rotates freely and that the speedometer and front brake work properly before taking the machine out on the road.

All other models

9 Remove the split pin (where fitted) and unscrew the spindle retaining nut. On KC100 and KH100 A models place the machine on its centre stand and place a block of wood or similar support under the crankcase so that the machine is supported securely with the front wheel clear of the ground. On KE100 models, lift the machine on to a strong wooden box or similar support so that it is supported securely with the front wheel clear of the ground.

10 Disconnect the front brake cable and the speedometer cable. On KH100 A models slacken fully the pinch bolt in the right-hand fork lower leg. Tap out the wheel spindle and withdraw the wheel, noting the arrangement of spacers and washers or sealing caps.

11 On reassembly, ensure that the tabs of the speedometer drive engage with the slots in the hub, as the brake backplate is refitted, then check that all spacers are correctly installed. Offer up the wheel, engaging the slot in the backplate with the lug on the left-hand fork lower leg. Smear grease over the spindle and refit it. Refit and tighten by hand only the spindle nut then connect the front brake cable.

12 On KH100 A models, push the machine off its stand, apply the front brake hard and pump the forks up and down to settle them on the spindle. Tighten the pinch bolt to a torque setting of 19 kg (14 lbf ft).

13 On all models, with the brake applied hard to centralise the shoes and backplate on the drum, tighten the spindle nut securely, to the specified torque setting, if given. On KH100 A models, fit a new split pin to secure the nut, spreading its end correctly. Connect the speedometer drive cable. Adjust the front brake as described in Routine Maintenance and check that the wheel revolves freely and that the brake works correctly.

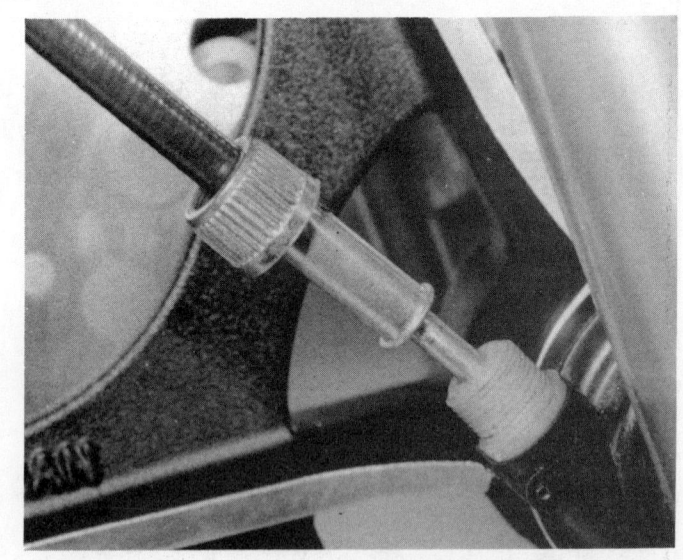

2.1 Remove knurled retaining ring and withdraw speedometer cable

2.4 Ensure speedometer gearbox drive ring tongues engage with hub slots

2.5a Lug on speedometer gearbox must fit against rear of lug on fork lower leg

2.5b Spacer must be refitted as shown – do not omit washer on refitting spindle nut

2.6 Tighten spindle nut securely (use torque settings where available) and secure with split pin (where fitted), as shown

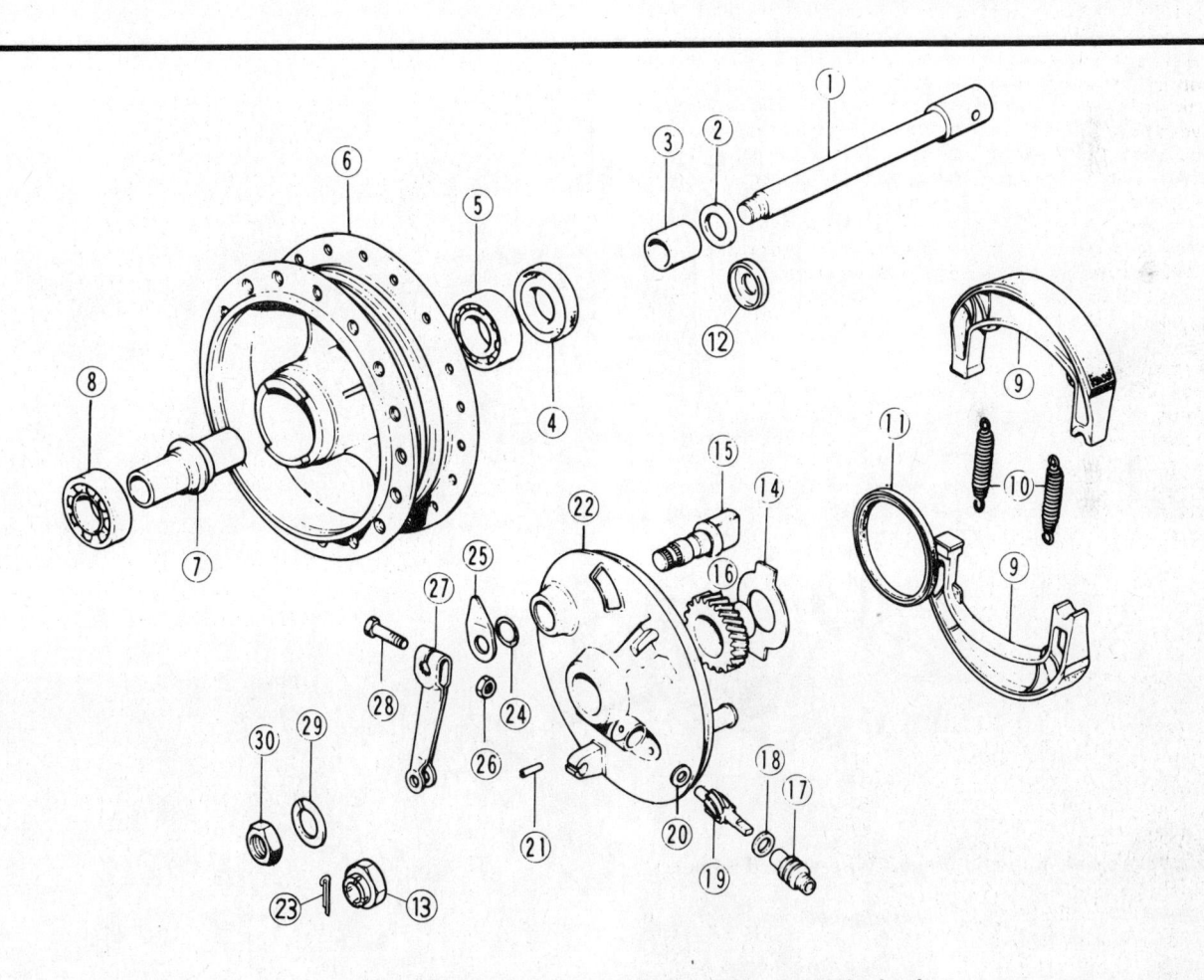

Fig. 5.2 Front wheel hub and brake – wire-spoked wheels

1	Spindle	9	Brake shoe – 2 off	17	Cable insert	25	Wear indicator pointer
2	Washer □	10	Return spring – 2 off	18	Washer	26	Nut
3	Spacer	11	Oil seal	19	Speedometer driven gear	27	Brake operating arm
4	Oil seal	12	Sealing cap △	20	Washer	28	Bolt
5	Right-hand bearing	13	Nut △	21	Roll pin	29	Washer □
6	Wheel hub	14	Drive plate	22	Brake backplate	30	Nut □
7	Spacer	15	Brake cam	23	Split pin △		□ KE and KC100 models
8	Left-hand bearing	16	Speedometer drive gear	24	O-ring		△ KH100 A model

3 Rear wheel: removal and refitting

1 On KC and KH100 models, place the machine on its centre stand. On KE100 models, lift the machine on to a strong wooden box or similar support so that the machine is securely supported with the rear wheel clear of the ground.

2 Pull the split pin (where fitted) from the spindle nut and remove the nut. Remove the split pin or R-clip, nut, and spring washer which retain the torque arm to the brake backplate and detach the arm.

3 Remove the nut from the brake rod end. Depress the brake pedal so the rod leaves the trunnion of the operating arm. Retain the nut, trunnion, spring and washer to prevent loss.

4 On KH100 models, lay a length of clean rag beneath the final drive chain and rotate the chain adjusters so that the wheel can be pushed forward and the chain lifted off the sprocket and looped over the swinging arm fork end. Support the wheel and pull out the spindle. If necessary, use a soft-metal drift and hammer to tap it from position. Remove the wheel, noting the fitted position of the adjusters and spacers.

5 On all other models pull out the spindle, displace the hub right-hand spacer, pull the wheel to the right to disconnect it from the cush drive vanes and withdraw it from the machine. If necessary, use a hammer and soft metal drift to tap out the spindle. Note that on some machines it may be necessary to slacken the sprocket carrier sleeve nut to provide extra working clearance so that the wheel can be manoeuvred out; in normal circumstances, however, there should be no need for this and the chain adjustment can remain undisturbed.

6 On KH100 models check that both spacers are correctly fitted before lifting the wheel into position and engaging the chain on the sprocket. Grease the spindle before insertion and tap lightly to seat it. Fit the retaining nut, finger-tight.

7 On all other models spray the cush drive rubber block with WD 40 or a similar lubricant to ease the operation, then insert the wheel and brake backplate into the swinging arm. Work it on the sprocket carrier vanes, then refit the spacer. Thoroughly clean and grease the spindle and refit it, complete with the right-hand chain adjuster and plain washer. Refit the spindle nut.

8 On all models, reconnect the brake rod and torque arm. Renew the torque arm retaining washer if flattened, tighten the nut to the specified torque setting and fit a new split pin, where applicable.

9 Check the chain (if disturbed) rear brake and stop lamp rear switch adjustments as described in Routine Maintenance. Do not forget to tighten the spindle nut to the specified torque setting (where available), and, on KH100 models only, to secure it with a new split pin.

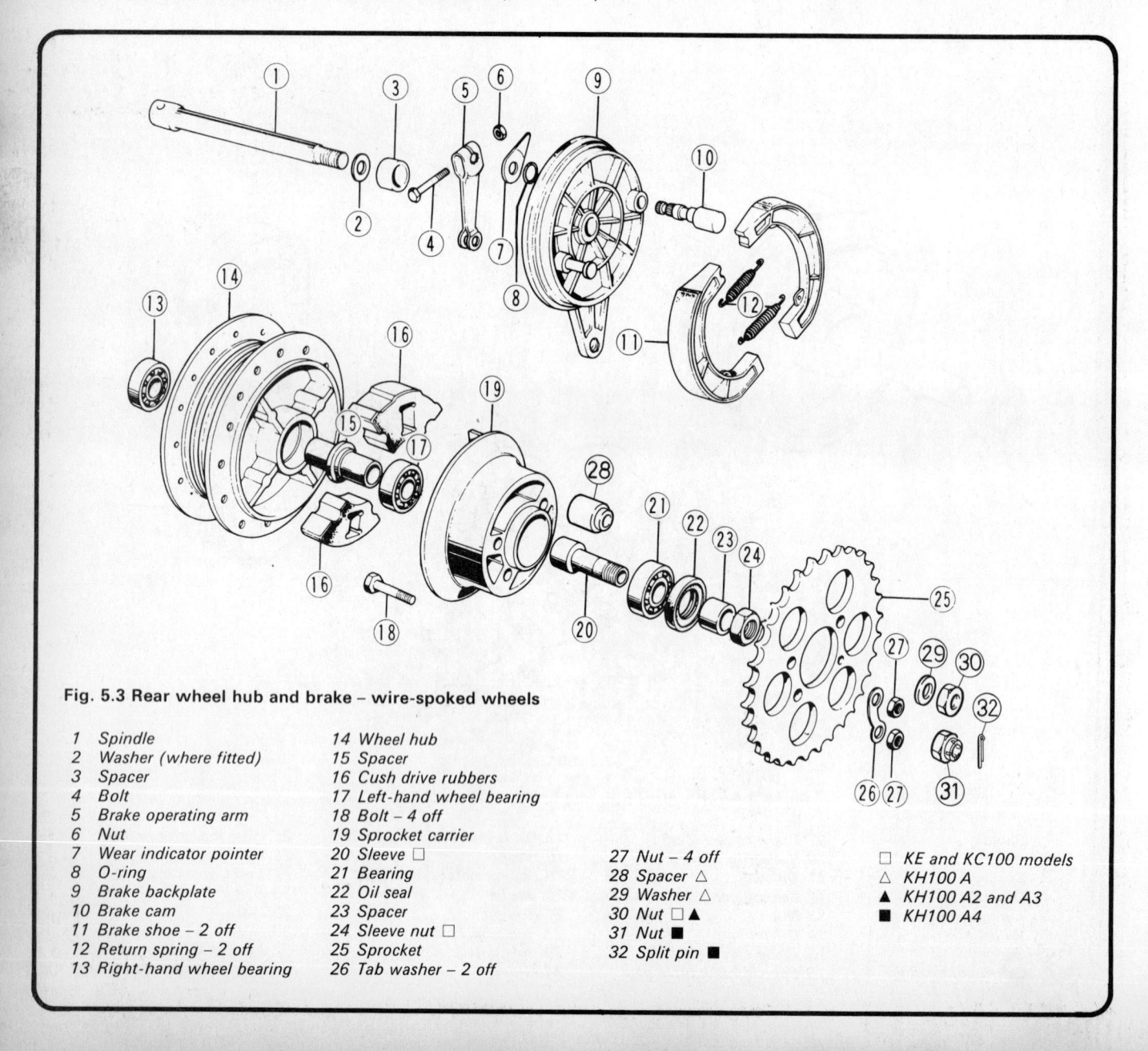

Fig. 5.3 Rear wheel hub and brake – wire-spoked wheels

1 Spindle
2 Washer (where fitted)
3 Spacer
4 Bolt
5 Brake operating arm
6 Nut
7 Wear indicator pointer
8 O-ring
9 Brake backplate
10 Brake cam
11 Brake shoe – 2 off
12 Return spring – 2 off
13 Right-hand wheel bearing
14 Wheel hub
15 Spacer
16 Cush drive rubbers
17 Left-hand wheel bearing
18 Bolt – 4 off
19 Sprocket carrier
20 Sleeve □
21 Bearing
22 Oil seal
23 Spacer
24 Sleeve nut □
25 Sprocket
26 Tab washer – 2 off
27 Nut – 4 off
28 Spacer △
29 Washer △
30 Nut □ ▲
31 Nut ■
32 Split pin ■

□ KE and KC100 models
△ KH100 A
▲ KH100 A2 and A3
■ KH100 A4

3.2a Remove split pin (where fitted) and unscrew spindle nut

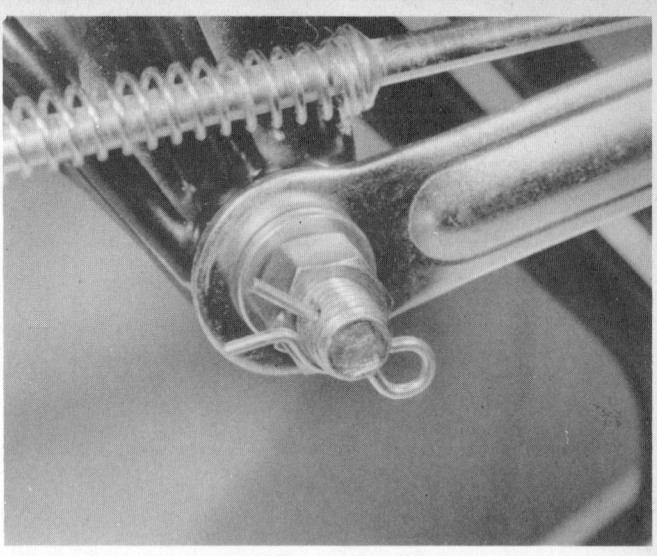

3.2b Remove split pin or R-clip and disconnect brake torque arm

3.7 Do not omit chain adjuster when refitting spindle

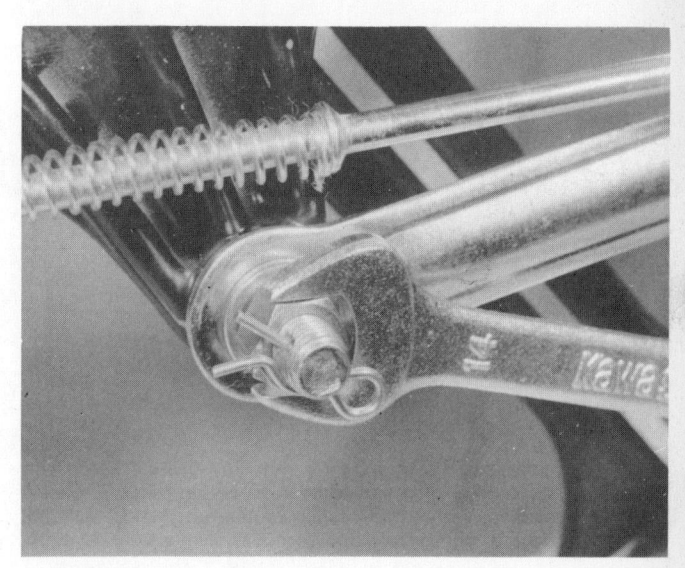

3.8 Tighten torque arm nut securely and refit R-clip or split pin

3.9 Tighten spindle nut securely – where fitted, split pin should be secured as shown

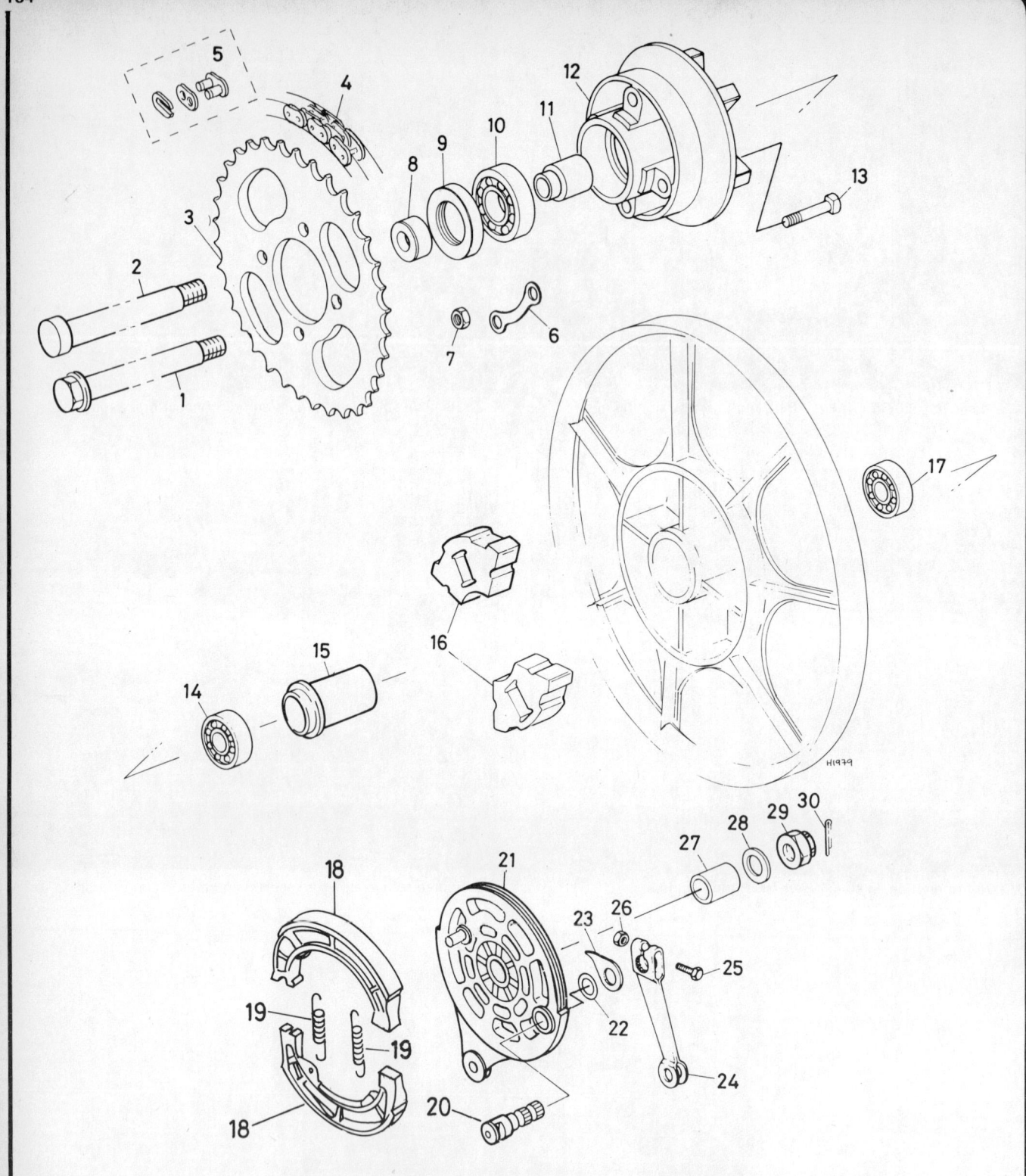

Fig. 5.4 Rear wheel – KH100 G2, G3 and G4 models

1 Spindle – KH100 G2 and G3	9 Oil seal	17 Right-hand bearing	25 Bolt
2 Spindle – KH100 G4	10 Bearing	18 Brake shoe – 2 off	26 Nut
3 Sprocket	11 Spacer	19 Return spring – 2 off	27 Spacer
4 Final drive chain	12 Sprocket carrier	20 Brake cam	28 Washer
5 Joining link	13 Bolt – 4 off	21 Brake backplate	29 Nut
6 Tab washer – 2 off	14 Left-hand bearing	22 O-ring	30 Split pin
7 Nut – 4 off	15 Spacer	23 Wear indicator pointer	
8 Spacer	16 Cush drive rubbers	24 Brake operating arm	

Chapter 5 Wheels, brakes and tyres 135

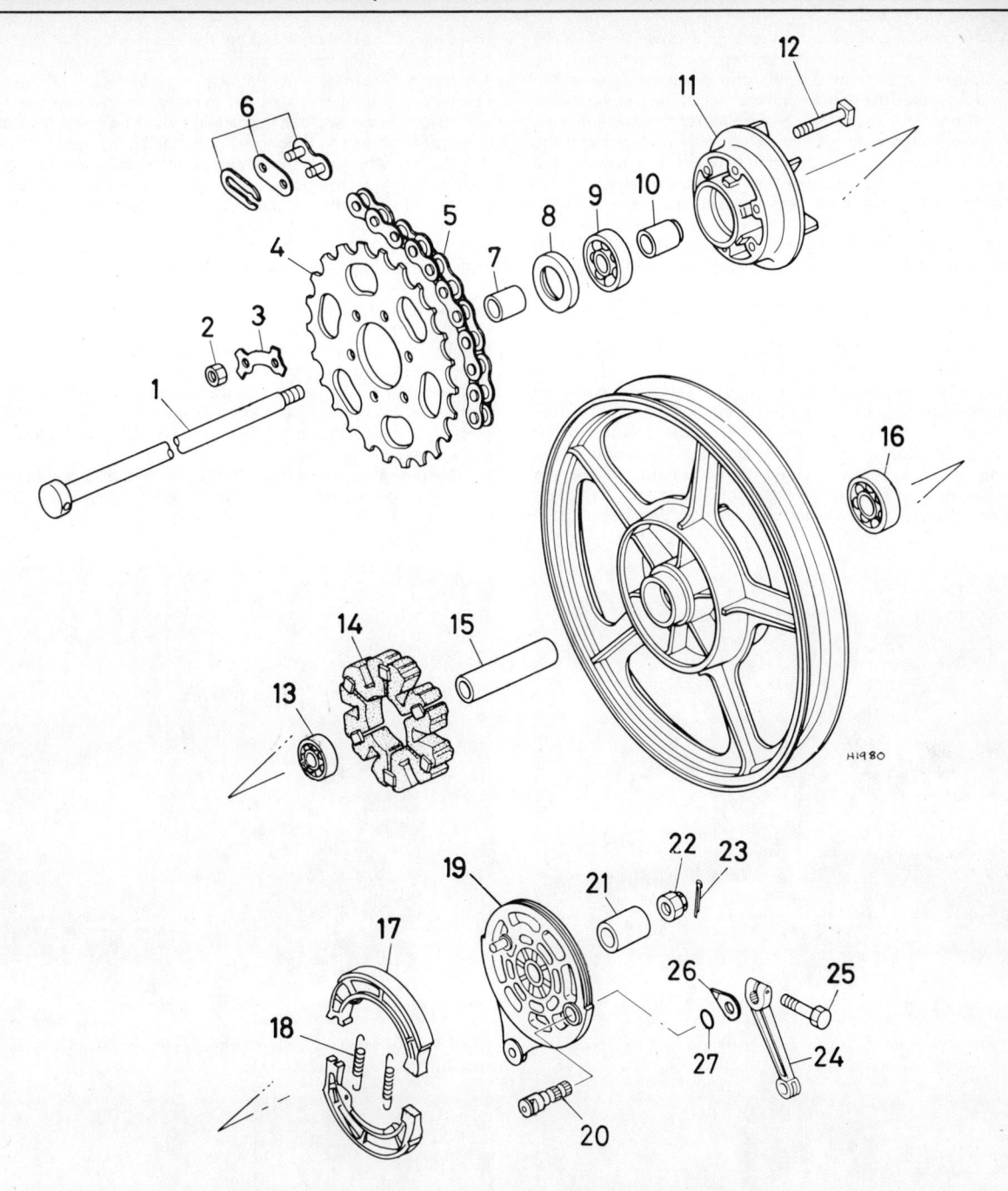

Fig. 5.5 Rear wheel – KH100 G5, G6, G7 and G8 models

1 Spindle	8 Oil seal	15 Spacer	22 Nut
2 Nut – 6 off	9 Bearing	16 Right-hand wheel bearing	23 Split pin
3 Tab washer – 3 off	10 Spacer	17 Brake shoe – 2 off	24 Brake operating arm
4 Sprocket	11 Sprocket carrier	18 Return spring – 2 off	25 Bolt
5 Final drive chain	12 Bolt – 6 off	19 Brake backplate	26 Wear indicator pointer
6 Joining link	13 Left-hand wheel bearing	20 Brake cam	27 O-ring
7 Spacer	14 Cush drive rubbers	21 Spacer	

4 Wheel bearings: removal, examination and refitting

1 Referring to Sections 2 or 3 of this Chapter, as appropriate, remove the wheel from the machine. On the front wheel, remove the hub spacer and the brake backplate or speedometer gearbox, as appropriate. Also, on KH100 G2, G3 and G4 models, carefully lever out the oil seal and remove from the hub left-hand side the circlip. On the rear wheel, remove the sprocket carrier (KH100 only) and the brake backplate.

2 Position the wheel on a work surface with its hub well supported by wooden blocks so that enough clearance is left beneath the wheel to drive out the bearing; place the wheel with the brake side (disc or drum) upwards. Ensure the blocks are placed as close to the bearing as possible, to lessen the risk of distortion of the hub casting whilst the bearings are being removed or refitted.

3 Place the end of a long-handled drift against the upper face of the

first bearing to be removed and tap the bearing downwards out of the wheel hub. The spacer located between the two bearings may be moved sideways slightly to allow the drift to be positioned against the face of the bearing. Move the drift around the face of the bearing whilst drifting it out of position, so that the bearing leaves the hub squarely.

4 With the one bearing removed, the wheel may be lifted and the spacer withdrawn from the hub. Invert the wheel and remove the second bearing, using a similar procedure. The dust seal which fits against the right-hand (front wheel) bearing will be driven out as the bearing is removed. This seal should be closely inspected for any indication of damage, hardening or perishing and renewed if necessary. It is advisable to renew this seal as a matter of course if the bearings are found to be defective.

5 Remove all the old grease from the hub and bearings, giving the latter a final wash in petrol. Check the bearings for signs of play or roughness when they are turned. If there is any doubt about the condition of a bearing, it should be renewed.

6 If the original bearings are to be refitted, they should be repacked with the recommended grease before being fitted into the hub. New bearings must also be packed with the recommended grease. Ensure that the bearing recesses in the hub are clean and the bearings and recess mating surfaces lightly greased to aid fitting. Check the condition of the hub recesses for evidence of abnormal wear which may have been caused by the outer race of a bearing spinning. If evidence of this is found, and the bearing is a loose fit in the hub, it is best to seek advice from an authorised Kawasaki dealer or a competent motorcycle engineer. Alternatively, a proprietary product such as Loctite Bearing Fit may be used to retain the bearing outer race; this will mean, however, that the bearing housing must be cleaned and degreased before the locking compound can be used.

7 With the wheel hub and bearing thus prepared, fit the bearings and central spacer as follows. With the hub again well supported by the wooden blocks, drift the first of the two bearings into position. Use a soft-faced hammer in conjunction with a socket or length of metal tube which has an overall diameter which is slightly less than that of the outer race of the bearing, but which does not bear at any point on the bearing sealed surface or inner race. Tap the bearing into place against the locating shoulder machined in the hub, remembering that the sealed surface of the bearing must always face outwards. With the first bearing in place, invert the wheel, insert the central spacer with its flange on the right and pack the hub centre no more than $2/3$ full with high melting-point grease. Fit the second bearing, using the same procedure. Take great care to ensure that each of the bearings enters its housing correctly, that is, square to the housing, otherwise the housing surface may be broached.

8 Use the same method to refit the seal and do not forget to refit all disturbed components to the front wheel. Refit the wheel to the machine.

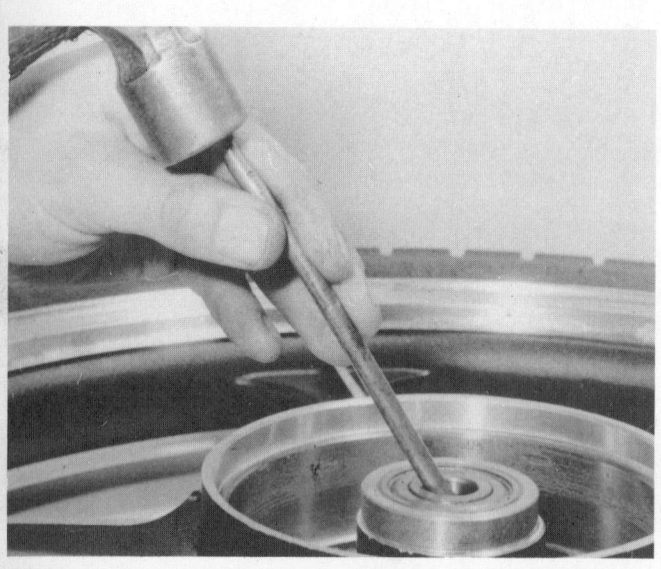

4.3 Starting from brake drum or disc side, use long drift to tap out first bearing

4.7a Fit first bearing then turn wheel over, fit central spacer and pack hub with grease

4.7b Bearings must always be refitted with sealed surface outwards ...

4.7c ... and must be tapped into place as shown

Chapter 5 Wheels, brakes and tyres

5 Rear wheel cush drive: examination and renovation

1 Remove the rear wheel from the machine as described in Section 3 of this Chapter. On KH100 models withdraw the sprocket carrier from the wheel and remove both spacers from the carrier. On all other models, unscrew the sleeve nut, remove the chain adjuster, unhook the chain from the sprocket and withdraw the carrier assembly from the machine; note that this will mean that the chaincase on KC100 models must be dismantled. Remove the single rear clamp bolt and nut, withdraw the six mounting bolts and separate both halves of the chaincase. Remove the left-hand spacer and tap out to the right the carrier sleeve.

2 Depending on the model, the sprocket is retained by six or four bolts; these are easiest to slacken while the assembly is on the wheel. Flatten the raised edges of the tab washers and slacken the retaining nuts, then remove the carrier assembly from the wheel. Remove the nuts (and tab washers) and withdraw the sprocket, then lever out the oil seal taking care not to scratch or damage the casing.

3 Remove the bearing, wash it and check it for wear as described in Section 4 of this Chapter, then repack it with grease and refit it. The oil seal should be renewed whenever it is disturbed.

4 If the rear sprocket teeth are hooked, chipped, missing or worn, the sprocket must be renewed, but this should be done only in conjunction with a new gearbox sprocket and chain. Refit the sprocket on the carrier, followed by the tab washers (where fitted) and the nuts, applying thread locking compound to their threads. Tighten the nuts to the recommended torque setting and secure each nut by bending up an unused portion of the tab washer (where fitted). Note that the tab washers should be renewed as soon as all their locking tabs have been used once. Insert the carrier sleeve and/or spacers into each side of the carrier assembly.

5 If the cush drive rubber block is perished, split, damaged or compressed to the extent that there is excessive movement between the sprocket carrier and wheel, it must be renewed. It can be pulled out of the hub by hand, but if a new one is a tight fit it should be lubricated using a very small amount of soapy water.

5.1a KH100 – remove sprocket carrier from wheel and withdraw left-hand spacer ...

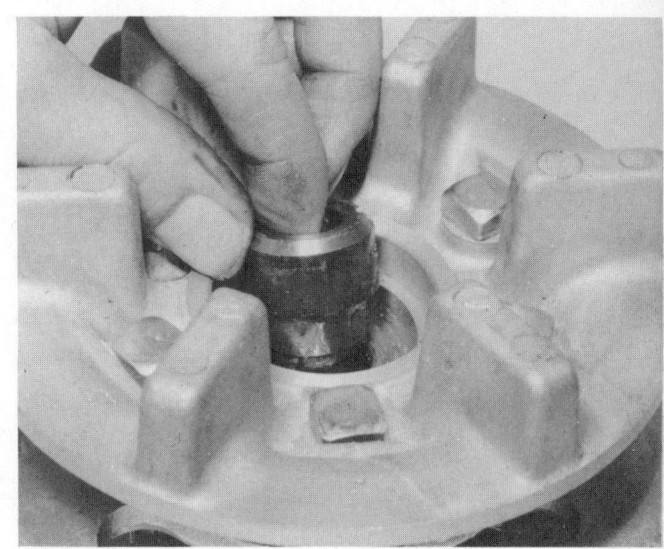

5.1b ... followed by right-hand spacer (carrier sleeve on all other models)

5.4 Tighten sprocket nuts securely and bend up locking tabs to secure

5.5 Check condition of cush drive block(s) and renew if necessary

6 Brake master cylinder: removal, overhaul and refittiing

1 Connect a length of tubing from the front brake caliper bleed nipple to a suitable container, unscrew the bleed nipple by 1 – 2 full turns and slowly pump the brake lever until all fluid is expelled. Tighten the bleed nipple and repeat the procedure on all remaining nipples in the system, where applicable. Remove the right hand mirror, then free the front brake light switch (KH100 G4) by depressing its locking pin, using a small electrical screwdriver via the hole in the underside of the switch housing; on KH100 G5, G6, G7 and G8 models the switch is a rectangular unit retained by a single screw.
2 Pull back the banjo union dust cover and remove the union bolt and washers. Wipe up any residual hydraulic fluid. Slacken the two master cylinder clamp bolts, remove the clamp half and lift the master cylinder away.
3 Remove the screws which retain the reservoir cover, remove the cover and diaphragm and empty out any remaining fluid. Remove the brake lever locknut and pivot bolt and remove the lever. On KH100 G2 and G3 models, remove the plastic liner by pushing in the locking tabs which retain it. On all other models displace the rubber dust cover and release the circlip beneath it. Pull out the piston assembly, separating the spring and, where applicable, the dust seal and piston stop from the piston.
4 Examine the piston surface and master cylinder bore for signs of wear or corrosion. Renew both components if damaged in any way; new seals will not compensate for scoring and will wear out quickly. Check the primary and secondary seals for damage or swelling renewing them unless in perfect condition. The cups are sold as a kit together with the piston and spring. Renew the dust seal at the same time to preclude road dirt entering the caliper body. Ensure that the supply port and the smaller relief port between the cylinder and reservoir are clear, especially where swollen or damaged cups have been noted.
5 In most cases, the components can be measured for wear if required, and the readings checked against the figures given in Specifications; renew any component that is found to be worn at any point to the specified wear limits or beyond.
6 Reassemble the master cylinder by reversing the dismantling sequence, ensuring that all components are kept spotlessly clean. Lubricate the piston, cups and cylinder bore with new hydraulic fluid during installation. Use new sealing washers on the banjo unions and tighten the union bolts to 3.0 kgf m (22 lbf ft). Refill and bleed the hydraulic system following the procedure described in Section 10 of this Chapter. Check brake operation before using the machine.

1 Pivot bolt
2 Brake lever
3 Nut
4 Union bolt
5 Sealing washer – 2 off
6 Master cylinder body
7 Brake hose
8 Diaphragm
9 Spring
10 Primary cup
11 Secondary cup
12 Piston
13 Circlip
14 Dust cover
15 Front brake light switch □
16 Bolt and washer – 2 off
17 Reservoir cover
18 Screw – 2 off
19 Dust cover
20 Front brake light switch △
21 Screw △

□ KH100 G4 model
△ KH100 G5, G6, G7 and G8 model

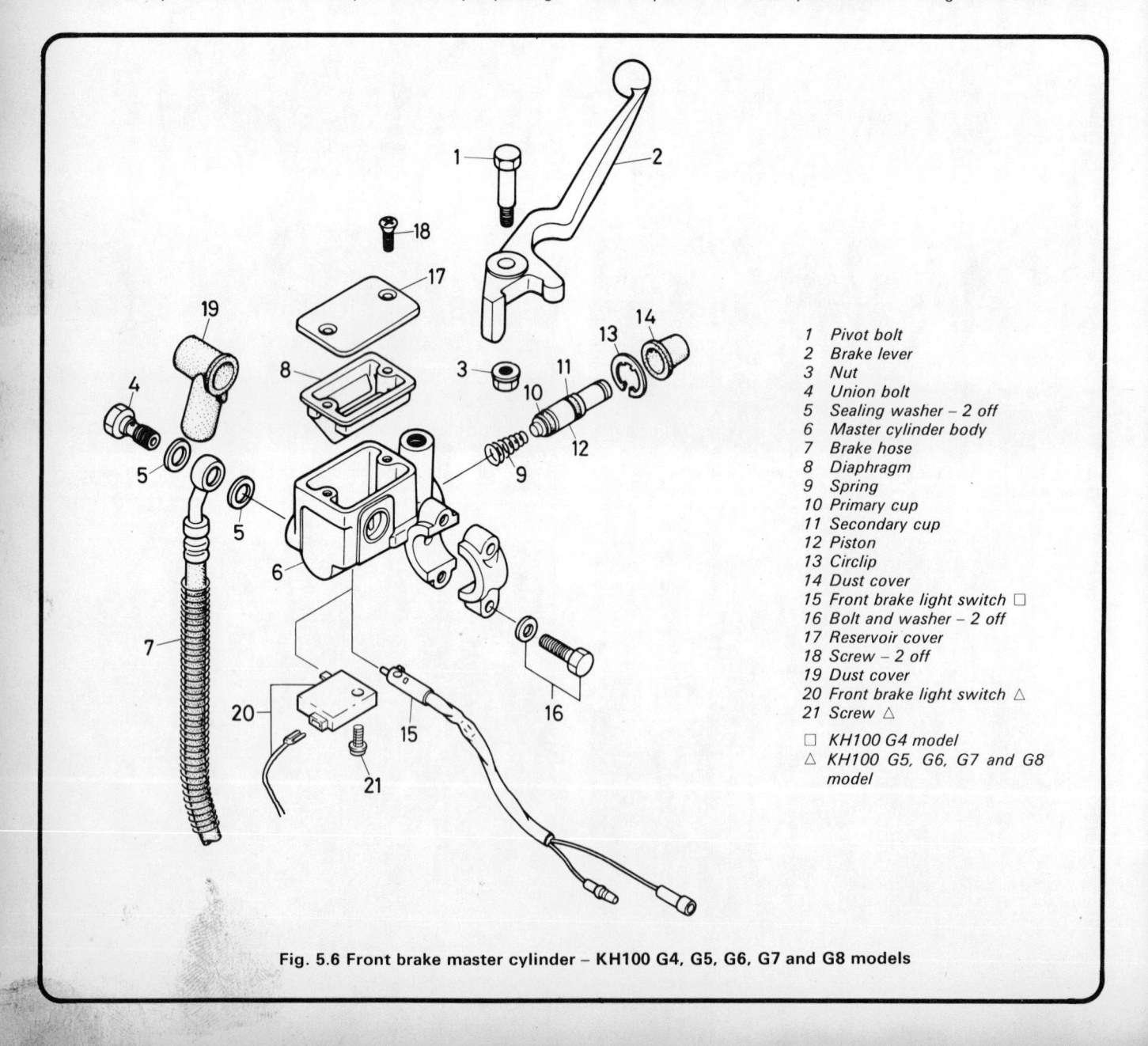

Fig. 5.6 Front brake master cylinder – KH100 G4, G5, G6, G7 and G8 models

Chapter 5 Wheels, brakes and tyres

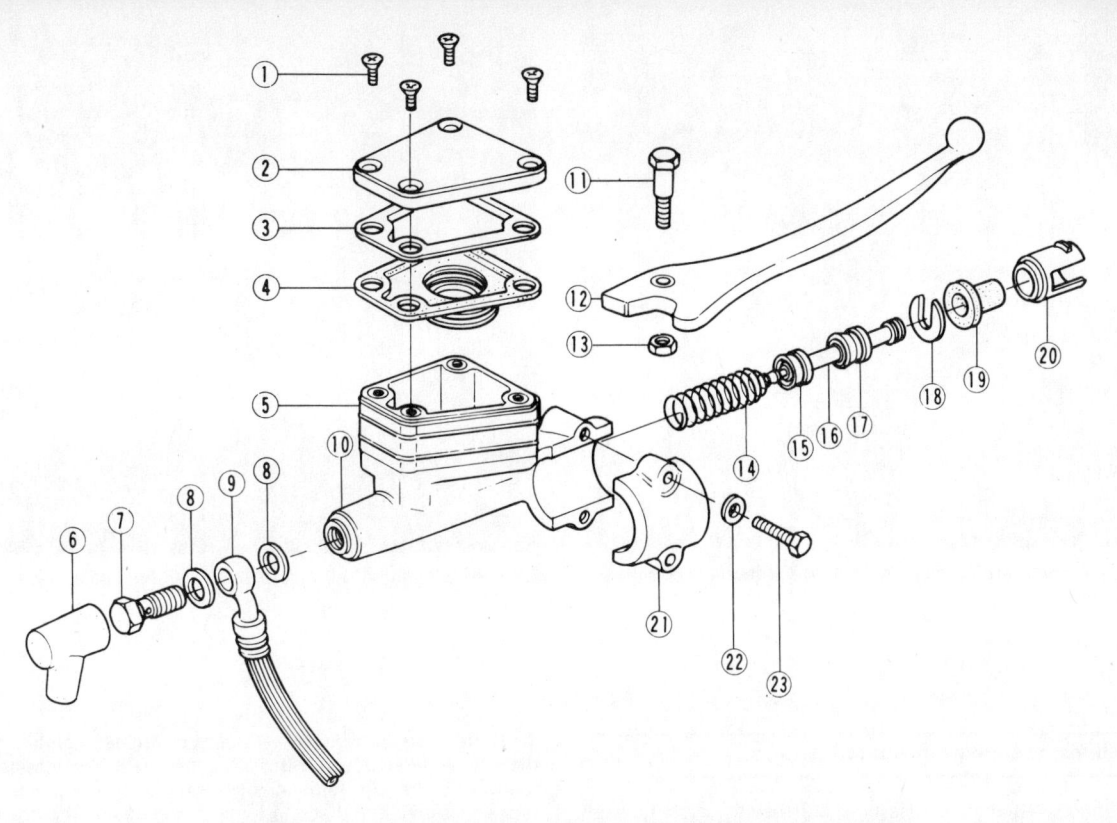

Fig. 5.7 Front brake master cylinder – KH100 G2 and G3 models

1 Screw – 4 off
2 Reservoir cover
3 Diaphragm plate
4 Diaphragm
5 Master cylinder reservoir
6 Dust cover
7 Union bolt
8 Sealing washer – 2 off
9 Brake hose
10 Master cylinder body
11 Pivot bolt
12 Brake lever
13 Nut
14 Spring
15 Primary cup
16 Piston
17 Secondary cup
18 Piston stop
19 Dust seal
20 Plastic liner
21 Handlebar clamp
22 Washer – 2 off
23 Bolt – 2 off

6.1 Remove mirror and handlebar clamp bolts to release master cylinder – note clamp 'Up' mark and refit accordingly

7 Hydraulic hoses: examination

Whenever the brake assembly is overhauled, check the condition of the hose for signs of leakage or scuffing. The union connections at each end must also be in good condition, with no stripped threads or damaged sealing washers. Renew immediately any hose found to be worn or damaged, referring to the torque settings given in the Specifications Section of this Chapter. Also renew any sealing washers whenever they are disturbed. Note that the manufacturer specifies that the brake hoses must be renewed as a matter of course at regular intervals. See Routine Maintenance.

8 Brake disc: examination and renovation

1 The disc can be examined while in place on the machine. If it is heavily scored or worn at any point to less than the limit specified, it must be renewed. If signs of warpage can be seen by the naked eye on spinning the front wheel, a dial gauge must be obtained and mounted on the fork lower leg to measure accurately the amount of distortion. If this proves excessive, the disc must be renewed. The only alternative is to find an engineering company who will skim the disc, but this must not reduce it to less than the minimum thickness specified, or braking efficiency will be greatly reduced, quite apart from the risk of sudden brake failure.

2 If the disc is to be renewed, remove the front wheel, and unscrew the disc mounting bolts or nuts. On reassembly on G2, G3 and G4 models tighten the disc retaining nuts to a torque setting of 1.0 kgf m (7 lbf ft) and then use a centre punch to stake the threads of each nut and bolt to prevent slackening. On G5, G6, G7 and G8 models tighten the mounting bolts to a torque setting of 2.0 kgf m (14.5 lbf ft).

3 Remember that the disc and hub mating surfaces must be absolutely clean whenever the disc is disturbed.

140   Chapter 5 Wheels, brakes and tyres

7.1 Ensure brake hoses are correctly routed and properly secured by clamps at all times

8.2 Disc is retained by nuts or bolts – tighten securely, as described

9 Brake caliper: examination and renovation

1 The procedures required to dismantle and rebuild each type of caliper are given below under separate headings, with general notes applicable to both types given at the end of the Section. In both cases, commence work by removing the brake pads as described in Routine Maintenance. The simplest way of removing the piston is to place the caliper in a plastic bag, tying its neck around the brake hose to prevent a shower of fluid, then pump the brake lever to expel the piston by hydraulic pressue.

2 If the above method does not displace the piston, waste no further time on the caliper; it must be renewed as a complete assembly as it is too badly damaged or corroded to be of any further use. If the hydraulic system has been drained or disconnected, apply a jet of compressed air to the fluid passage or bleed nipple orifice (having removed the nipple) to displace the piston, but wrap the caliper first in a thick layer of rag to prevent the piston flying out.

KH100 G2, G3

3 With the caliper removed from the fork leg and the piston expelled as described above, carefully displace the dust seal and piston seal.

4 Straighten the lock washer tab that is bent over the stopper clip, withdraw the clip and the lockwasher and slide the caliper body off its pivot, noting the presence of the two sealing O-rings. Discard the lock washer; this must be renewed whenever it is disturbed. Check for wear at the pivot as described below.

5 On reassembly, refit the seals and piston as described below. Fitting new sealing O-rings and greasing the pivot bearing surfaces as described below, refit the caliper body to the mounting bracket/pivot. Fit the second O-ring, a **new** lock washer and the stopper clip. Slide the clip into the pivot groove and rotate it so that its open end fits around the raised tang of the lock washer. Secure the clip by bending the lock washer tab right over it so that it cannot move. Refit the pads to the caliper and the caliper to the machine, then fill the system with fresh hydraulic fluid and bleed the system to remove all traces of air as described in the following Section.

All later models

6 Pushing the mounting bracket away from the piston, remove it from the caliper, then displace the two rubber dust covers and the anti-rattle spring. Displace the piston cap and expel the piston, as described above, then remove the piston seals and disconnect the brake hose.

7 On reassembly, refit the seals and piston as described below, press the rubber dust covers into place and refit the mounting bracket, followed by the anti-rattle spring. Using new sealing washers, refit the brake hose ensuring that it is correctly routed and that the banjo bolt is tightened to a torque setting of 3.0 kgf m (22 lbf ft). Refit the pads to the caliper and the caliper to the machine then fill the system with fresh hydraulic fluid and carry out the bleeding procedure to remove all traces of air.

Caliper overhaul – general

8 Before any repair work is undertaken, check with a local Kawasaki dealer to establish what is available for the caliper to be worked on. In some cases nearly all components are individually available, in other cases only complete assemblies can be obtained as replacement parts.

9 Clean all components carefully, removing all traces of road dirt, friction material and corrosion. Note that only clean hydraulic fluid (or ethyl or isopropyl alcohol) should be used to clean hydraulic components; petrol, paraffin or other normal cleaning solvents will attack the rubber seals. It is permissible to use a wire brush gently to remove dirt and corrosion except in the caliper bores and piston skirt.

10 Renew all piston seals (both fluid and dust seals) and all sealing O-rings as a matter of course. Never re-use a hydraulic seal after it has been disturbed and note that the piston seal must be in excellent condition as its secondary role is to return the piston when lever or pedal pressure is released, thus preventing brake drag. Carefully examine the axle bolt dust covers and any other rubber seals, renewing any that are perished, split or otherwise damaged. Similarly, discard the sealing washers fitted at the brake hose union; these should be renewed as a matter of course.

11 Examine the piston surface and caliper bore for signs of wear or scoring, normally caused by the presence of road dirt or corrosion. If wear is found, or deep scoring or scratches which might cause fluid leaks, the component concerned must be renewed. Where measurements are given in the Specifications Section of this Chapter, check that neither has worn to beyond the service limits.

12 Where applicable, check that there is no free play between the caliper body and its axle shafts or mounting bracket pivot (as appropriate). Renew any component that is found to be worn. As mentioned in Routine Maintenance, it is essential that single-piston brake calipers can move smoothly on their mountings. Make a final check that there are no signs of damage on any other part of the caliper assembly.

Chapter 5 Wheels, brakes and tyres

13 On reassembly, soak the new piston (fluid) seal in clean hydraulic fluid then refit it to the caliper bore, taking great care that it is seated correctly in its groove and that the bore is not scratched. Smear hydraulic fluid over the caliper bore and piston surface and refit the piston, rotating it slightly while keeping it square to the caliper bore so that it does not stick or displace the piston seal. Press the piston fully into the caliper, wipe off any surplus fluid, then refit the new dust seal ensuring that it locates correctly on the caliper lip.

14 Where applicable, apply PBC (Poly Butyl Cuprysil) grease to all sliding surfaces on the caliper body and mounting bracket or on the axle shafts, and pack grease into the recesses in the mounting bracket or caliper body. Be careful to wipe off all surplus grease once the caliper assembly is rebuilt. Check that the caliper body moves smoothly and fully from side to side before refitting the pads.

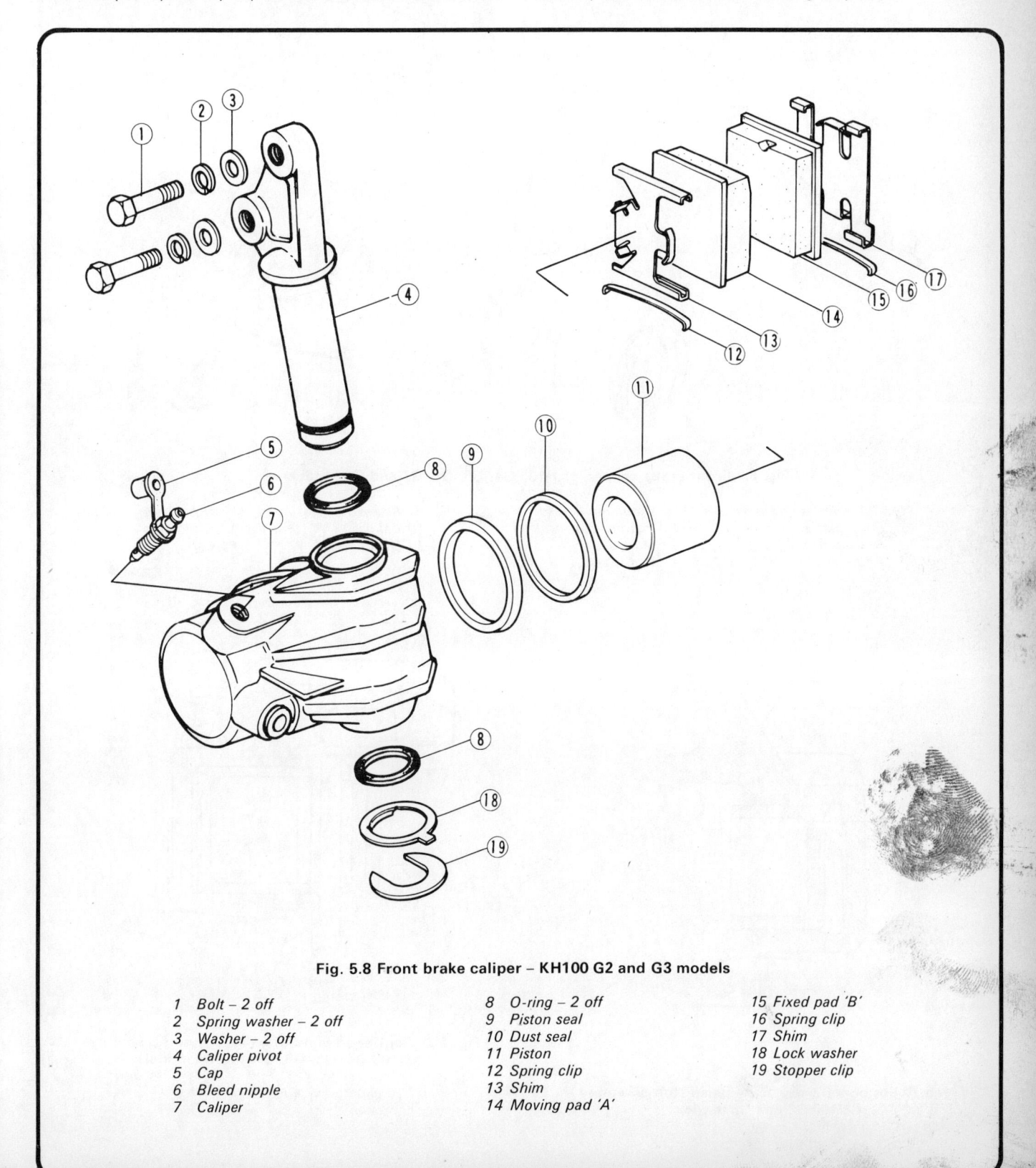

Fig. 5.8 Front brake caliper – KH100 G2 and G3 models

1 Bolt – 2 off
2 Spring washer – 2 off
3 Washer – 2 off
4 Caliper pivot
5 Cap
6 Bleed nipple
7 Caliper
8 O-ring – 2 off
9 Piston seal
10 Dust seal
11 Piston
12 Spring clip
13 Shim
14 Moving pad 'A'
15 Fixed pad 'B'
16 Spring clip
17 Shim
18 Lock washer
19 Stopper clip

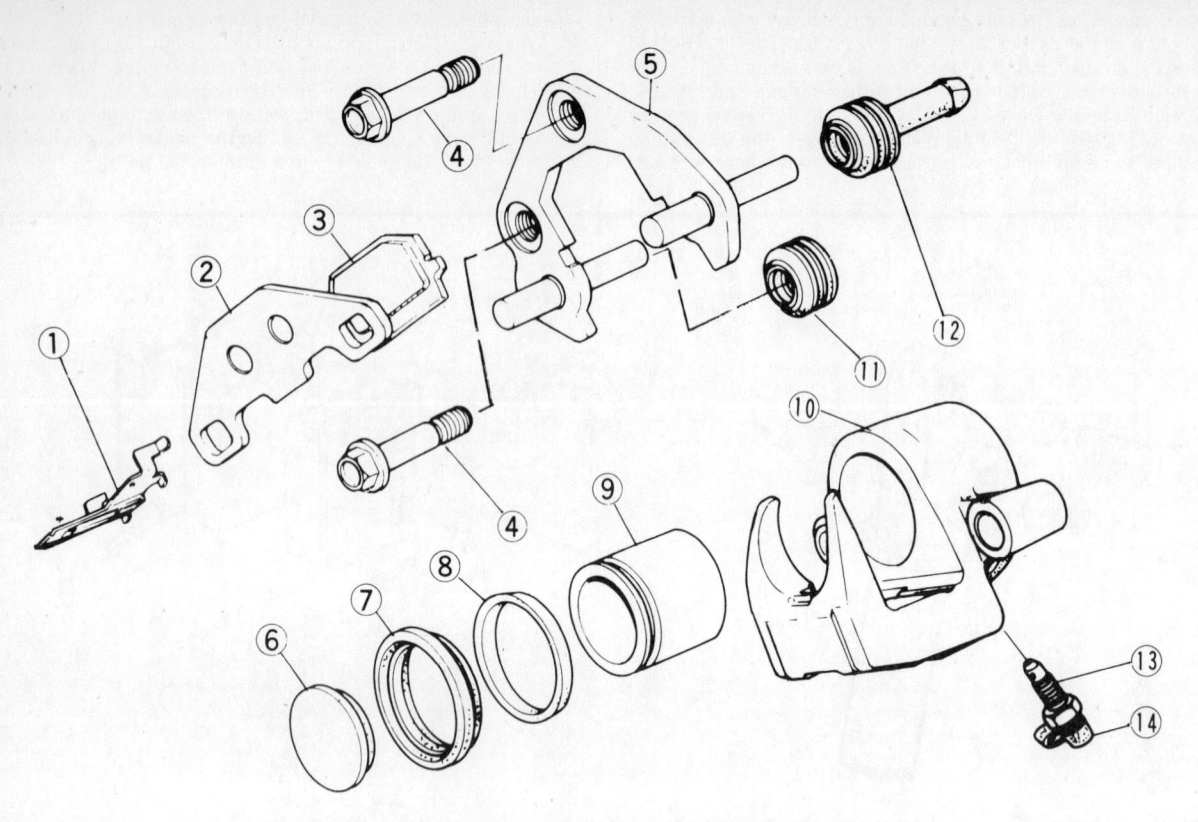

Fig. 5.9 Front brake caliper – KH100 G4, G5, G6, G7 and G8 models

1 Anti-rattle spring	5 Mounting bracket	9 Piston	12 Dust cover
2 Fixed pad 'B'	6 Piston head/insulator	10 Caliper	13 Bleed nipple
3 Moving pad 'A'	7 Dust seal	11 Dust cover	14 Cap
4 Bolt – 2 off	8 Piston seal		

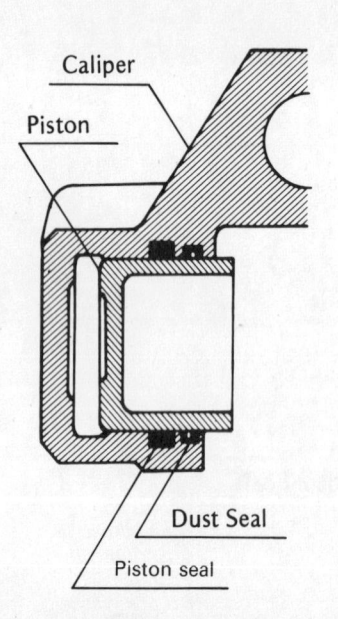

Fig. 5.10 Sectioned view of caliper piston assembly – KH100 G2 and G3 models

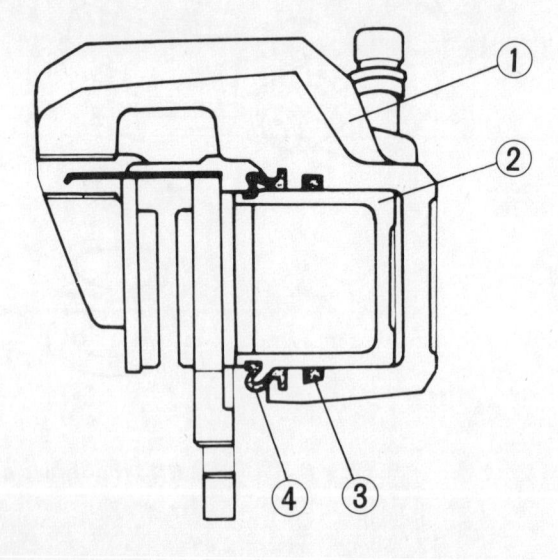

Fig. 5.11 Sectioned view of caliper piston assembly – KH100 G4, G5, G6, G7 and G8 models

1 Caliper
2 Piston
3 Piston seal
4 Dust seal

10 Bleeding the hydraulic brake system

1 If the brake action becomes spongy, or if any part of the hydraulic system is dismantled, it is necessary to bleed the system to remove all traces of air. The procedure for bleeding the hydraulic system is best carried out by two people.
2 Check the fluid level in the reservoir and top up with new fluid of the specified type if required. Keep the reservoir at least half full during the bleeding procedure; if the level is allowed to fall too far air will enter the system requiring that the procedure be started again from scratch. Refit the reservoir cover to prevent the ingress of dust or the ejection of a spout of fluid.
3 Remove the dust cap from the caliper bleed nipple and clean the area with a rag. Place a clean glass jar below the caliper and connect a pipe from the bleed nipple to the jar. A clear plastic tube should be used so that air bubbles can be more easily seen. Place some clean hydraulic fluid in the glass jar so that the pipe is immersed below the fluid surface throughout the operation.
4 If the system has to be filled, open the bleed nipple about one turn and pump the brake lever until fluid starts to issue from the clear tube. Tighten the bleed nipple and then continue the normal bleeding operation as described in the following paragraphs. Keep a close check on the reservoir level whilst the system is being filled.
5 Operate the brake lever as far as it will go and hold it in this position against the fluid pressure. If spongy brake operation has occurred, it may be necessary to pump the brake lever rapidly a number of times until pressure is achieved. With pressure applied, loosen the bleed nipple about half a turn. Tighten the nipple as soon as the lever has reached its full travel and then release the lever. Repeat this operation until no more air bubbles are expelled with the fluid into the glass jar. When this condition is reached, the air bleeding operation should be complete, resulting in a firm feel to the brake operation. If sponginess is still evident, continue the bleeding operation; it may be that an air bubble trapped at the top of the system has yet to work down through the caliper.
6 When all traces of air have been removed from the system, top up the reservoir and refit the diaphragm and cover. Check the entire system for leaks, and check also that the brake system in general is functioning efficiently before using the machine on the road.
7 Brake fluid drained from the system will almost certainly be contaminated, either by foreign matter or more commonly by the absorption of water from the air. All hydraulic fluids are to some degree hygroscopic, that is, they are capable of drawing water from the atmosphere which degrades their specification. In view of this, and the relative cheapness of the fluid, old fluid should always be discarded.
8 Great care should be taken not to spill hydraulic fluid on any painted cycle parts; it is a very effective paint stripper. Also, the plastic glasses in the instrument heads, and most other plastic parts, will be damaged by contact with this fluid.

10.3 Brake bleeding equipment ready for use

11 Drum brake: examination and renovation

1 The brake assembly can be withdrawn from its hub after removal of the wheel from the machine.
2 Examine the condition of the brake linings. If they are worn beyond the specified limit, the brake shoes should be renewed. The linings are bonded on and cannot be supplied separately.
3 If oil or grease from the wheel bearings has badly contaminated the linings, the brake shoes should be renewed. There is no satisfactory way of degreasing the lining material. Any surface dirt on the linings can be removed with a stiff-bristled brush. High spots on the linings should be carefully eased down with emery cloth.
4 Examine the drum surface for signs of scoring, wear beyond the service limit or oil contamination. All of these conditions will impair braking efficiency. Remove all traces of dust, preferably using a brass wire brush, taking care not to inhale any of it as it is of an asbestos nature, and consequently harmful. Remove oil or grease deposits, using a petrol soaked rag.
5 If deep scoring is evident, due to the linings having worn through to the shoe at some time, the drum must be skimmed on a lathe, or renewed. Whilst there are firms who will skim the drum, it should be borne in mind that excessive skimming will change the radius of the drum in relation to the brake shoes, thereby reducing the friction area until extensive bedding in has taken place. Also, full adjustment of the shoes may not be possible. If in doubt about this point, the advice of one of the specialist engineering firms who undertake this work should be sought.
6 It is false economy to try to cut corners with brake components, the whole safety of both machine and rider being dependent on their good condition.
7 Removal of the brake shoes is accomplished by folding the shoes together so that they form a 'V'. With the spring tension relaxed, both shoes and springs may be removed from the brake backplate as an assembly. Detach the springs from the shoes and carefully inspect them for any signs of fatigue or failure. If in doubt measure their free lengths and renew them if they have stretched to the specified service limit or beyond.
8 Before fitting the brake shoes, check that the brake operating cam is working smoothly and is not binding in its pivot. The cam can be removed by withdrawing the retaining bolt on the operating arm and pulling the arm off the shaft. Before removing the arm, mark its position in relation to the shaft so that it can be located correctly, also that of the wear indicator pointer.
9 Remove any deposits of hardened grease or corrosion from the bearing surface of the brake cam and shoe by rubbing it lightly with a strip of fine emery paper or by applying a solvent with a piece of rag. Check the cam and backplate bearing surfaces for wear by measuring them. Renew any component that is worn to the specified service limit or beyond, and check the sealing O-ring for wear or damage. Lightly grease the length of the shaft and the face of the operating cam prior to reassembly. Clean and grease the pivot stud which is set in the backplate then refit the camshaft and sealing O-ring.
10 Before refitting the original shoes, roughen the lining surface sufficiently to break the glaze which will have formed in use. Glasspaper or emery cloth is ideal for this purpose but take care not to inhale any of the asbestos dust that may come from the lining surface.
11 Fitting the brake shoes and springs to the brake backplate is a reversal of the removal procedure. Some patience will be needed to align the assembly with the pivot and operating cam whilst still retaining the springs in position; once the brake shoes are correctly aligned, they can be pushed back into position by pressing downwards to snap them into position. Do not use excessive force, as there is risk of distorting them permanently.
12 If the original shoes have been refitted, place the wear indicator pointer in its original position. If new shoes have been fitted, place the pointer on the shaft splines so that it aligns with the extreme right-hand (towards the backplate centre) end of the 'Usable Range' arc. The operating arm should be refitted in its original position if the same shoes are being refitted, but if new shoes have been installed, the arm's final position can only be set when in place on the machine. The arm must be set so that the angle between it and the operating cable or rod is between 80° – 90° (ie just less than a right angle) when the brake is firmly applied.
13 When the arm position is correct, tap it fully on to the shaft splines then refit and tighten securely the pinch bolt and nut.

144 Chapter 5 Wheels, brakes and tyres

11.1 Remove wheel from machine and withdraw brake assembly from hub

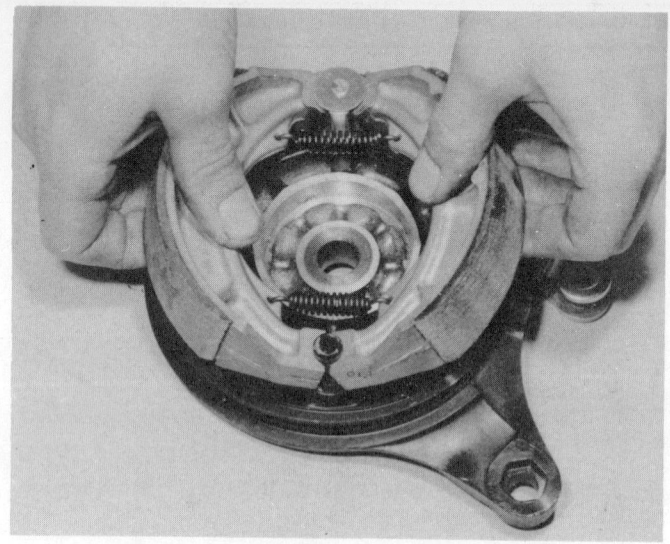

11.7 Removing brake shoes – take care to prevent damage or injury

11.12 If brake shoes are renewed, wear indicator pointer must be refitted to align as shown with 'usable range' arc

12 Tyres: removal, repair and refitting

1 To remove the tyre from either wheel, first detach the wheel from the machine. Deflate the tyre by removing the valve core, and when the tyre is fully deflated, push the bead away from the wheel rim on both sides so that the bead enters the centre well of the rim. Remove the locking ring and push the tyre valve into the tyre itself.
2 Insert a tyre lever close to the valve and lever the edge of the tyre over the outside of the rim. Very little force should be neessary; if resistance is encountered it is probable that the tyre beads have not entered the well of the rim all the way round. Damage to alloy wheel rims by tyre levers can be prevented by the use of plastic rim protectors.
3 Once the tyre has been edged over the wheel rim, it is easy to work round the rim, so that the tyre is completely free from one side. At this stage the inner tube can be removed.
4 Working from the other side of the wheel, ease the other edge of the tyre over the outside of the wheel rim that is furthest away. Continue to work around the rim until the tyre is completely free from the rim.
5 If a puncture has necessitated the removal of the tyre, reinflate the inner tube and immerse it in a bowl of water to trace the source of the leak. Mark the position of the leak, and deflate the tube. Dry the tube, and clean the area around the puncture with a petrol soaked rag. When the surface has dried, apply rubber solution and allow this to dry before removing the backing from the patch, and applying the patch to the surface.
6 It is best to use a patch of self vulcanizing type, which will form a permanent repair. Note that it may be necessary to remove a protective covering from the top surface of the patch after it has sealed into position. Inner tubes made from a special synthetic rubber may require a special type of patch and adhesive, if a satisfactory bond is to be achieved.
7 Before replacing the tyre, check the inside to make sure that the article that caused the puncture is not still trapped inside the tyre. Check the outside of the tyre, particularly the tread area to make sure nothing is trapped that may cause a further puncture.
8 If the inner tube has been patched on a number of past occasions, or if there is a tear or large hole, it is preferable to discard it and fit a replacement. Sudden deflation may cause an accident, particularly if it occurs with the rear wheel.
9 When fitting the tyre note first the arrows indicating the direction of rotation. If only one arrow is found this is to be fitted pointing in the direction of rotation on rear wheels, but *opposite* to the direction of rotation, to accept braking loads, on *front* wheels. Where two arrows (marked 'Front wheel' and 'Rear wheel') are provided, fit the tyre as indicated by them.
10 To ensure correct wheel balance, check the tyre sidewall for a spot of paint. On most German makes of tyre this indicates the lightest point of the tyre (check with the tyre manufacturer, to be safe) and must be fitted next to the valve.
11 To replace the tyre, inflate the inner tube for it just to assume a circular shape but only to that amount, and then push the tube into the tyre so that it is enclosed completely. Lay the tyre on the wheel at an angle, and insert the valve through the rim tape and the hole in the wheel rim. Attach the locking ring on the first few threads, sufficient to hold the valve captive in its correct location.
12 Starting at the point furthest from the valve, push the tyre bead over the edge of the wheel rim until it is located in the centre well. Continue to work around the tyre in this fashion until the whole of one side of the tyre is on the rim. It may be necessary to use a tyre lever during the final stages.
13 Make sure there is no pull on the tyre valve and again commencing with the area furthest from the valve, ease the other bead of the tyre over the edge of the rim. Finish with the area close to the valve, pushing the valve up into the tyre until the locking ring touches the rim. This will ensure that the inner tube is not trapped when the last section of bead is edged over the rim with a tyre lever.

Tyre changing sequence — tubed tyres

A — Deflate tyre. After pushing tyre beads away from rim flanges push tyre bead into well of rim at point opposite valve. Insert tyre lever adjacent to valve and work bead over edge of rim.

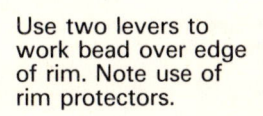

B — Use two levers to work bead over edge of rim. Note use of rim protectors.

C — Remove inner tube from tyre.

D — When first bead is clear, remove tyre as shown.

E — When fitting, partially inflate inner tube and insert in tyre.

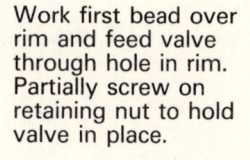

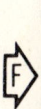

F — Work first bead over rim and feed valve through hole in rim. Partially screw on retaining nut to hold valve in place.

G — Check that inner tube is positioned correctly and work second bead over rim using tyre levers. Start at a point opposite valve.

H — Work final area of bead over rim whilst pushing valve inwards to ensure that inner tube is not trapped.

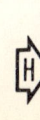

14 Check that the inner tube is not trapped at any point. Reinflate the inner tube, and check that the tyre is seating correctly around the wheel rim. There should be a thin rib moulded around the wall of the tyre on both sides, which should be an equal distance from the wheel rim at all points. If the tyre is unevenly located on the rim, try bouncing the wheel when the tyre is at the recommended pressure. It is probable that one of the beads has not pulled clear of the centre well.

15 Always run the tyres at the recommended pressures and never under or over inflate. The correct pressures for original equipment tyres are given in the Specifications Section of this Chapter; if non-standard tyres are fitted, check with the manufacturer or supplier for recommended pressures.

16 Tyre replacement is aided by dusting the side walls, particularly in the vicinity of the beads, with a liberal coating of french chalk or a proprietary tyre-fitting lubricant.

17 Never replace the inner tube and tyre without the rim tape in position. If this precaution is overlooked there is a good chance of the ends of the spoke nipples chafing the inner tube and causing a crop of punctures.

18 Never fit a tyre that has a damaged tread or sidewalls. Apart from legal aspects, there is a very great risk of a blowout, which can have very serious consequences on a two wheeled vehicle.

13 Valve cores and caps: general

1 Valve cores seldom give trouble, but do not last indefinitely. Dirt under the seating will cause a puzzling 'slow-puncture'. Check that they are not leaking by applying spittle to the end of the valve and watching for air bubbles.

2 A valve cap is a safety device, and should always be fitted. Apart from keeping dirt out of the valve, it provides a second seal in case of valve failure, and may prevent an accident resulting from sudden deflation.

Chapter 6 Electrical system

Contents

General description	1
Electrical system: general information and preliminary checks	2
Battery: examination and maintenance	3
Charging/lighting circuit components: testing – late US models	4
Charging/lighting circuit components: testing – UK models and early US models	5
Charging system: boosting the charging rate – UK models	6
Voltage regulator/rectifier unit: location and testing – late US models	7
Rectifier: location and testing – UK models and early US models	8
Fuse: location and renewal	9
Switches: general	10
Horn: location and testing	11
Turn signal relay: location and testing	12
Bulbs: renewal	13

Specifications

Electrical system
Voltage 6
Earth .... Negative

Battery
Manufacturer .... Furukawa or Yuasa
Electrolyte specific gravity .... 1.26 @ 20° (68°F)

Type:

	Battery code	Capacity
KC100, KE100 (UK), KE100 A5, A6, A7 (US)	6N4-2A-5	6V, 4Ah
KE100 A8, A9, A10 (US), KE100 B (US)	6N6-1D-2	6V, 6Ah
KH100	6N6-3B-1	6V, 6Ah

Fuse rating

	Main circuit	Lighting circuit
KE100 A8, A9, A10 (US), KE100 B (US)	15A	10A
All other models	10A	N/App

Bulbs

	KE100 (UK)	KE100 (US)
Headlamp	6V, 25/25W	6V, 30/30W
Pilot lamp	6V, 4W	Not applicable
Stop/tail lamp	6V, 21/5W	6V, 25/5.3W (6V, 32/3 cp)
Turn signal lamp	6V, 21W	6V, 17W (6V, 21 cp)
Instrument illuminating lamp and neutral indicator lamp	6V, 3W	6V, 3W
Turn signal warning lamp and main beam warning lamp:		
KE100 A	6V, 1.5W	6V, 1.5W
KE100 B	6V, 1.7W	6V, 1.7W

	KC100, KH100
Headlamp	6V, 25/25W
Pilot lamp	6V, 4W
Stop/tail lamp	6V, 21/5W
Turn signal lamp:	
KC100	6V, 8W
KH100 A and G2 to G6	6V, 17W
KH100 G7 and G8	6V, 21W
Instrument illuminating lamp(s):	
KC100 and KH100 G	6V, 1.7W
KH100 A	6V, 3W
Neutral indicator lamp:	
KC100 C1, C2	6V, 1.5W
KC100 C3, C4	6V, 1.7W
KH100	6V, 3W
Turn signal warning lamp:	
KH100 A	6V, 1.5W
KH100 G	6V, 1.7W
Main beam warning lamp:	
KC100 C1, C2	6V, 1.5W
KC100 C3, C4	6V, 1.7W

Chapter 6 Electrical system

1 General description

The system is powered by the charging and lighting coils of the flywheel generator, a third coil being fitted to supply power for the ignition system only.

Part of the output on most models is converted to direct current by a silicon diode rectifier and used to charge the battery which then supplies most of the ancillary electrical components. The greater part of the generator output is controlled by careful switching and by exact matching of the load (ie bulb wattages) to output. Reference to the appropriate wiring diagram will show the exact details of each model's system.

Later US models have a more sophisticated system in which all the generator's output is fed to a solid state regulator/rectifier unit which converts the supply into a stable direct current feed, used to charge the battery and also to power other electrical components.

2 Electrical system: general information and preliminary checks

1 In the event of an electrical system fault, always check the physical condition of the wiring and connectors before attempting any of the test procedures described here and in subsequent Sections. Look for chafed, trapped or broken electrical leads and repair or renew these as necessary. Leads which have broken internally are not easily spotted, but may be checked using a multimeter or a simple battery and bulb circuit as a continuity tester. This arrangement is shown in the accompanying illustration. The various multi-pin connectors are generally trouble-free but may corrode if exposed to water. Clean them carefully, scraping off any surface deposits, and pack with silicone grease during assembly to avoid recurrent problems. The same technique can be applied to the handlebar switches.
2 The wiring harness is colour-coded and will correspond with the accompanying wiring diagrams. When socket connections are used, they are designed so that reconnection can be made only in the correct position.
3 Visual inspection will usually show whether there are any breaks or frayed outer coverings which will give rise to short circuits. Occasionally a wire may become trapped between two components, breaking the inner core but leaving the more resilient outer cover intact. This can give rise to mysterious intermittent or total circuit failure. Another source of trouble may be the snap connectors and sockets, where the connector has not been pushed fully home in the outer housing, or where corrosion has occurred.
4 Intermittent short circuits can often be traced to a chafed wire that passes through or is close to a metal component such as a frame member. Avoid tight bends in the lead or situations where a lead can become trapped between casings.
5 A sound, fully charged battery, is essential to the normal operation of the system. There is no point in attempting to locate a fault if the battery is partly discharged or worn out. Check battery condition and recharge or renew the battery before proceeding further.
6 Many of the test procedures described in this Chapter require voltages or resistances to be checked. This necessitates the use of some form of test equipment such as a simple and inexpensive multimeter of the type sold by electronics or motor accessory shops.
7 If you doubt your ability to check the electrical system, entrust the work to an authorised Kawasaki dealer. In any event have your findings double-checked before consigning expensive components to the scrap bin.

3 Battery: examination and maintenance

1 Details of the regular checks needed to maintain the battery in good condition are given in Routine Maintenance, together with the instructions on removal, refitting and general battery care. Batteries can be dangerous if mishandled; read carefully the 'Safety First' section at the front of this Manual before starting work, and always wear overalls or old clothing in case of accidental acid spillage. If acid is ever allowed to splash into your eyes or onto your skin, flush it away with copious quantities of fresh water and seek medical advice immediately.
2 When new, the battery is filled with an electrolyte of dilute sulphuric acid having a specific gravity of 1.260 at 20°C (68°F). Subsequent evaporation, which occurs in normal use, can be compensated for by topping up with distilled or demineralised water only. Never use tap water as a substitute and do not add fresh electrolyte unless spillage has occurred.
3 The state of charge of a battery can be checked using an hydrometer.
4 The normal charge rate for a battery is $1/10$ of its rated capacity, thus for a 4 ampere hour unit charging should take place at 0.4 amp. Exceeding this figure could cause the battery to overheat, buckling the plates and rendering it useless. Few owners will have access to an expensive current controlled charger, so if a normal domestic charger is used check that after a possible initial peak, the charge rate falls to a safe level. If the battery becomes hot during charging **stop**. Further charging will cause damage. Note that cell caps should be loosened and vents unobstructed during charging to avoid a build-up of pressure and risk of explosion.
5 After charging, top up with distilled water as required, then check the specific gravity and battery voltage. Specific gravity should be above 1.270 and a sound, fully charged battery should produce 6 – 7 volts. If the recharged battery discharges rapidly if left disconnected it is likely that an internal short caused by physical damage or sulphation has occurred. A new battery will be required. A sound item will tend to lose its charge at about 1% per day.

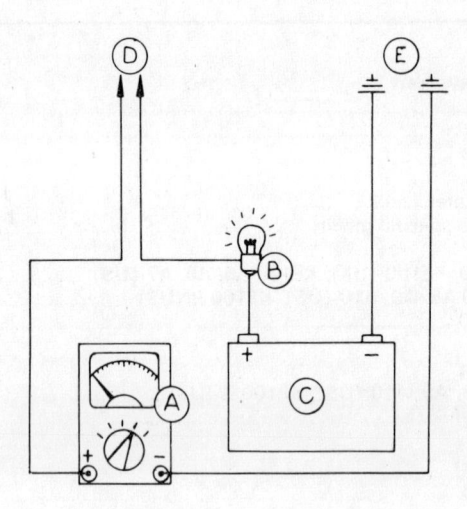

Fig. 6.1 Simple testing apparatus

A   Multimeter   D   Positive probe
B   Bulb   E   Negative probe
C   Battery

4 Charging/lighting circuit components: testing – late US models

1 This Section refers to the **US** KE100 A8, A9, A10 and KE100 B models. In addition to the preliminary checks described in Section 2, any investigation of the charging system in the search for a possible fault must start with a check of the battery and the rectifier/regulator unit. Look for obvious signs of physical damage to the unit concerned or its wiring and renew or repair as necessary.
2 Try to ascertain why a unit has failed, if a fault is discovered, and resolve the problem before a new component is fitted. Note also that faults may be found in more than one component of the system. To check the system, proceed as follows.

Regulator output test – voltage
3 Referring to Fig. 6.2 connect into the charging circuit as shown a meter set to the 10 volts dc scale, start the engine and note the readings obtained at various engine speeds with the lights on. Repeat the test with the lights switched off by disconnecting the white/blue

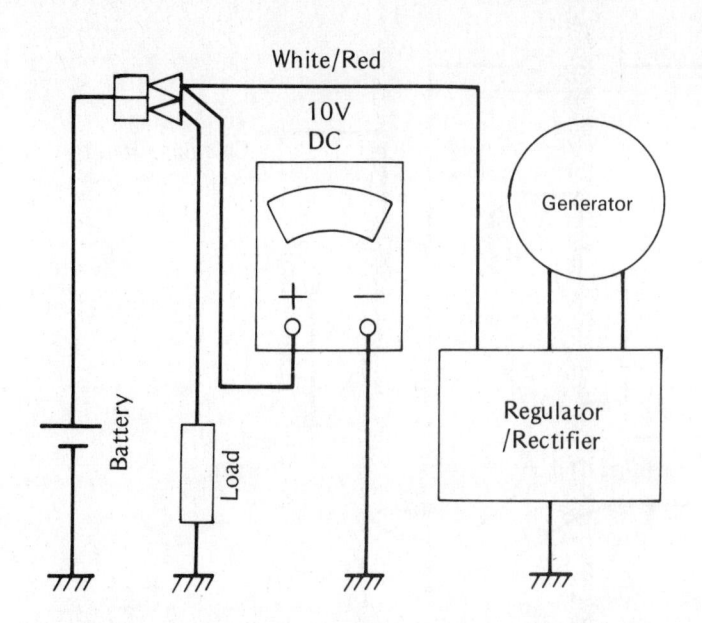

Fig. 6.2 Regulator output voltage measurement – late US models

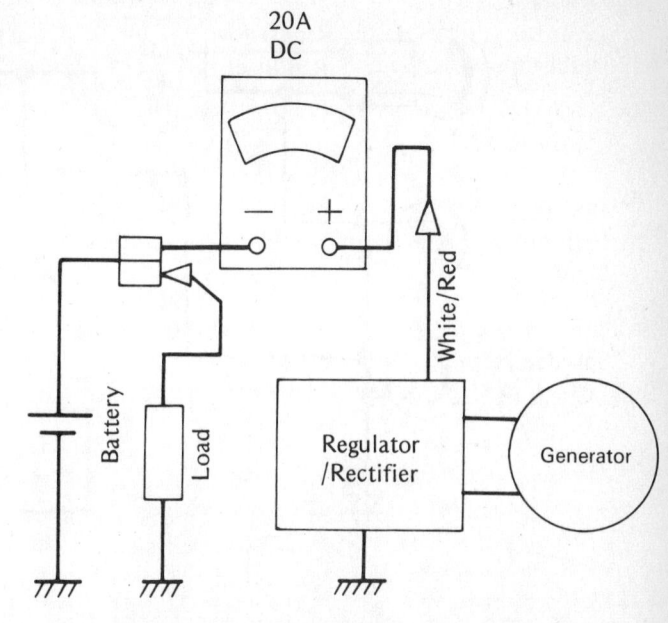

Fig. 6.3 Regulator output amperage measurement – late US models

wire at its connector next to the fuse casing. Readings should be obtained between normal battery voltage (approx 6 – 7 volts) and 8 volts; they should be near battery voltage at low speed and should increase with the engine speed but should never exceed 8 volts.

Regulator output test – amperage
4 Referring to Fig. 6.3 connect into the charging circuit as shown a meter set to the 20 amps dc scale, thus putting the meter in series with the regulator output lead. With the lights on, start the engine and increase speed to just over 4000 rpm, then take a reading. Stop the engine and disconnect the white/blue wire at its connector next to the fuse casing, thus switching off the lights. Repeat the test at the same engine speed and note the reading. With a load reduced by some 40 watts the difference in output should be approximately 5.3 amps. Note that the exact figure is not that important, the essential fact being that the regulator/rectifier unit increases output to meet demand.

5 **Note:** If the tests are satisfactory, the system is in good order. However, if the output voltage is significantly higher than that specified, either the regulator/rectifier unit is defective or there is a fault in the brown wire from the unit. Check the wire first for insecure connections or other damage; the only way of testing the unit is to substitute a new component. If the output amperage does not alter to suit demand, either the regulator/rectifier unit is faulty or the generator output is insufficient. The generator can be eliminated as described below; if it is found to be in good condition the regulator/rectifier unit is faulty and must be renewed.

Generator output test
6 Disconnect at the two-pin connector the lead (yellow wires) which runs from the generator to the main wiring loom. Connect across the two yellow wire terminals a meter set to the 250 volts ac scale, start the engine and increase speed to 4000 rpm; a reading of approximately 25 volts should be obtained.

7 If the reading is close to that specified, the generator is in good working order. If not it must be checked further. Warm the engine up to normal operating temperature before making the test.

Generator coil test
8 The condition of the coils can be tested only by measuring their resistance. First, check the resistance between the two yellow wire terminals; a figure of 0.15 – 0.22 ohm should be obtained. Switch the meter to its highest resistance range and check for continuity between each yellow wire and a good earth point on the crankcase; any reading less than infinity indicates a short circuit. If the first test revealed no reading at all, or one significantly higher than that specified, an open circuit (broken wire) is indicated. Unless the owner can repair the damage or find an auto-electrician who will attempt a repair, the coil(s) must be renewed.

9 If the output test revealed a generator fault but the coils are apparently in good condition, the fault must be in the generator rotor. It is possible for the rotor magnets to lose their magnetism if struck violently or through old age. The only solution is the renewal of the rotor.

5 Charging/lighting circuit components: testing – UK models and early US models

1 This Section refers to the US KE100 A5, A6 and A7 models and to all UK KE and KH100 models. In addition to the preliminary checks described in Section 2, any investigation of the charging system in the search for a possible fault must start with a check of the battery, the rectifier and of all circuit loads ie disconnect any electrical accessories which may have been fitted and check that all bulbs are of the correct wattage.

2 Try to ascertain why a unit has failed, if a fault is discovered, and resolve the problem before a new component is fitted. Remember that faults may be found in more than one component of the system. To check the system, proceed as follows:

AC lighting voltage check
3 To measure the AC voltage produced by the lighting coil connect a meter set to the 12 or 30 volts ac scale, in parallel across, the ac circuit load, making the meter connections as shown in the appropriate accompanying illustration; for KE100 B models refer to Fig. 6.4 and for all other KE100 A and KH100 models refer to Fig. 6.5.

4 Start the engine and switch on all lights (ignition switch on 'Night' position) with the headlamp on main beam; check that all bulbs are working. Increase speed to 4000 rpm and note the reading obtained. This should be 6.5 – 7.1 volts on KE100 models, approximately 5.8 volts on KH100 models. Stop the engine and disconnect the meter leads.

150 Chapter 6 Electrical system

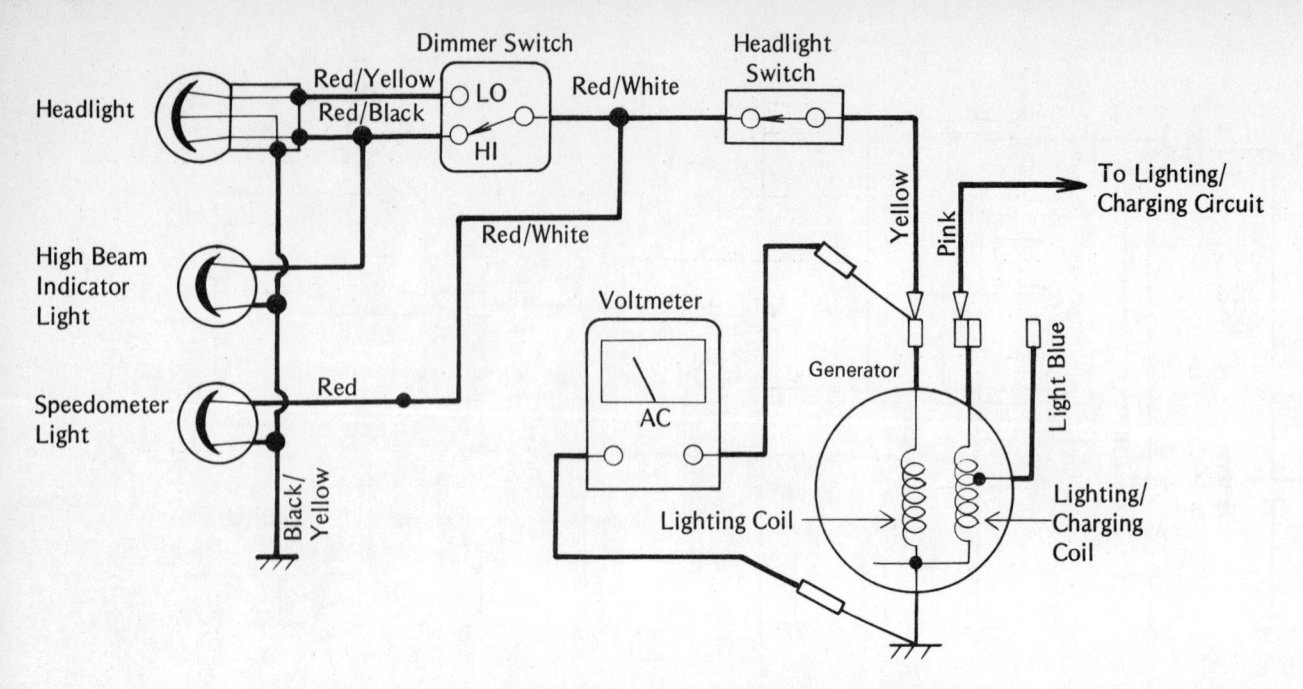

Fig. 6.4 AC lighting voltage check – KE100 B (UK) model

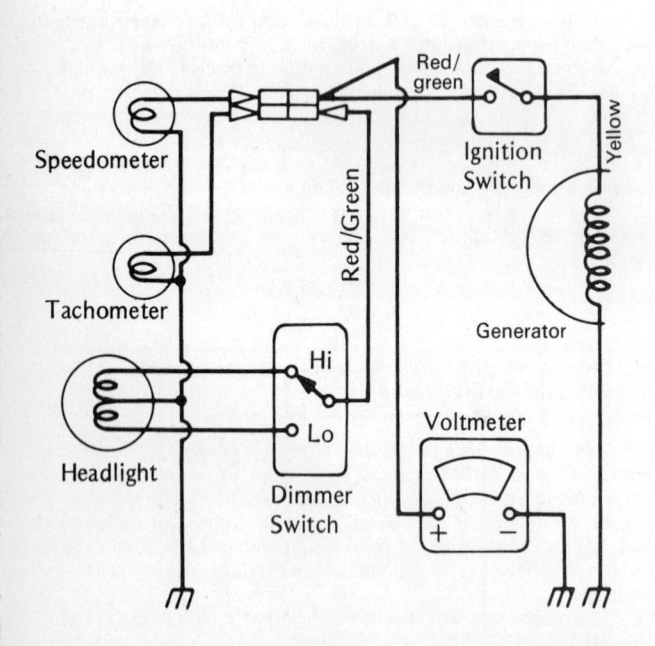

Fig. 6.5 AC lighting voltage check – KE100 A and KH100 models

DC charging voltage check

5 Connect a meter, set to the 20 volts dc scale, across the battery as shown in the appropriate accompanying illustration. For KE100 A models refer to Fig 6.6, for KE100 B models refer to Fig. 6.7 and for KH100 models refer to Fig. 6.8. Switch on all the lights as described for the previous test and note the reading obtained at 4000 rpm. For all KE100 models a reading of between 7.3 – 8.9 volts should be obtained, for KH100 A and KH100 G2 models the reading should be 7.7 volts, and for all later KH100 G models a reading of 8.0 volts should be obtained.

DC charging amperage check

6 For this test connect a meter, set to the 20 amps dc range, in series with the battery and rectifier to measure the charging current. Make the meter connections as shown in the appropriate accompanying illustration; for KE100 A models refer to Fig. 6.9, for KE100 B models refer to Fig. 6.10 and for KH100 models refer to Fig. 6.11. Make two tests, each at 4000 rpm, one with the lights switched off (Day) and one with the lights switched fully on to main beam (Night) as described in paragraph 4 above. Results should be as follows:

Model	Day	Night
KE100 A	0.95 – 1.14 amp	0.40 – 0.60 amp
KE100 B	0.95 – 2.00 amp	0.40 – 0.60 amp
KH100 A, KH100 G2	0.80 amp (approx)	0.60 amp (approx)
KH100 G3 to G8	0.80 amp (approx)	1.00 amp (approx)

7 Note: If any of the tests shows a low reading there must be a fault in the system. Since the preliminary checks will have eliminated all other components, the fault must lie either with the generator coils or with the rotor (see paragraph 9 of Section 4). To eliminate the coils, measure their resistances as follows:

8 Trace the generator lead from the crankcase upwards and disconnect the coil wires at the connectors joining them to the main wiring loom. Note: the engine must be warmed up to normal operating temperature for the tests to be accurate. Using a meter set to the 1 ohm resistance range, measure the resistance between each coil terminal and a good earth point on the crankcase. Results should be as follows:

	KE100 A	KE100 B
Lighting coil – Yellow to earth	0.41 – 0.61 ohm	0.4 – 0.8 ohm
Charging coil – Pink to earth	0.96 – 1.40 ohm	1.0 – 1.6 ohm
Charging coil – Light blue to earth	0.18 – 0.28 ohm	N/App

	KH100 A, G2	KH100 G3-G8
Lighting coil – Yellow to earth	0.4 – 0.6 ohm	0.4 – 0.6 ohm
Charging coil – Pink to earth	2.4 – 3.6 ohm	1.0 – 1.5 ohm
Charging coil – Light blue to earth	4.0 – 6.0 ohm	1.0 – 1.5 ohm

9 If any reading obtained is significantly lower than that specified, it indicates a trapped or otherwise short-circuited wire. A much higher reading indicates a broken wire. Unless the owner can repair the damage or find an auto-electrician who will attempt a repair, a faulty coil must be renewed.

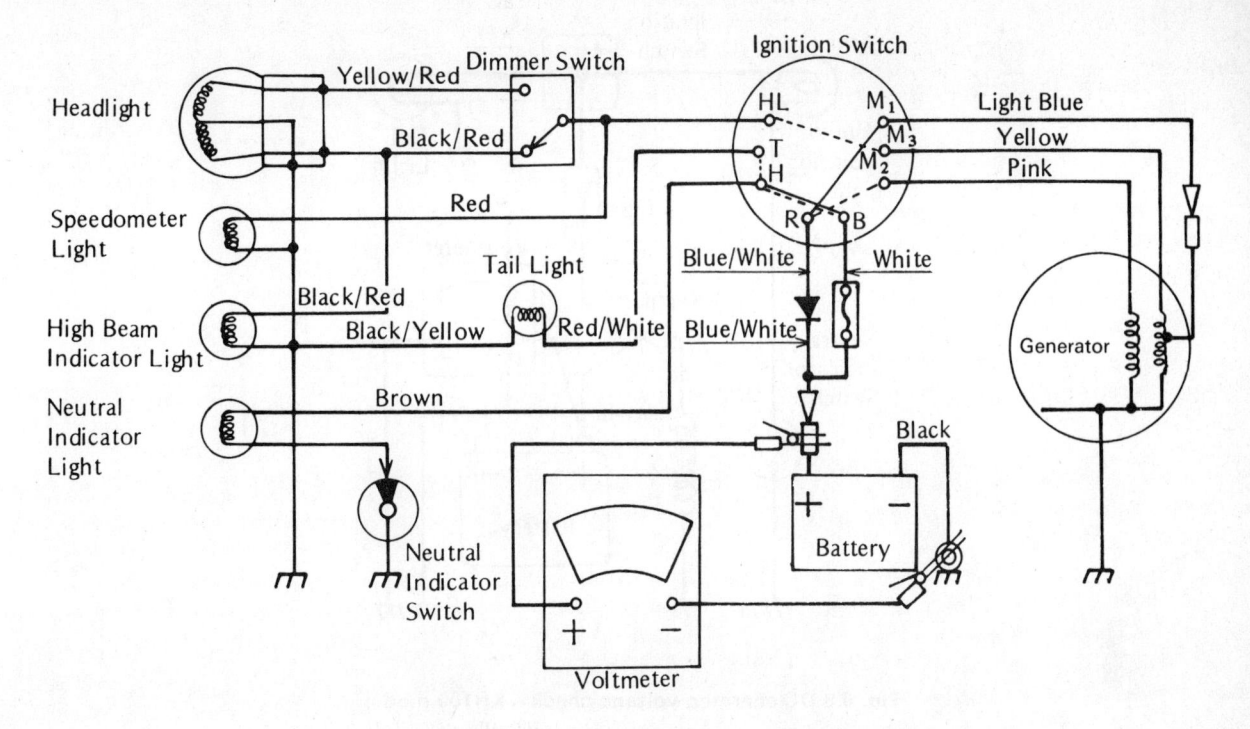

Fig. 6.6 DC charging voltage check – KE100 A model

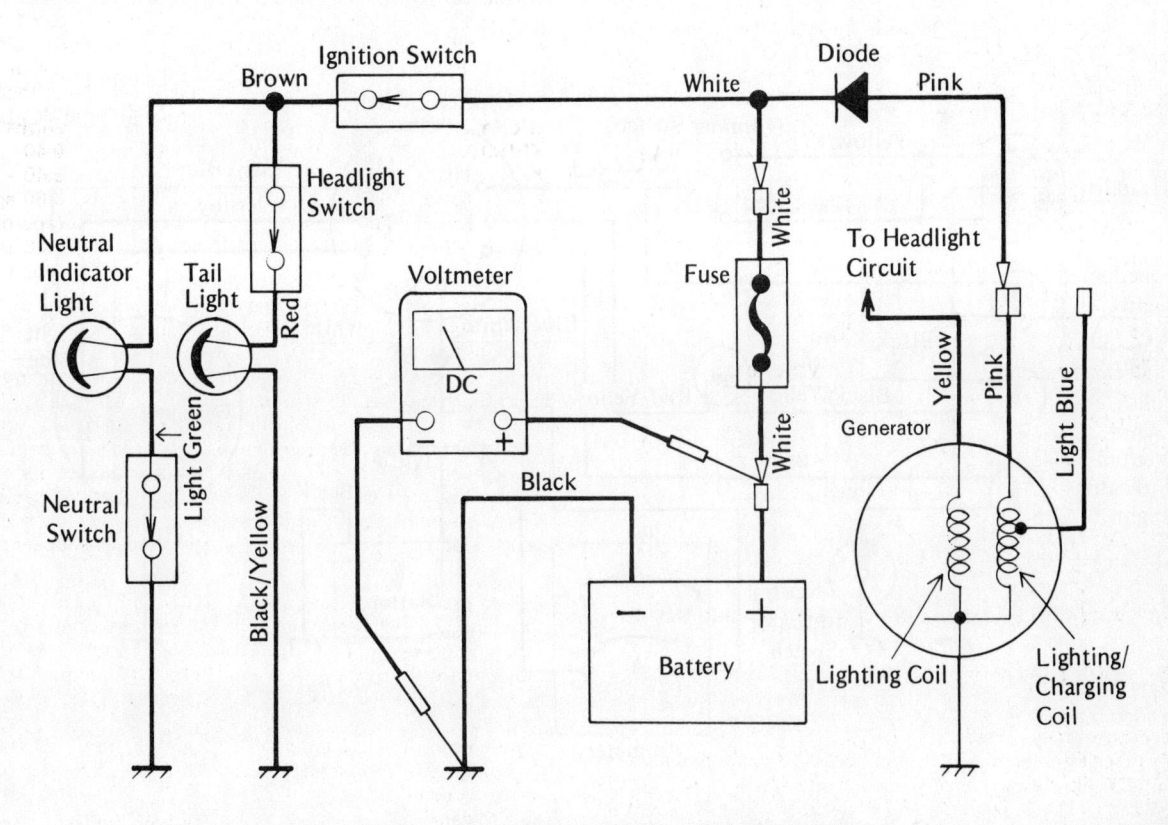

Fig. 6.7 DC charging voltage check – KE100 B model

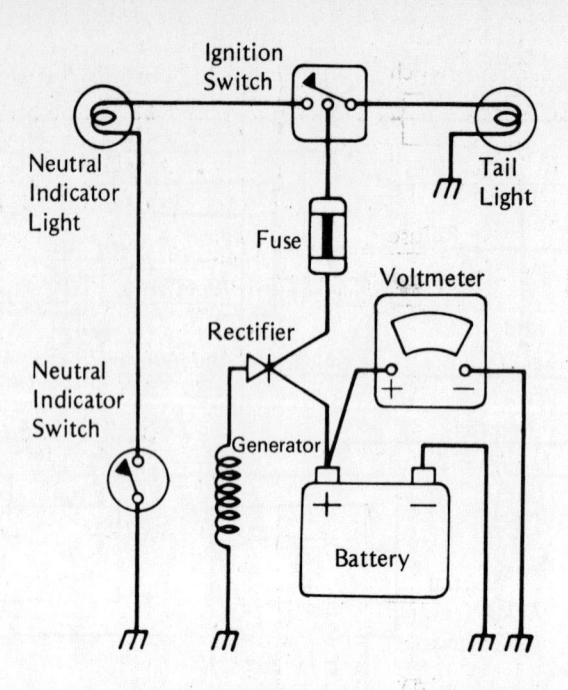

Fig. 6.8 DC charging voltage check – KH100 model

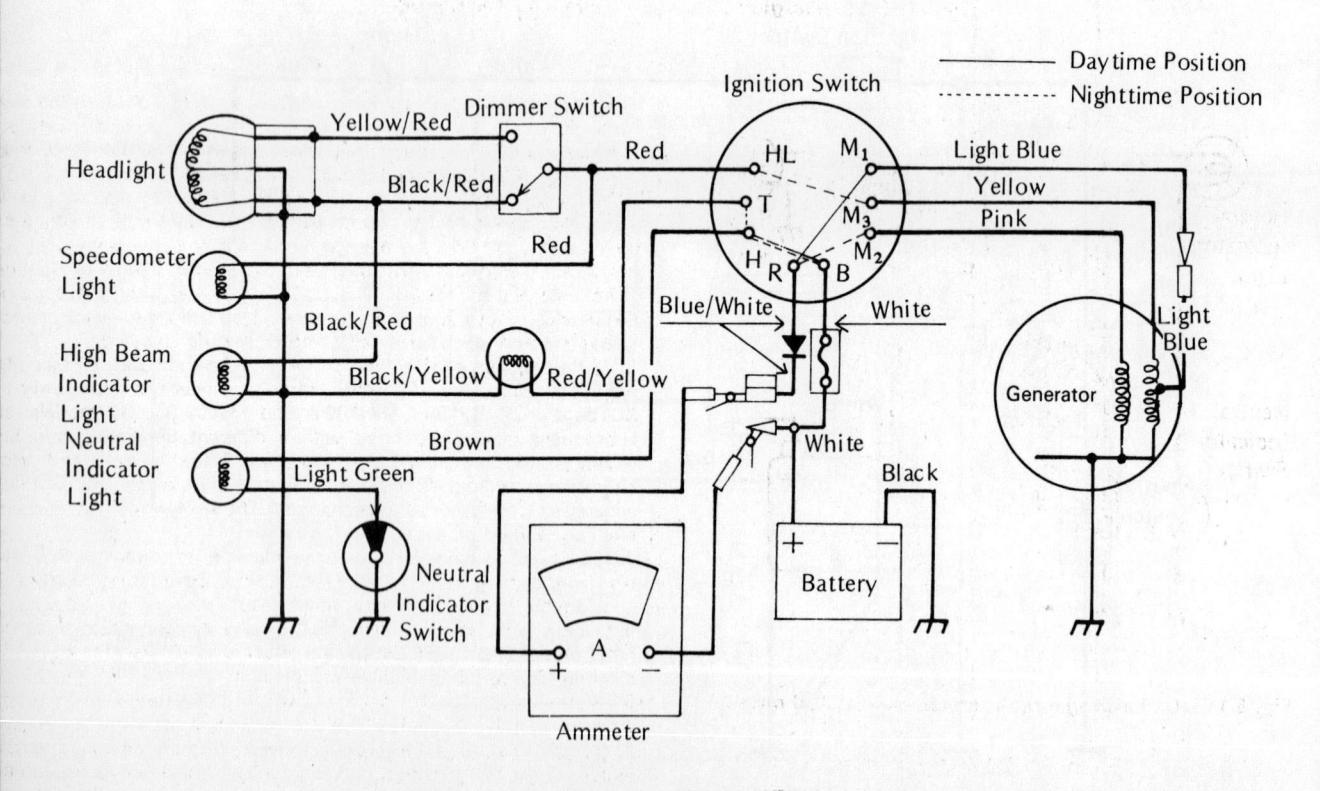

Fig. 6.9 DC charging amperage check – KE100 A model

Chapter 6 Electrical system

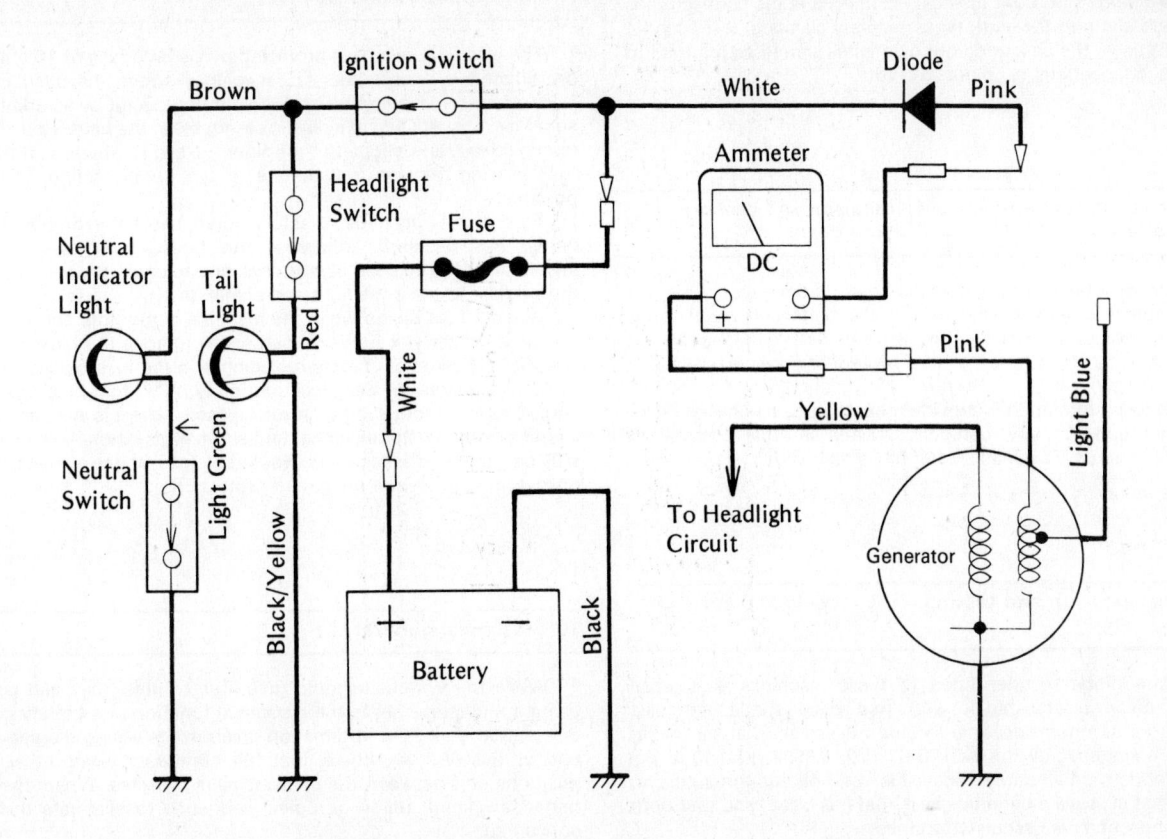

Fig. 6.10 DC charging amperage check – KE100 B model

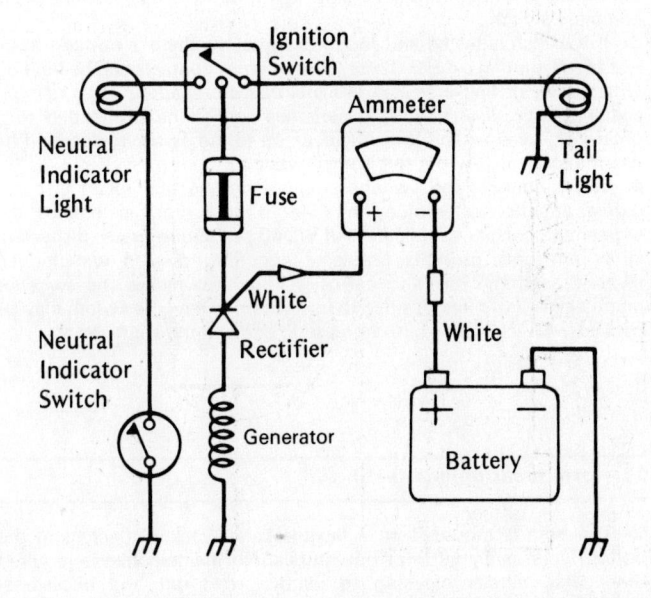

Fig. 6.11 DC charging amperage check – KH100 model

6 Charging system: boosting the charging rate – UK models

1 It is possible that certain earlier UK models may have difficulty in keeping the battery sufficiently charged to keep pace with demand. This is particularly true of short winter journeys when turn signals, horn, lights etc are used much more frequently, or in the case of machines which are infrequently used, allowing the battery to lose charge at a faster rate than normal.

2 In most cases the problem can be cured by altering the riding style slightly to make the minimum use of the electrical components concerned, or by ensuring that the machine is correctly maintained eg giving the battery a regular refresher charge, as described in Routine Maintenance, if the machine is out of service for any length of time, or by ensuring that components such as the stop lamp rear switch are not consuming too much power by being incorrectly adjusted.

3 If the problem persists, one or two modifications may be made to provide a permanent cure. **Note:** these modifications apply **only** to KC100 C1, C2, KE100 A, KH100 A and KH100 G1, G2 models; all subsequent models are fitted with a different generator assembly which produces improved charging rates. Since the coils and rotor only function efficiently as a set, components from the early and late generators should not be interchanged. The later models should not require modification at all.

4 For the early models the first modification is to transfer the light blue wire coming from the main loom to the spare fitting of the pink wire double female connector which leads from the generator. The light blue wire from the generator remains disconnected; since no other connections are touched, the effect is to convert the day-time charging rate to full night-time specification.

5 Note that on KE100 A models it will be necessary to make up an adaptor lead with a single male connector and a double female connector. The male connector must be fitted to the pink wire female terminal from the generator and the double female connector should have the pink and light blue wires (from the main loom) fitted to it.

6 If this modification is not sufficient, a higher output coil (Part Number 21047–022) is available from any authorised Kawasaki dealer, but fitting it requires the removal of the generator assembly (see Chapter 1) so that the original coil can be detached from its location opposite the points. Its connections must be carefully unsoldered so

that both the existing pink and light blue wires can be soldered to the new coil's single output. Refit the coil, ensuring that the connection is fully insulated and that the earth tag is securely clamped. If the higher output coil is used, the pink and light blue wires should be returned to their original connections ie colour to colour.

7 Voltage regulator/rectifier unit: location and testing – late US models

1 This unit is a heavily-finned metal-cased component which is mounted under the seat, to the rear of the battery. It requires no maintenance except a regular check that its connections and mountings are securely fastened. It must be kept as clean as possible at all times.
2 Since no test information is provided the unit can be checked only by the substitution of a new component when all other possibilities have been eliminated. For details, refer to Section 4.

8 Rectifier: location and testing – UK models and early US models

1 The silicon diode rectifier fitted to these machines is a small rectangular black plastic block with two male spade terminals projecting from its underside. It is located next to the battery, being retained by a single screw on KC100 C1, C2, KE100, KH100 A and KH100 G2 models; on all other models it is free-floating. It requires no maintenance at all, save a periodic check that it is clean, and that both it and its connections are securely fastened.
2 The rectifier consists of a small diode which converts the ac output of the flywheel generator into dc to charge the battery. It should be thought of as a one-way valve, that will allow the current to flow in one direction only, thus blocking half of the output wave from the generator.
3 Using a multimeter set to the resistance mode, check for continuity between the two terminals. There should only be continuity in one direction. This direction of flow may be shown by an arrow on the surface of the unit. If there is continuity in the reverse direction as well, or if resistance is measured in both directions, the rectifier is faulty and must be renewed. No repair is possible.

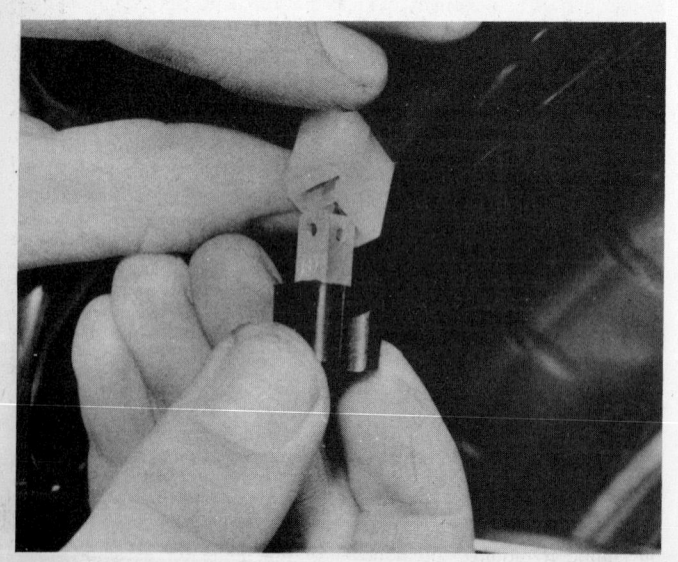

8.1 Rectifier is not directly mounted on later models, but is next to battery

9 Fuse: location and renewal

1 The electrical system is protected by a single fuse of 10 amp rating on all models except later US machines where the main circuit is protected by a 15 amp fuse and the lighting circuit by an additonal 10 amp fuse. It is retained in a plastic casing set in the battery positive (+) terminal lead, and clipped to a holder next to the battery. If the spare fuse is used, replace it with one of the correct rating as soon as possible.
2 Before renewing a fuse that has blown, check that no obvious short circuit has occurred, otherwise the replacement fuse will blow immediately it is inserted. It is always wise to check the electrical circuit thoroughly, to trace the fault and eliminate it.
3 When a fuse blows while the machine is running, and no spare is available, a 'get you home' remedy is to remove the blown fuse and wrap it in silver paper before replacing it in the fuse holder. The silver paper will restore the electrical continuity by bridging the broken fuse wire. This expedient should **never** be used if there is evidence of short circuit or other major electrical faults, otherwise more serious damage will be caused. Replace the 'doctored' fuse at the earliest possible opportunity, to restore full circuit protection.

10 Switches: general

1 While the switches should give little trouble, they can be tested using a multimeter set to the resistance function or a battery and bulb test circuit. Using the information given in the wiring diagrams at the end of this Manual, check that full continuity exists in all switch positions and between the relevant pairs of wires. When checking a particular circuit, follow a logical sequence to eliminate the switch concerned.
2 As a simple precaution always disconnect the battery before removing any of the switches, to prevent the possibility of a short circuit. Most troubles are caused by dirty contacts, but in the event of the breakage of some internal part, it will be necessary to renew the complete switch.
3 If a switch is tested and found to be faulty, there is nothing to be lost by attempting a repair. It may be that worn contacts can be built up with solder, or that a broken terminal can be repaired, again using a soldering iron. The handlebar switches can all be dismantled to a greater or lesser extent. It is, however, up to the owner to decide if he has the skill to carry out this sort of work.
4 While none of the switches require routine maintenance, some regular attention will prolong their life. In the author's experience, the regular and constant application of WD40 or a similar water-dispersant spray not only prevents problems occurring due to waterlogged switches and the resulting corrosion, but also makes the switches much easier and more positive to use. Alternatively, the switch may be packed with a silicone-based grease to achieve the same result.

11 Horn: location and testing

1 The horn is mounted on a flexible steel bracket attached to the bottom yoke or frame front downtubes. No maintenance is required other than regular cleaning to remove road dirt and occasional spraying with WD40, or a similar water dispersant lubricant, to minimise internal corrosion.
2 Different types of horn may be fitted; if a screw and locknut is provided on the outside of the horn, the internal contacts may be adjusted to compensate for wear and to cure a weak or intermittent horn note. Slacken the locknut and slowly rotate the screw until the clearest and loudest note is obtained, then retighten the locknut. If no means of adjustment is provided on the horn fitted, it must be renewed.
3 If the horn fails to work, first check that the power is reaching it by disconnecting the wires. Substitute a 6 volt bulb, switch on the ignition and press the horn button. If the bulb lights, the circuit is

proved good and the horn is at fault; if the bulb does not light, there is a fault in the circuit which must be found and rectified.

4 To test the horn itself, connect a fully-charged 6 volt battery directly to the horn. If it does not sound, a gentle tap on the outside may free the internal contacts. If this fails, the horn must be renewed as repairs are not possible.

12 Turn signal relay: location and testing

1 The relay is a cylindrical sealed metal unit, rubber-mounted under the seat, next to the battery.

2 If the turn signal lamps cease to function correctly, there may be any one of several possible faults responsible which should be checked before the relay is suspected. First check that the turn signal lamps are correctly mounted and that all the earth connections are clean and tight. Check that the bulbs are of the correct wattage and that corrosion has not developed on the bulbs or in their holders. Any such corrosion must be thoroughly cleaned off to ensure proper bulb contact. Also check that the turn signal switch is functioning correctly and that the wiring is in good order. Finally, ensure that the battery is fully charged.

3 Faults in any one or more of the above items will produce symptoms for which the turn signal relay may be blamed unfairly. If the fault persists even after the preliminary checks have been made, the relay must be at fault. Unfortunately the only practical method of testing the relay is to substitute a known good one.

13 Bulbs: renewal

1 To renew the headlamp or pilot lamp bulb on UK models, remove the single screw which retains the headlamp assembly. Withdraw the assembly from its shell, then disengage the spring clip, withdraw the bulb holder and release the bulb by pressing it in and twisting it anti-clockwise. If the reflector unit is separated from the rim at any time, note the 'Top' marking on the reflector showing which way up it is to be refitted.

2 On US models, remove the three screws which secure the headlamp assembly and pull the rim and reflector unit out of the shell. Withdraw the connector plug from the back of the unit, dismantle the horizontal adjuster screw assembly and remove the two screws to separate the reflector from the rim. On reassembly, ensure that the reflector is refitted with the 'Top' mark uppermost and adjust the headlamp aim.

3 The pilot lamp bulb assembly (UK models only) can be pulled out of its rubber grommet in the reflector; the bulb should be pressed in and twisted anti-clockwise to release it.

4 All instrument panel bulbs are of the bayonet type, fitted into bulbholders which are rubber-mounted in the base of the instrument or the instrument panel.

5 All turn signal lamp bulbs and the stop/tail lamp bulb are of the bayonet type and can be released by pushing in, turning anticlockwise and pulling from the holder. The stop/tail lamp is fitted with a twin-filament bulb which has offset pins to prevent accidental reversal in its holder.

6 All lamp lenses are retained by two screws; take care not to tear the lens seal on removal . On refitting, clean away any moisture or corrosion from the bulb and holder, check that the holder contacts are free to move against spring pressure, and that the lens seal is correctly fitted and remains in place as the lens is refitted. Do not overtighten the lens retaining screws(s), the lens will crack if over-stressed.

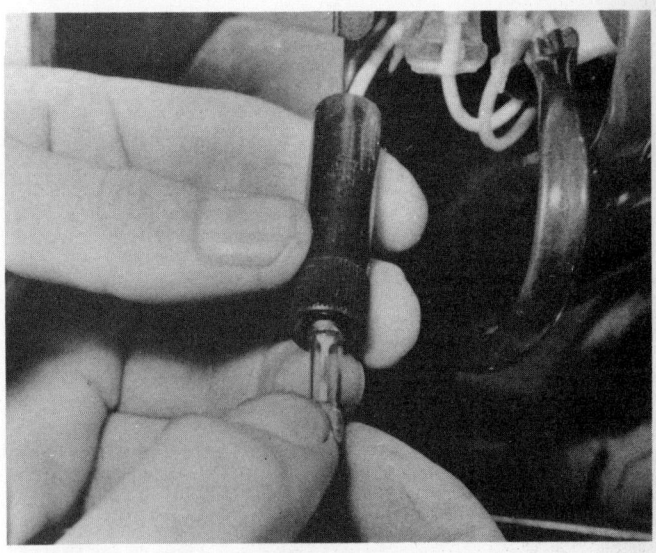

9.1 Fuse is contained in plastic casing clipped next to battery

10.4 Spray switches regularly with lubricant to prevent corrosion

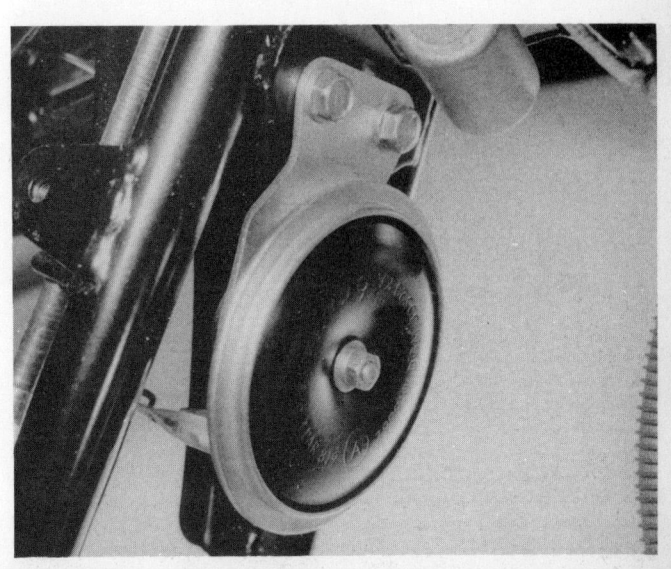

11.1 Location of horn – KH100

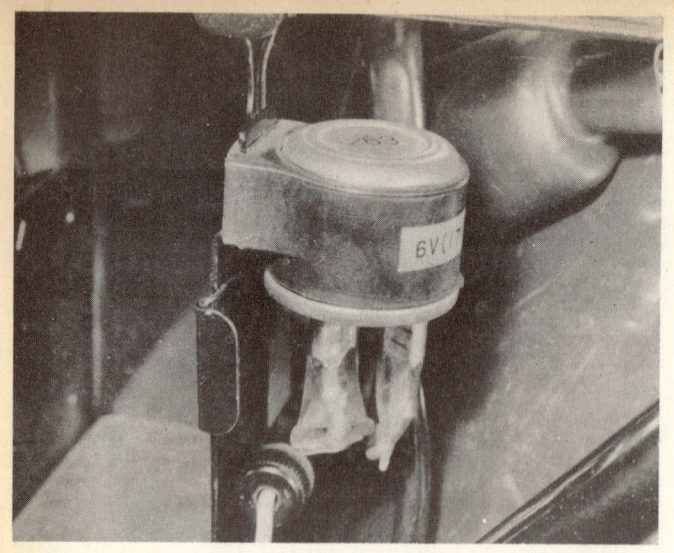

12.1 Location of turn signal relay

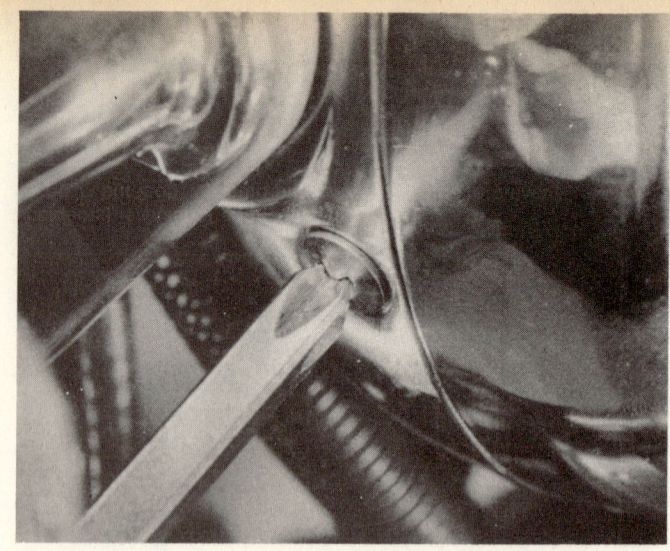

13.1a Remove single screw to release headlamp – UK models ...

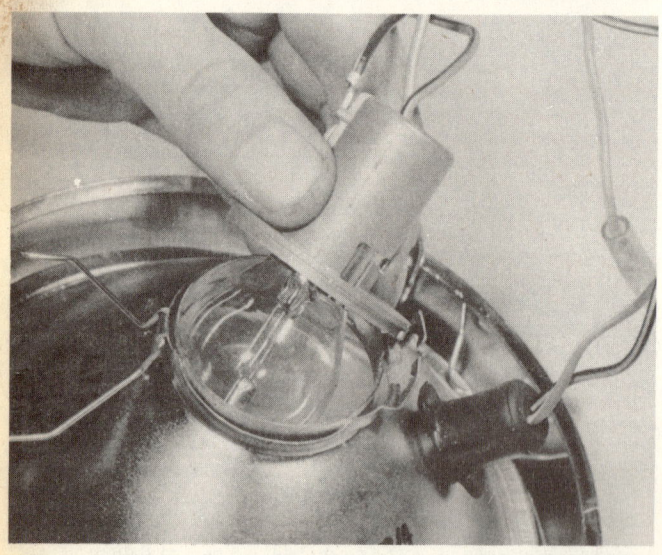

13.1b ... release spring clip and withdraw bulb holder to gain access to bulb

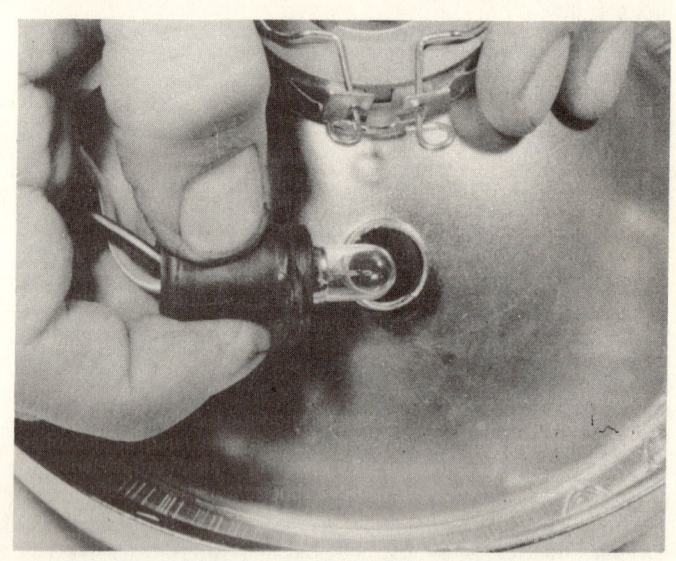

13.3 Pilot lamp bulb is retained in separate bulb holder

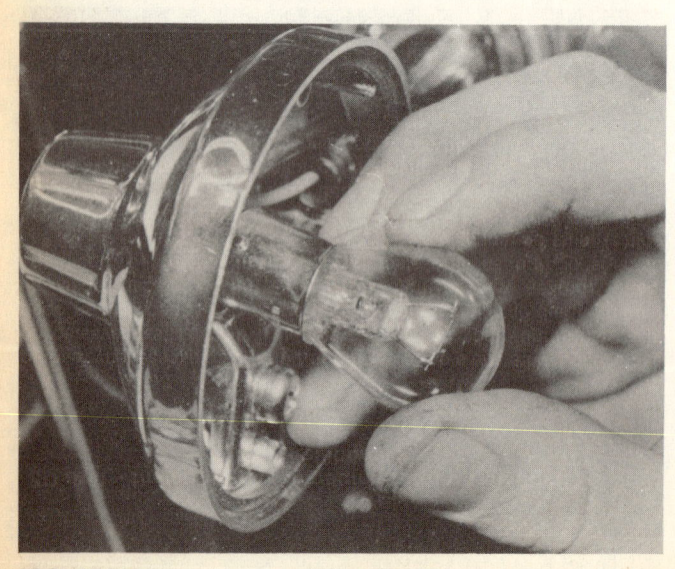

13.5a All bulbs are of bayonet fitting ...

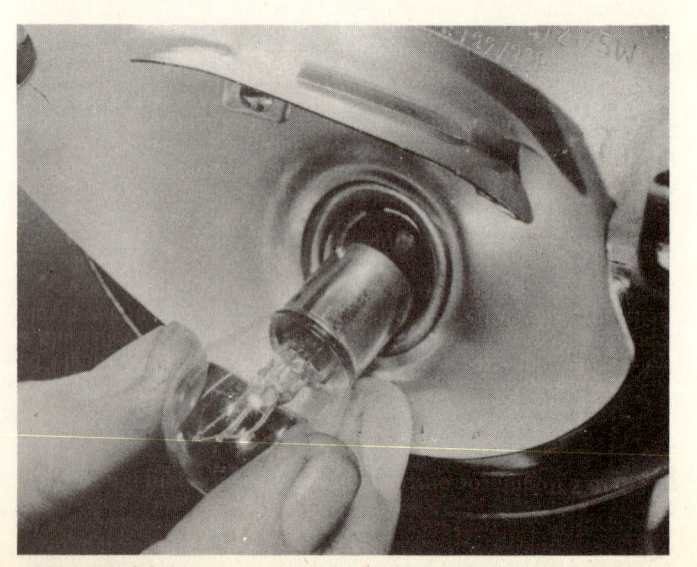

13.5b ... but stop/tail lamp bulb has offset pins to prevent accidental reversal of connections

13.6a Thoroughly clean lens and bulb holder, check gasket for tears ...

13.6b ... do not overtighten retaining screws or lens will crack

Wiring diagram – KC100 models

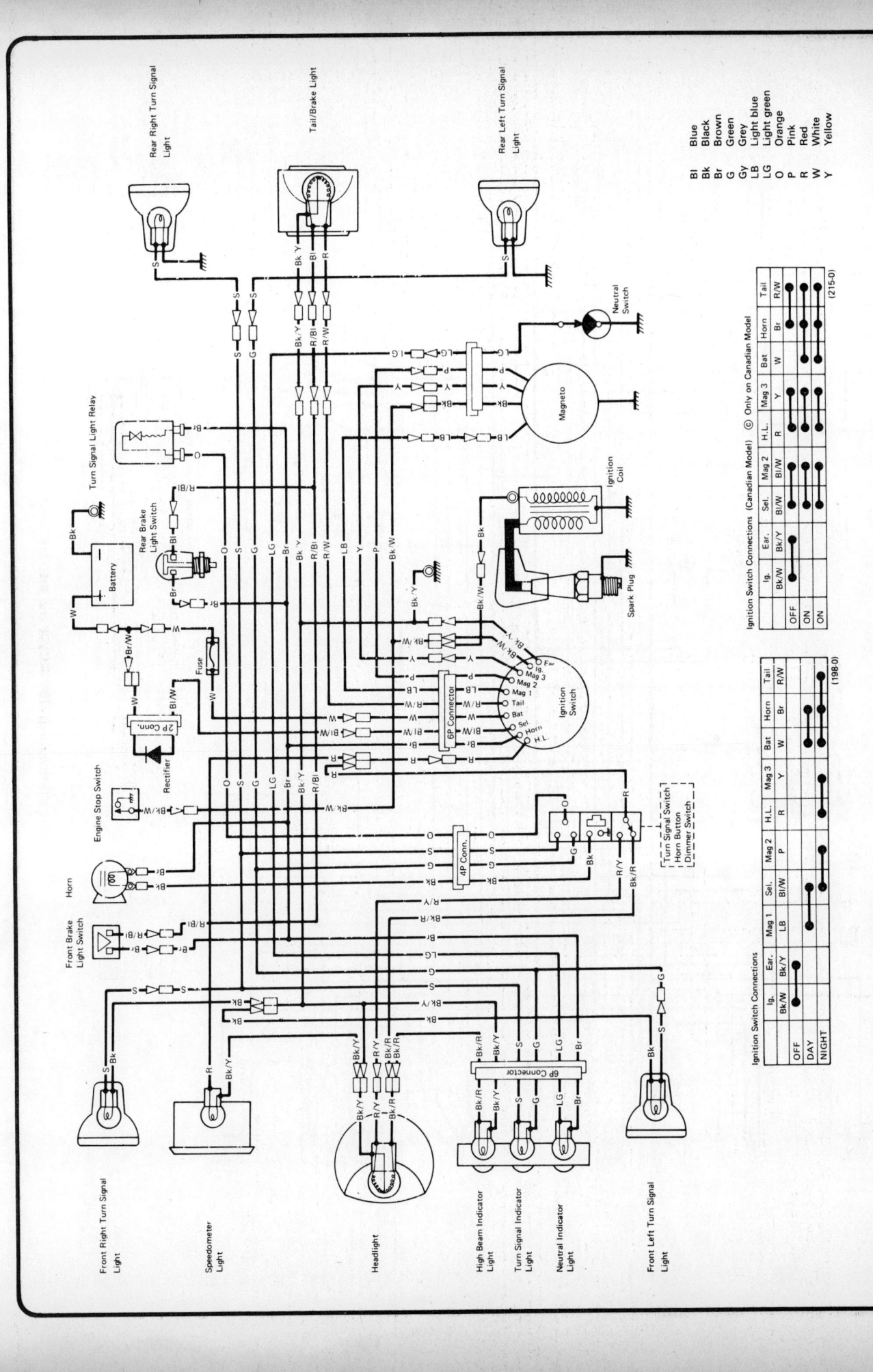

Wiring diagram – KE100 A5, A6 and A7 models

Wiring diagram – KE100 A8, A9 and A10 UK models

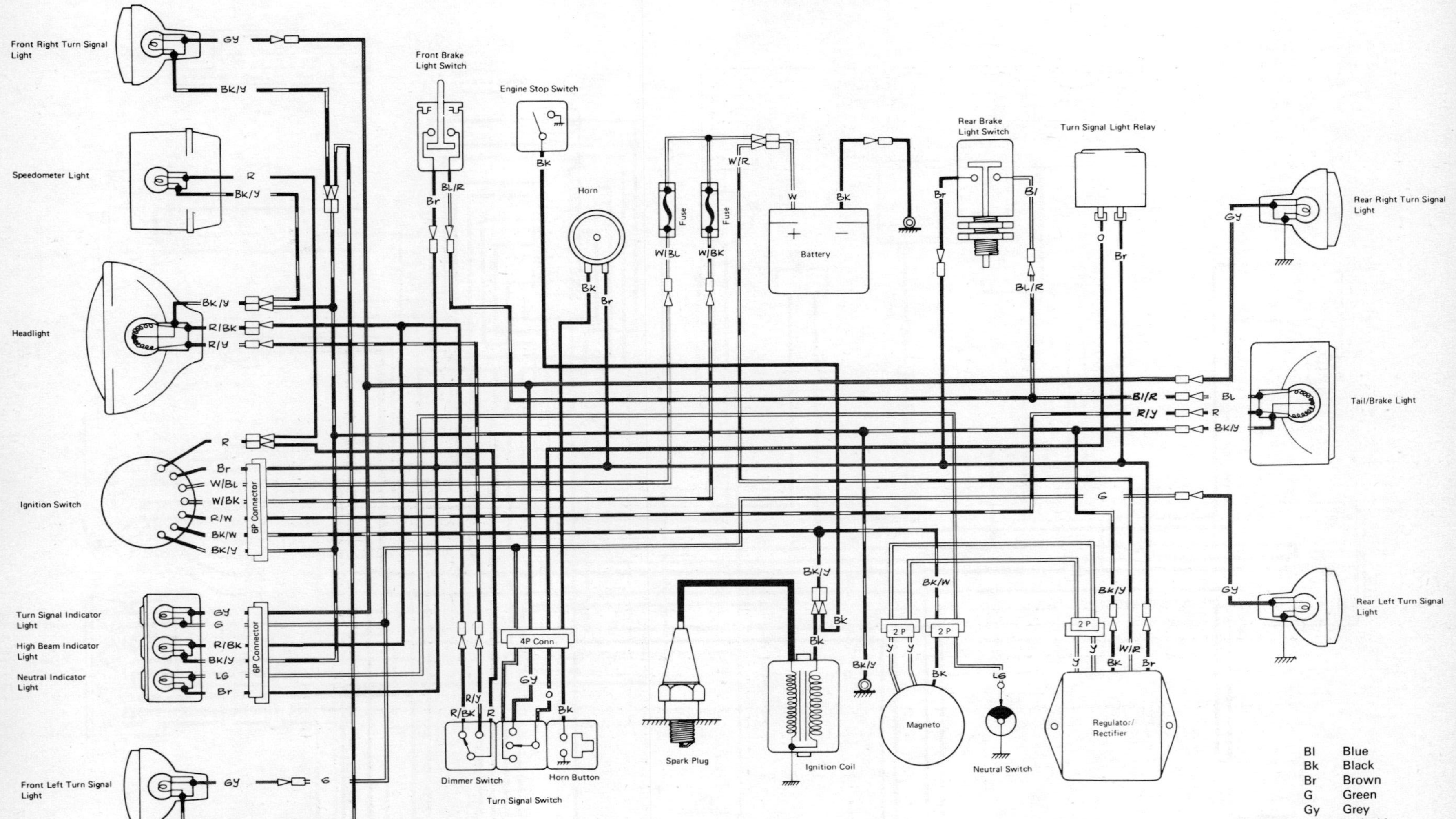

IGNITION SWITCH CONNECTIONS							
	Ignition	Earth	Battery 1	Horn	Battery 2	Headlight	Tail Light
	Bk/W	Bk/Y	White	Brown	W/Bl	Red	R/W
OFF	o——	——o					
ON			o——	——o	o——	——o——	——o

(1074A)

Bl	Blue
Bk	Black
Br	Brown
G	Green
Gy	Grey
LB	Light blue
LG	Light green
O	Orange
P	Pink
R	Red
W	White
Y	Yellow

Wiring diagram – KE100 A8, A9 and A10 US models

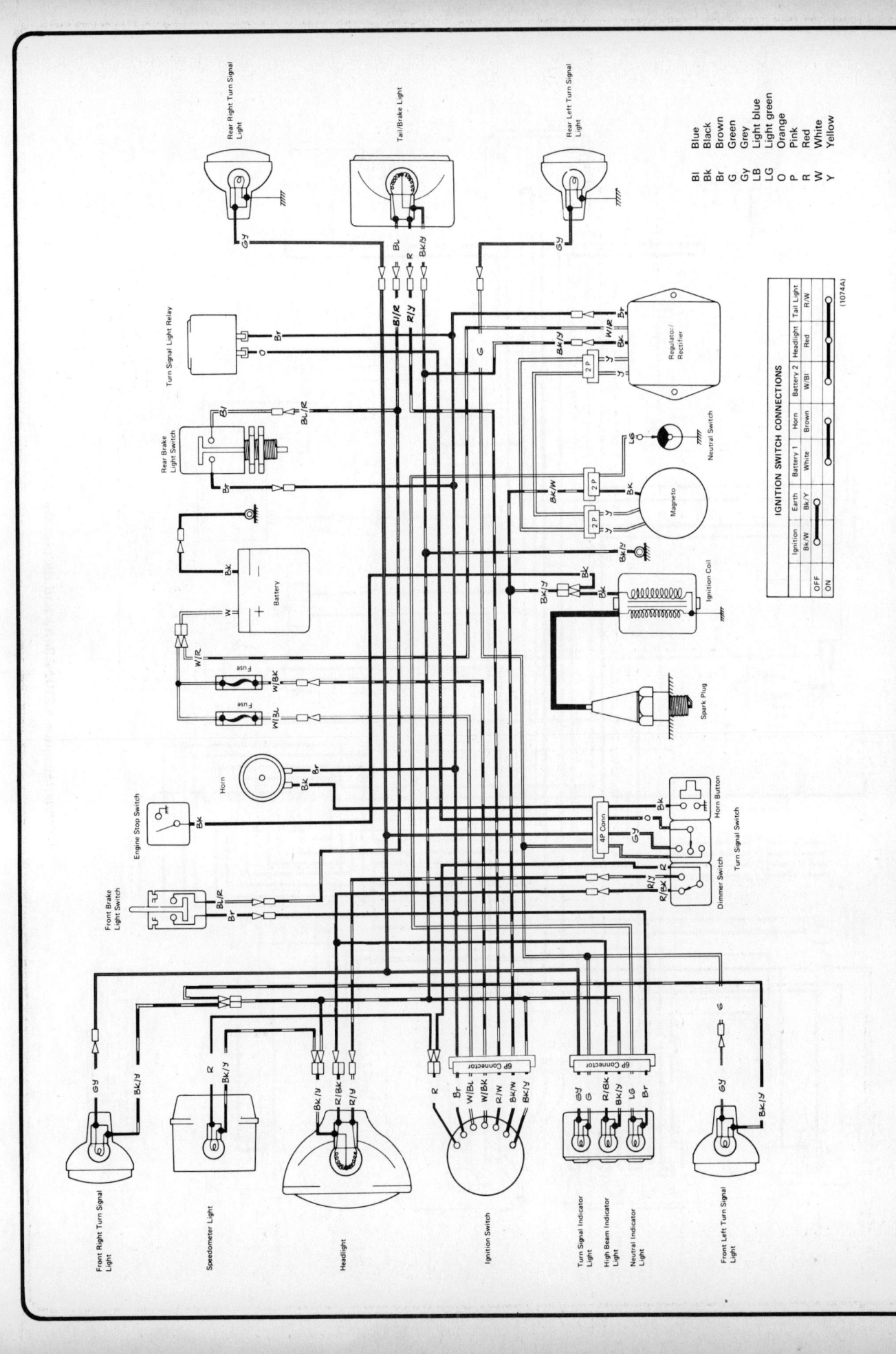

Wiring diagram – KE100 B1 to B10 UK models

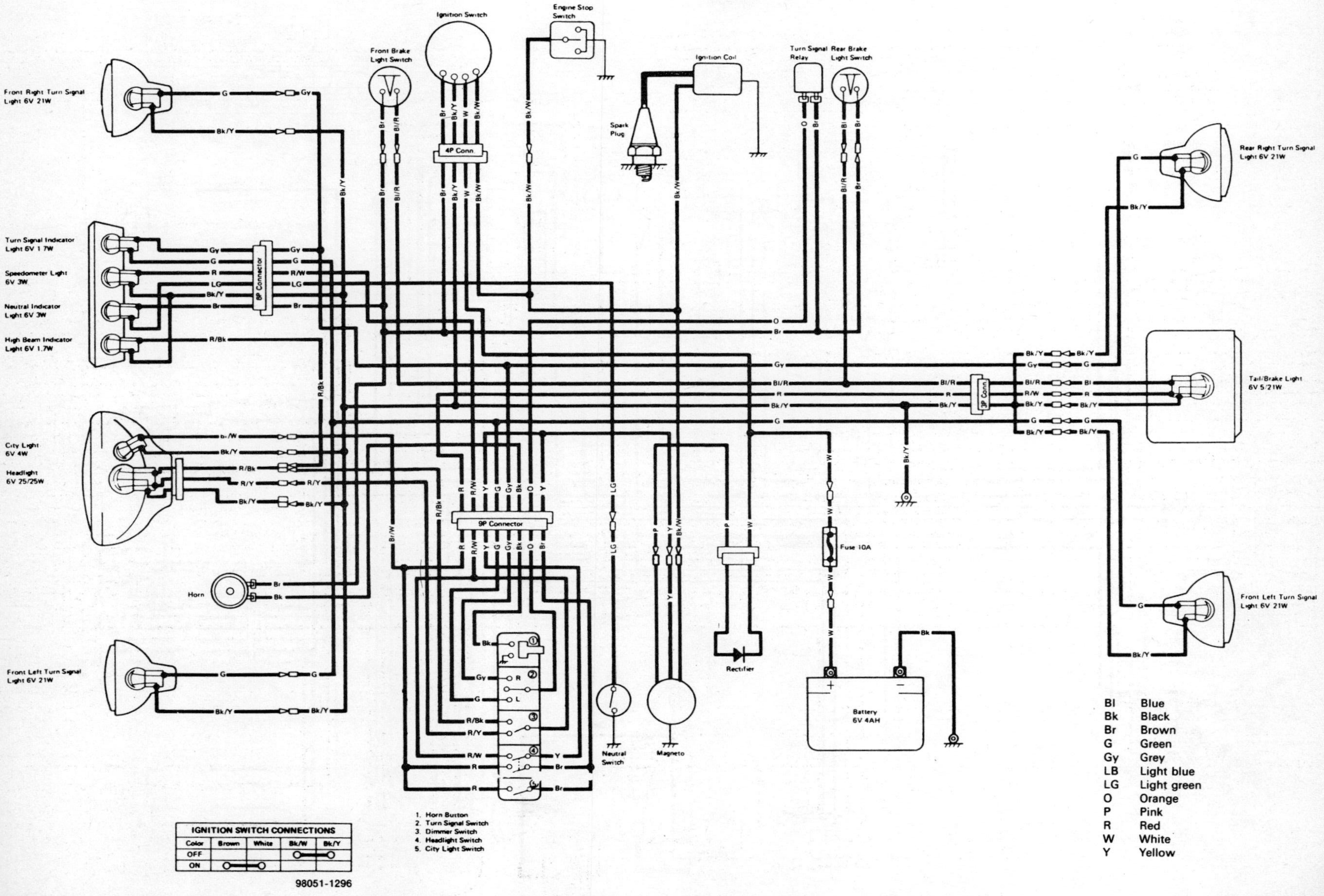

Wiring diagram – KE100 B11 UK model

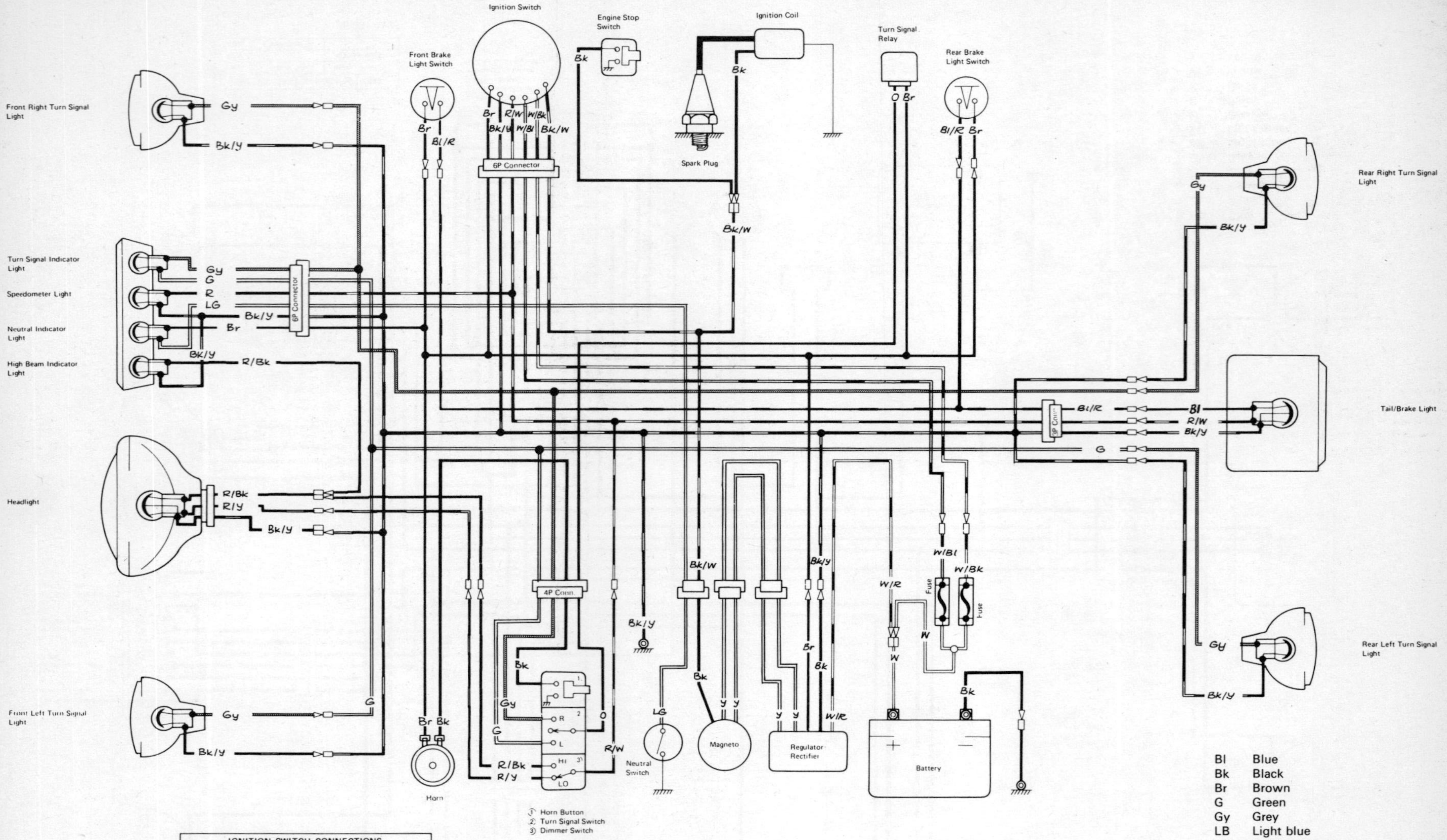

IGNITION SWITCH CONNECTIONS						
Color	BK/W	BK/Y	W/BK	Brown	W/Bl	R/W
OFF	○	○				
ON			○	○	○	○

(1466A)

Bl	Blue
Bk	Black
Br	Brown
G	Green
Gy	Grey
LB	Light blue
LG	Light green
O	Orange
P	Pink
R	Red
W	White
Y	Yellow

Wiring diagram – KE100 B US model

Wiring diagram – KH100 A model

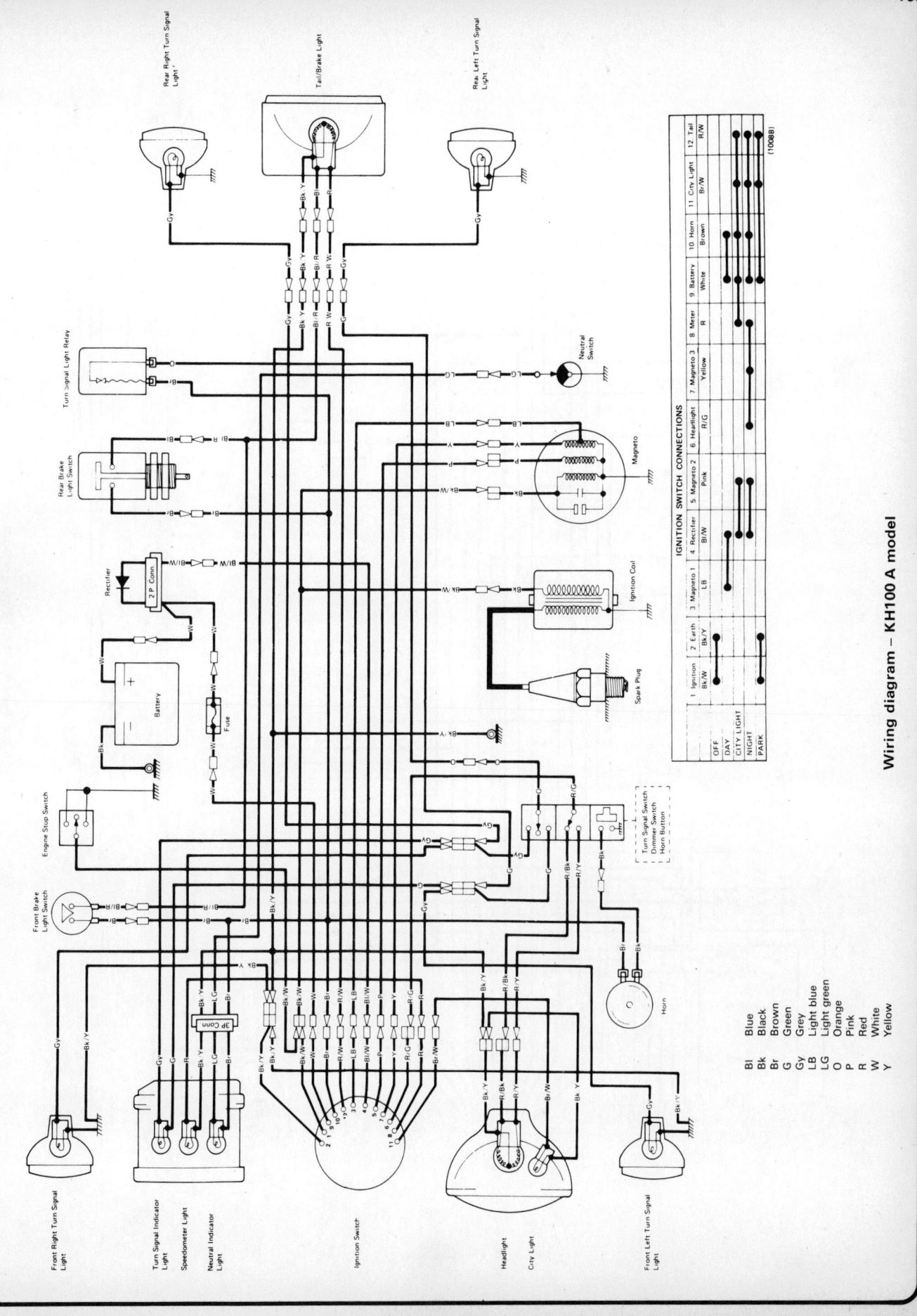

Wiring diagram – KH100 G2 model

Wiring diagram – KH100 G3 and G4 models

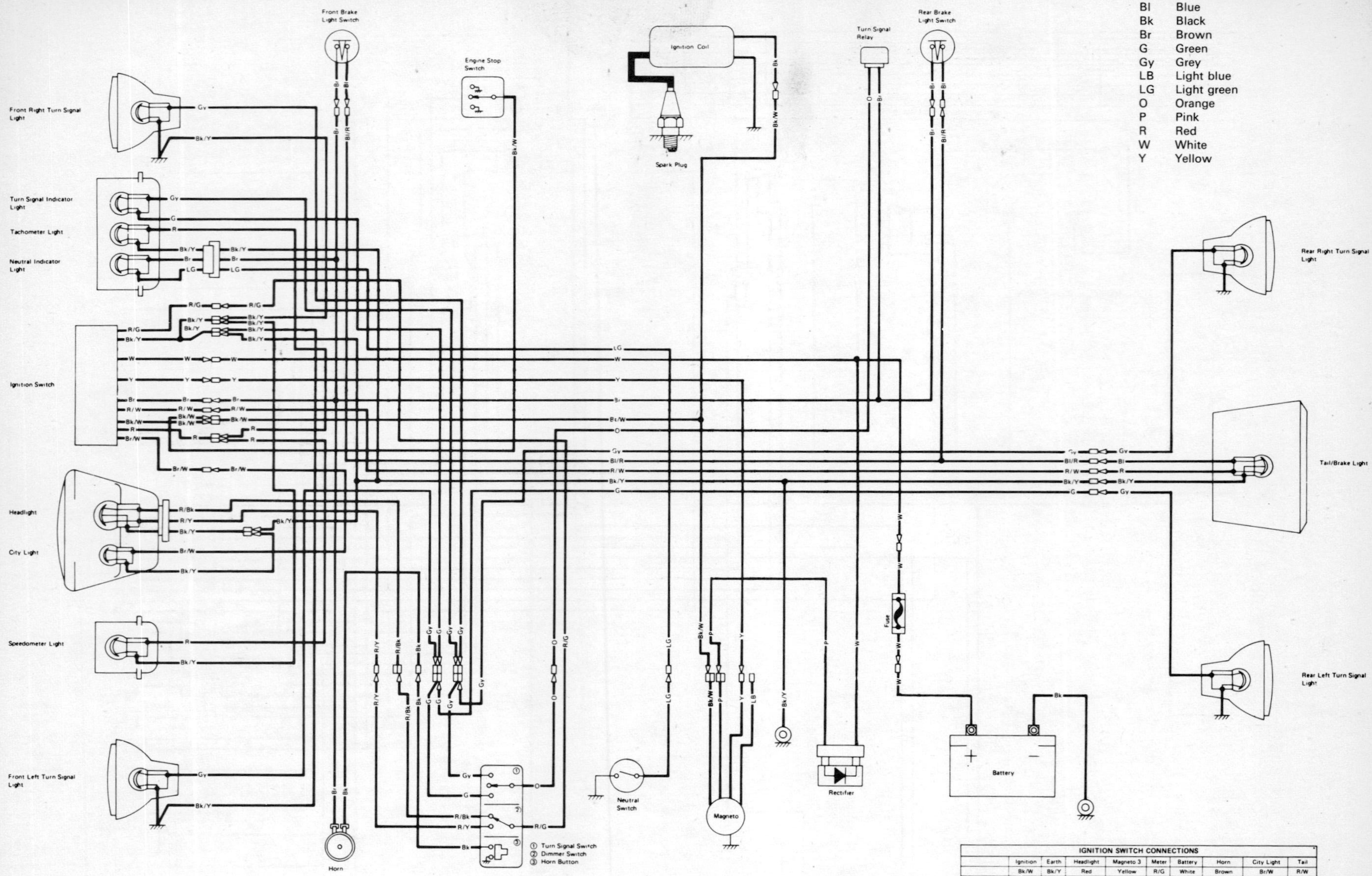

IGNITION SWITCH CONNECTIONS									
	Ignition	Earth	Headlight	Magneto 3	Meter	Battery	Horn	City Light	Tail
	Bk/W	Bk/Y	Red	Yellow	R/G	White	Brown	Br/W	R/W
OFF	○	○							
DAY						○	○		
CITY LIGHT			○			○	○	○	○
NIGHT			○	○	○	○	○	○	○
PARK	○	○				○		○	○

(1970A)

Wiring diagram – KH100 G5, G6, G7 and G8 models

English/American terminology

Because this book has been written in England, British English component names, phrases and spellings have been used throughout. American English usage is quite often different and whereas normally no confusion should occur, a list of equivalent terminology is given below.

English	American	English	American
Air filter	Air cleaner	Number plate	License plate
Alignment (headlamp)	Aim	Output or layshaft	Countershaft
Allen screw/key	Socket screw/wrench	Panniers	Side cases
Anticlockwise	Counterclockwise	Paraffin	Kerosene
Bottom/top gear	Low/high gear	Petrol	Gasoline
Bottom/top yoke	Bottom/top triple clamp	Petrol/fuel tank	Gas tank
Bush	Bushing	Pinking	Pinging
Carburettor	Carburetor	Rear suspension unit	Rear shock absorber
Catch	Latch	Rocker cover	Valve cover
Circlip	Snap ring	Selector	Shifter
Clutch drum	Clutch housing	Self-locking pliers	Vise-grips
Dip switch	Dimmer switch	Side or parking lamp	Parking or auxiliary light
Disulphide	Disulfide	Side or prop stand	Kick stand
Dynamo	DC generator	Silencer	Muffler
Earth	Ground	Spanner	Wrench
End float	End play	Split pin	Cotter pin
Engineer's blue	Machinist's dye	Stanchion	Tube
Exhaust pipe	Header	Sulphuric	Sulfuric
Fault diagnosis	Trouble shooting	Sump	Oil pan
Float chamber	Float bowl	Swinging arm	Swingarm
Footrest	Footpeg	Tab washer	Lock washer
Fuel/petrol tap	Petcock	Top box	Trunk
Gaiter	Boot	Torch	Flashlight
Gearbox	Transmission	Two/four stroke	Two/four cycle
Gearchange	Shift	Tyre	Tire
Gudgeon pin	Wrist/piston pin	Valve collar	Valve retainer
Indicator	Turn signal	Valve collets	Valve cotters
Inlet	Intake	Vice	Vise
Input shaft or mainshaft	Mainshaft	Wheel spindle	Axle
Kickstart	Kickstarter	White spirit	Stoddard solvent
Lower leg	Slider	Windscreen	Windshield
Mudguard	Fender		

Index

A

Accessories 13
Adjustment:-
   brakes 36
   carburettor 101
   clutch 39
   contact breaker points 32
   final drive chain 29
   headlamp beam 27
   ignition timing 32
   oil pump 36
   spark plug 32
   steering head bearings 44, 122
Air filter 31, 43, 103

B

Battery:-
   check 28
   examination and renovation 148
Bearings:-
   engine 60, 62
   steering head 40, 44, 122
   wheel 44, 135
Bleeding:-
   hydraulic brake 143
   oil pump 107
Brakes:-
   cable operated disc:
      cable 44
      caliper 140
      disc 139
      pad renewal 36
   check 27, 36
   drum brake:
      adjustment 36
      camshafts 44
      examination and renovation 143
   fault diagnosis 22, 23
   hydraulically operated disc:
      bleeding 143
      caliper 140
      disc 139
      fluid level check 28
      fluid renewal 43
      hose 44, 139
      master cylinder 138
      pad renewal 36
      seals 44
   specifications 127
Bulbs:-
   renewal 155
   wattages 147

C

Cables:-
   brake 36, 41, 44
   clutch 39, 41
   lubrication 41
   throttle/oil pump 35, 41
Carburettor:-
   adjustment 101
   check 35
   dismantling, examination and reassembly 94
   removal and refitting 93
   settings 101

Chain – final drive 27, 29
Clutch:-
   adjustment 39
   cable 41
   examination and renovation 64
   fault diagnosis 19, 20
   refitting 82
   removal 53
   specifications 47
Condenser 110
Contact breaker points 32
Crankcase:-
   examination and renovation 61
   joining 75
   left-hand cover:
      refitting 88
      removal 49
   reassembly 72
   right-hand cover:
      refitting 85
      removal 53
   separating 58
Crankshaft:-
   examination and renovation 64
   refitting 73
   removal 60
Cylinder barrel and head:-
   examination and renovation 62
   refitting 85
   removal 52

D

Decarbonisation 44
Dimensions – model 7
Disc brake see Front brake
Disc valve – rotary 55, 78, 103

E

Electrical system:-
   battery 28, 148
   bulbs 155
   charging system 148 – 150, 153
   checking 148
   fault diagnosis 23, 24
   fuse 154
   headlamp 27, 155
   horn 154
   lighting system 148 – 150
   rectifier 154
   specifications 147
   switches 154
   tail lamp 155
   turn signals 155
   voltage regulator/rectifier 154
   wiring diagrams 158 – 168
Engine:-
   bearings 60, 62
   crankcase 58, 61, 72, 75
   crankcase covers:
      left-hand 49, 88
      right-hand 53, 85
   crankshaft 60, 64, 73
   cylinder barrel and head 52, 62, 85
   decarbonisation 44

Index

dismantling – general 50
examination and renovation 61
fault diagnosis 17 – 21
oil level check 26
oil pump 36, 106, 107
oil seals 60, 62
piston 52, 63, 85
reassembly – general 72
refitting into frame 88
removal from frame 49
road testing 89
rotary disc valve 55, 78, 103
specifications 46
starting and running a rebuilt engine 88
Exhaust:
decarbonisation 44
removal, examination and refitting 103

F

Fault diagnosis:-
brakes 22, 23
clutch 19, 20
electrical 23, 24
engine 17 – 21
frame and suspension 21, 22
gearbox 20, 21
Filter-air 31, 43, 103
Final drive chain:-
check 27
adjustment and lubrication 29
Flywheel generator:-
refitting 75
removal 57
testing 111, 148 – 150
Frame 123
Front disc brake – cable operated:-
cable 44
caliper 140
disc 139
pad renewal 36
Front disc brake – hydraulically operated:-
bleeding 143
caliper 140
disc 139
fluid level check 28
fluid renewal 43
hose 44, 139
master cylinder 138
pad renewal 36
seals 44
Front drum brake:-
adjustment 36
camshafts 44
check 27, 36
examination and renovation 143
Front forks:-
dismantling 114
examination and renovation 119
oil change 43
reassembly 120
refitting 121
removal 113
Front wheel:-
bearings 44, 135
check 41
removal and refitting 129
Fuel system:-
carburettor 35, 93 – 101
cleaning 43
pipe 44
specifications 90
tank 91
tap 92
Fuse 154

G

Gearbox:-
examination and renovation 67
fault diagnosis 20, 21
oil change 43
oil level check 26
refitting 73
removal 60
shafts 68
specifications 48
Gear selector:-
examination and renovation 67
external components:
refitting 78
removal 56
internal components:
refitting 73
removal 60
Generator – flywheel:-
refitting 75
removal 57
testing 111, 148 – 150

H

Headlamp:-
beam adjustment 27
bulb renewal 155
HT coil 110
HT lead 111
Horn 154

I

Ignition system:-
condenser 110
contact breaker points 32
fault diagnosis 17 – 19
HT coil 110
HT lead 111
ignition source coil 111
ignition timing 32
spark plug 32, 33
specifications 109
Instruments:-
speedometer 44, 125, 126
tachometer 125, 126

K

Kickstart:-
lever:
refitting 88
removal 49
mechanism:
examination and renovation 66
refitting 73
removal 60

L

Lamps 147, 155
Legal check 27
Lubrication:-
cables 41
engine oil level check 26
final drive chain 29
general 41
oil pump 36, 106, 107
specifications 25, 91
transmission 26, 43

M

Main bearings 60, 62
Maintenance – routine 25 – 45
Master cylinder 138

N

Neutral indicator switch 57, 154

O

Oil – engine:-
  level check 26
  pump:
    bleeding 107
    removal, examination and refitting 106
    settings 36
Oil – front forks 43
Oil seals – engine 60, 62
Oil – transmission:-
  change 43
  level check 26

P

Parts – ordering 8
Piston:-
  examination and renovation 63
  refitting 85
  removal 52
  rings 63
Primary drive:-
  examination and renovation 64
  refitting 82
  removal 53
Pump – oil 36, 106, 107

R

Rear brake:-
  adjustment 36
  camshaft 44
  check 27, 36
  examination and renovation 143
Rear suspension units 124
Rear wheel:-
  bearings 44, 135
  check 41
  cush drive 137
  removal and refitting 132
Rectifier 154
Rectifier/regulator 154
Rev-counter *see* **Tachometer**
Rings – piston 63
Rotary disc valve:-
  examination 103
  refitting 78, 103
  removal 55, 103
Routine maintenance 25 – 45

S

Safety checks 9, 27
Source coil – ignition 111
Spare parts 8
Spark plug 32, 33
Specifications:-
  brakes 127
  clutch 47
  electrical 147
  engine 46
  frame and suspension 112
  fuel system 90
  gearbox 48
  ignition 109
  lubrication 25
  maintenance 25
  tyres 127
  wheels 127
Speedometer:-
  drive 44, 126
  head 125
Stands 40, 124
Steering:-
  check 40
  head:
    bearings 44, 122
    removal and refitting 122
Suspension:-
  check 40
  fault diagnosis 21, 22
  front 43, 113 – 121
  rear 123, 124
  specifications 112
Swinging arm 123, 124
Switches 155

T

Tachometer:-
  drive 126
  head 125
Tail lamp 155
Tank – fuel 91
Tap – fuel 92
Timing – ignition 32
Tools 10
Torque settings 12, 48, 91, 109, 112, 128
Transmission *see* **Gearbox**
Turn signals 155
Tyres:-
  check 27, 41
  pressures 128
  removal, repair and refitting 144
  specifications 128
  valves 146

V

Valve – rotary disc:-
  examination 103
  refitting 78, 103
  removal 55, 103
Valve – tyre 146
Voltage regulator/rectifier 154

W

Weights – model 7
Wheels:-
  bearings 44, 135
  check 41
  cush drive (rear) 137
  removal and refitting:
    front 129
    rear 132
  specifications 128
Wiring diagrams 158 – 168